Student Solutions

Intermediate Algebra

Second Edition

BARKER/ROGERS/VAN DYKE

by Grace Malaney
Donnelly College

SAUNDERS COLLEGE PUBLISHING
Harcourt Brace Jovanovich College Publishers
Fort Worth Philadelphia San Diego New York Orlando Austin
San Antonio Toronto Montreal London Sydney Tokyo

Printed in the United States of America.

Barker/Rogers/ Van Dyke: Student's Soloutions Manual to accompany
    INTERMEDIATE ALGEBRA, SECOND EDITION

ISBN 0–03–072858–4

234 017 987654321

PREFACE

This Student Solutions Manual is to accompany Intermediate Algebra, second edition, by Barker, Rogers and Van Dyke. This manual includes solutions for every other odd-numbered problem of each exercise set. Instead of solutions only, word explanations accompany the symbolic solutions, which are often written out in considerable detail. It should not be necessary to ask, "Now what did she do?" In most cases, what is offered is a "no steps left out" solutions manual.

Note to Students:

To use this manual I would recommend that it be placed to one side of a pad of paper with the text on the other side. If you are successful in solving the problem then you may or may not refer to the manual. However, if after trying to solve the problem you are unable to solve it, or if you are not sure of your solution and would like to check it by comparing your work to the manual solution, or if you would just like to look at another solution, then refer to the manual and follow through the solution there.

The manual may also be used to study for examinations by working your way through the solutions (don't just read or glance at them) and the word explanations.

Note to Instructors:

This manual may also be used by instructors as an aid in selecting homework assignments and examination items. By glancing through the manual and looking at the length of the solutions, it is possible to get an idea of the difficulty of the problems.

I have worked hard trying to make the manual free of both typing errors and conceptual errors. If any errors have gone undetected, I would appreciate it very much if you would bring them to my attention. Thank you.

Grace Malaney
Mathematics and Science Division
Donnelly College
608 North 18th Street
Kansas City, KS 66102

CONTENTS

Contents

CHAPTER 1

PROPERTIES OF REAL NUMBERS

EXERCISES 1.1   REAL NUMBERS: A REVIEW

A

True or false:

1. {3, 5, 9} is a subset of {0, 1, 2, 3, 4, 5, 6, 7, 8, 9}.

    True. Every element of the first set is an element of the second set.

5. {0, 3, 5 is the intersection of the sets {0, 1, 2, 3, 4, 5, 6} and {-1, 0, 2, 3, 4, 5, 10}.

    False. Since 2 and 4 are also common to both sets, the intersection of the two given sets is the set {0, 2, 3, 4, 5}.

9. {-12, 7, 0, -34, 681} is a subset of the integers.

    True. Every element of the given set is an integer.

13. 3x = 14 is a conditional equation.

    True. This statement is true for some but not all replacements of x.

17. -72 < -5

    True. -72 is to the left of -5 on the number line.

B

For exercises 21 and 25, given A = {6, 12, 15, 91, 106, 210} and B = {-5, 0, 6, 10, 91, 300}:

21. Is B a subset of A?

    No. Not every element of B is an element of A.

25. Is 6 ∈ A?

    Yes. 6 is an element of A.

For exercise 29, given
$$A = \left\{-8, \ -\sqrt{3}, \ -\frac{2}{3}, \ -0.16, \ 0, \ 5, \ \frac{16}{3}, \ \sqrt{39}, \ 42, \ 55.8\right\}:$$

29. List the subset of irrational numbers.

    The elements of A which are irrational numbers are: $\{-\sqrt{3}, \ \sqrt{39}\}$.

1

For exercise 33, given

$$B = \left\{-1.01, \ -\frac{21}{4}, \ 2.3, \ \sqrt[3]{9}, \ -0.21\overline{21}, \ \frac{4}{5}, \ 0.78, \ 62, \ -29\right\}:$$

33. List the subset of integers.   The elements of B which
                                            are integers are:
                                            {-29, 62}.

For exercise 37, use -, <, or > in the box to make the statement
true.

37. 5 ☐ 12   Use the symbol < since 5

                                            is to the left of 12 on
                                            the number line.

C

For exercises 41, 45, and 49, given
    A = {-3, -2, -1, 0, 1, 2, 3}, B = {0, 1, 2, 3},
    C = {-8, -4, -2} and D = {-3, -1, 0, 1, 3, 5, 7}, find:

41. A ∪ B
     A ∪ B = {-3, -2, -1, 0, 1, 2, 3}   The union of the sets
                                          contains all the elements
                                          that are in A or B.

42. A ∪ D
     A ∪ D = {-3,-2,-1,0,1,2,3,5,7}   The union of the sets
                                          contains all the elements
                                          that are in A or D.

49. Is C a subset of the integers?   Yes. Every element of C
                                          is an integer.

True or false:

53. 5x = 7x + 3 is a conditional   True. This statement is
     equation.   true for some but not all
                                          replacements of x.

57. {2, 10, 15, 20, 25} is a   True. Every element of
     subset of the rational numbers.   the given set is a
                                          rational number.

61. The symmetric property of equality states if a = b then b = a. State this in your own words.

If the left side of an equation is equal to the right side, then the right side equals the left side and can be rewritten in that order.

CHALLENGE EXERCISES

Given A = {0, 1, 2, 3} and B = {9, 10, 12}

65. B ∪ ∅ = ?
     B ∪ ∅ = B

Since the union of these two sets is the set that contains all the elements of B or ∅, the empty set, then the union is B, as ∅ has no elements.

EXERCISES 1.2 PROPERTIES OF REAL NUMBERS

A

The following are examples of what properties of real numbers?

1. $-8 + 4 = 4 + (-8)$

Commutative property of addition. The order in which the terms are added has been changed.

5. $-8 \cdot 1 = -8$

Multiplication property of one. When any real number is multiplied by one, the product is the real number.

9. $3 + x = x + 3$

Commutative property of addition. The order in which the terms are added has been changed.

13. $0 + (-32) = -32$

Addition property of zero. When zero is added to any real number, the sum is the real number.

True or false:

17. 4 is the reciprocal of $\frac{1}{4}$.

True. 4 and $\frac{1}{4}$ are reciprocals since their product is one.

B

True or false:

21. $2 \cdot 3 + 4 = 2 \cdot 3 + 2 \cdot 4$ is an example of the distributive property.

False. Multiplication is not being distributed over addition. A true statement would be: $2(3 + 4) = 2 \cdot 3 + 2 \cdot 4$.

25. $7 + (-7) = 0$ is an example of the addition property of opposites.

True. The sum of a number and its opposite is zero.

29. $3(x + y) = 3x + y$ is an example of the associative property of addition.

False. The statement is an <u>unsuccessful</u> example of the distributive property. It could be corrected by writing: $3(x + y) = 3x + 3y$.

33. $1(x - 2) = (x - 2)(1)$ is an example of the multiplication property of one.

False. The factors have been interchanged, so this is an example of the commutative property of multiplication. The multiplication property of one says $1(x - 2) = x - 2$.

The following are examples of what property of real numbers?

37. $(7 + 9)4 = 7 \cdot 4 + 9 \cdot 4$

Distributive property. Multiplication by 4 has been "distributed" over the addition of 7 and 9.

41. $\left(\frac{2}{x}\right)\left(\frac{x}{2}\right) = 1, \ x \neq 0$

Multiplication property of reciprocals. The product of reciprocals is one.

4

C

For exercises 45 through 65 justify the step used in the simplification of the expression:

45. 3 + 4(-8 + 8) = 3 + 4(0)

Addition property of opposites. The sum of a number and its opposite is zero.

49. 3 + 1(-3) + 1(8) = 3 + (-3) + 8

Multiplication property of one. The product of any real number and one is the real number itself.

53. -12 + 4(3 + 2) = -12 + 12 + 8

Distributive property. Multiplication was "distributed" over addition.

57. -1(2x + 2y) = -2x + (-2y)

Distributive property.

61. 5 + 2(3 + 2(1 + -1)) = 5 + 2(3 + 2(0))

Addition property of opposites. The sum of opposites, (1 + -1) equals zero.

65. 5 + 6 = 11

Closure property of addition. The sum of two integers is an integer.

CHALLENGE EXERCISES

69. Is the set {1, 2} closed with respect to addition?

No. A set is closed with respect to addition when the sum of any two of its members is also a member of the set. 1 + 2 = 3, but 3 is not a member of the set. So the set is not closed with respect to addition.

EXERCISES 1.3 OPPOSITES, ABSOLUTE VALUE, ADDITION OF REAL NUMBERS

A

Perform the indicated operations:

1. $|-6| = -(-6) = 6$

    The absolute value of a negative number is its opposite.

5. $-|-4| = -[-(-4)]$

    $\quad\quad = -(4)$

    $\quad\quad = -4$

    The absolute value of -4 is its opposite, 4.
    The opposite of 4 is -4.

9. $-8 + (-2) = -10$

    The sum of two negative numbers is negative.

13. $-(4 + 1) = (-4) + (-1)$

    $\quad\quad\quad = -5$

    The opposite of a sum is the sum of the opposites.
    The sum of two negative numbers is negative.

17. $|8 + 7| = |15| = 15$

    The absolute value of a positive quantity is itself.

B

21. $-(5 + 12) = (-5) + (-12)$

    $\quad\quad\quad = -17$

    The opposite of a sum is the sum of the opposites.
    The sum of two negative numbers is negative.

25. $\quad -8 + (-3) + 5$

    $= (-11) + 5$

    $= -(|-11| - |5|)$

    $= -6$

    Combine terms from left to right. The sum of the first two terms is negative since both are negative.
    To find the sum of a positive and a negative number, subtract their absolute values.
    Use the _sign_ of the larger absolute value.

29. $\quad -6 + (-8) + (-12)$

    $= (-14) + (-12)$

    $= -26$

    Combine like terms from left to right.
    The sum of negative numbers is negative.

33. $-[x + (-2)] = (-x) + (2)$ The opposite of a sum is
                                        the sum of the opposites.

              $= -x + 2$ or $2 - x$

37. $-[2x + 0(3y - 4)]$
      $= -[2x + 0]$ Multiplication property of
                                        zero.
      $= -[2x]$ Addition property of zero.
      $= -2x$

C

41. $-(15 + 12) + (-36)$
      $= (-15) + (-12) + (-36)$ The opposite of a sum is
                                        the sum of opposites:
                                        $-(15 + 12) = (-15) + (-12)$
      $= (-27) + (-36)$ Now combine terms from
                                        left to right.
      $= -63$

45. $-(3.2) + (-7.5) + (1.9)$ Combine like terms from
      $= (-3.2) + (-7.5) + (1.9)$ left to right.
      $= (-10.7) + (1.9)$ The sum of the first two
                                        terms is negative since
                                        both terms are negative.
      $= -|-10.7| - |1.9|$ Find the difference of
      $= -8.8$ absolute values and use
                                        the sign of the number
                                        with the larger absolute
                                        value.

49. $-[-|-4| + (-|3|)] + [3 + (-2)]$
      $= -[-|-4| + (-3)] + [3 + (-2)]$ Rewrite the absolute value
                                        inside the parentheses.
      $= [(-4) + (-3)] + [3 + (-2)]$ Do the indicated
                                        operations with the
      $= -[(-7)] + [3 + (-2)]$ brackets on the left, and
      $= [(-7)] + [1]$ then on the right.
      $= (7) + (1)$
      $= 8$

53. $-|-4 + 7| + |-16 + 12| + |-12 + 2|$

      $= -|3| + |-4| + |-10|$ Simplify within each
                                        absolute value from left
                                        to right.
      $= -(3) + (4) + (10)$ Rewrite each absolute
                                        value.
      $= (-3) + (4) + (10)$ Combine like terms from
      $= (1) + (10)$ left to right.
      $= 11$

7

57. $|5 + (-8)| + |-16 + (-4)| + (-|-5 + (-9)|)$

$= |-3| + |-20| + (-|-14|)$ Simplify within each absolute value from left to right.

$= (3) + (20) + (-14)$ Rewrite each absolute value.

$= 23 + (-14)$ Combine like terms from
$= 9$ left to right.

D

61. The checking account of the Square Hole Donut Company showed credits of \$894.72 and debits of \$674.72 (−\$674.72) for one day. What increase or decrease will the balance of the account show?

$\$894.72 + (-\$674.72)$ Find the result of the day's transactions by

$= |\$894.72| - |-\$674.72|$ finding the sum of the two given amounts.

$= \$220$ increase   Since the result is positive, the amount represents an increase in the balance of the account.

65. Jan Long has a checking account balance of \$641.32. She wrote checks for \$49.50 and \$36.47, then made a deposit of \$50.00. What is her current balance?

$\$641.32 + (-\$49.50) + (-\$36.47) + \$50.00$

To find the current balance, calculate the result of the given transactions. Remember that checks written on the account are written as negative values and deposites are written as positive values.

$= \$591.82 + (-\$36.47) + \$50.00$ Combine like terms from left to right.

$= \$555.35 + \$50.00$
$= \$605.35$ Jan's current balance.

8

69. Harry and Tom formed a partnership. They agree to split all earnings and expenses equally between each other. If Harry earned $172.50 and Tom earned $156.25, what is the net worth of the partnership if Harry's expenses were $33.15 and Tom's expenses were $28.35?

$172.50 + $156.25 + (-$33.15) + (-$28.35)
                           Find the sum of the transactions. Write earnings as positive values and expenses as negative values.

= $328.75 + (-$33.15) + (-$28.35) Work from left to right.
= $295.60 + (-$28.35)
= $267.25   Net worth of the partnership.

CHALLENGE EXERCISES

79. Is $|a + b| \leq |a| + |b|$ always true? If the answer is no, give an example when it is false.

Always true, because $|a + b|$ can represent the absolute value of the sum of a positive number and a negative number.

$$|(-5) + (2)| < |-5| + |2|$$

$$|-3| < 5 + 2$$

$$3 < 7$$

Note in the example how $|a + b| < |a| + |b|$, when replacing $\underline{a}$ with a negative number while replacing $\underline{b}$ with a positive number. When a and b are either both positive or both negative, $|a + b| = |a| + |b|$ from exercise 71.

EXERCISES 1.4   SUBTRACTION OF REAL NUMBERS AND COMBINING TERMS

A

Perform the indicated operations (subtract or combine like terms):

1. $12 - 6 = 12 + (-6)$
      $= 6$ Convert the subtracion to addition. Add the opposite of 6.

5. $-8 - (-4)$
  $= -8 + (4)$ Add the opposite of -4.
  $= -4$

9. $7a + 4a = 11a$

Mentally add the coefficients.

13. $-8abc - 9abc$
    $= -8abc + (-9abc)$

Convert the subtraction to addition.
Simplify, combining the coefficients.

    $= [(-8) + (-9)]abc$
    $= -17abc$

17. $115 - 291$
    $= 115 + (-291)$
    $= -176$

Add the opposite of 291.

B

21. $\dfrac{5}{8} - \dfrac{1}{4} = \dfrac{5}{8} - \dfrac{2}{8}$

Find a common denominator.

    $= \dfrac{5}{8} + \left(-\dfrac{2}{8}\right)$

Convert subtraction to addition.

    $= \dfrac{3}{8}$

25. $-0.5 - (-0.4)$
    $= -0.5 + (0.4)$
    $= -0.1$

Add the opposite of -0.4.

29. $6x + 5x - 7x$
    $= 6x + 5x + (-7x)$

Convert the subtraction to addition.
Combine the coefficients.

    $= [6 + 5 + (-7)]x$
    $= 4x$

33. $6x + 8 + 12x = (6x + 12x) + 8$

Group like terms using the associative and commutative properties.
Combine like terms.

    $= 18x + 8$

37. $-8.7 - 14.6 = -8.7 + (-14.6)$

Convert subtraction to addition by adding the opposite of 14.6

    $= -23.3$

C

41. $22.5 - (-16.3) = 22.5 + 16.3$
    $= 38.8$

Add the opposite of -16.3.

45.   4x + 7y - 8x - 4y
    = 4x + 7y + (-8x) + (-4y)   Convert subtractions to
                                        additions.

    = [4x + (-8x)] + [7y + (-4y)]   Regroup using associative
                                        and commutative
                                        properties.
    = -4x + 3y   Combine like terms.

49. $-\frac{1}{3}a - \frac{1}{4}b - \frac{1}{12}a - \frac{1}{2}b$

    $= -\frac{1}{3}a + \left(-\frac{1}{4}b\right) + \left(-\frac{1}{12}a\right) + \left(-\frac{1}{2}b\right)$ Convert subtractions
                                                                                                              to additions.

    $= -\frac{4}{12}a + \left(-\frac{3}{12}b\right) + \left(-\frac{1}{12}a\right) + \left(-\frac{6}{12}b\right)$ Rewrite with common
                                                                                                              denominator.

    $= \left[\left(-\frac{4}{12}a\right) + \left(-\frac{1}{12}a\right)\right] + \left[\left(-\frac{3}{12}b\right) + \left(-\frac{6}{12}b\right)\right]$ Regroup like
                                                                                                                                                      terms
                                                                                                                                                      together.

    $= \left(-\frac{5}{12}a\right) + \left(-\frac{9}{12}b\right)$ Combine like terms.

    $= -\frac{5}{12}a - \frac{3}{4}b$ Simplify.

53.   3.52y - 12 + 6.3z + 14
    = 3.52y + 6.3z + [(-12) + (14)]   Convert subtraction to
                                          addition and group like
                                          terms.
    = 3.52y + 6.3z + 2   Simplify.

57.   -8.3x + 5.9x - 3.7x - 10.3x + 15.2x
    = (-8.3x) + (5.9x) + (-3.7x) + (-10.3x) + (15.2x)
                                          Convert subtractions to
                                          additions.
    = [(-8.3) + (5.9) + (-3.7) + (-10.3) + (15.2)]x
                                          Combine the coefficients.
    = -1.2x

D

61. What was the change in temperature (difference) from a low
     of -8° F to a high of 40° F?

        40° F - (-8° F)   Find the difference.
     = 40° F + (8° F)
     = 48° F   Change in temperature.

11

65. What is the change in altitude from 250 ft below sea level (-250) to 600 ft above sea level?

    600 - (-250)   Subtract to find the difference

= 600 + 250

= 850   Feet above sea level.

69. If a check is written for the amount of $85.42 on an account with a balance of $96.57, what is the new balance?

    $96.57 - ($85.42)   Subtract the amount of the

= $96.57 + (-$85.42)   check from the beginning

= $11.15   balance.

CHALLENGE EXERCISES

73. Is (a - b) - c = a - (b - c) ever true? If the answer is yes, state when it is true.

    (a - b) - c = a - (b - c)   Yes, sometimes true.

    (a - b) - 0 = a - (b - 0)   If c = 0, then the statement is true.

[a + (-b)] + 0 = a + [-(b + 0)]   Convert subtractions to additions. Zero is neither positive nor negative.

    [a + (-b)] = a + (-b)   Addition property of zero.

        a + b = a + b

EXERCISES 1.5 RECIPROCALS, MULTIPLICATION, AND DIVISION OF REAL NUMBERS

A

Find the reciprocal of:

1. 6 $6 = \frac{6}{1}$, so its

    The reciprocal of 6 is $\frac{1}{6}$.   reciprocal is $\frac{1}{6}$.

Perform the indicated operations:

5. (-7)(-5) = 35   The product of two negative numbers is positive.

9. $-12(0) = 0$

Multiplication property of zero.

13. $-64 \div 8 = \dfrac{-64}{8} = -8$

Write the division in fraction form. Simplify. The quotient of a negative number and a positive number is negative.

17. $(-10x) \div (-2) = \dfrac{-10x}{-2}$
$= 5x$

Write the division as a fraction. Divide the coefficient -10 by -2. The quotient of two negative numbers is positive.

B

Find the reciprocal of:

21. $-\dfrac{11}{12}$

The reciprocal of $-\dfrac{11}{12}$ is $-\dfrac{12}{11}$.

The reciprocal of $\dfrac{a}{b}$ is $\dfrac{b}{a}$. Also, the product of reciprocals is 1.
$\left(-\dfrac{11}{12}\right) \cdot \left(-\dfrac{12}{11}\right) = 1$

Perform the indicated operations:

25. $(-2)(-3)4 = 6(4)$

$= 24$

The product of two negative numbers is positive.

29. $\dfrac{-50}{-5} = 10$

The quotient of two negative numbers is positive.

33. $(-3a)(4)(-2) = [(-3)(4)(-2)]a$
$= 24a$

Group the coefficients. Simplify.

37. $3x \div 5 = \dfrac{3x}{5}$ or $\dfrac{3}{5}x$

Write the division as a fraction.

41. $-(3)^2(-2)(5) = -(9)(-2)(5)$

$= (-9)(-2)(5)$
$= 90$

Perform the exponentiation. Write the opposite of 9. Simplify.

C

Find the reciprocal of:

45. $0.25 = \dfrac{25}{100} = \dfrac{1}{4}$

So the reciprocal of 0.25 is 4.

Rewrite the decimal as a fraction, and reduce.

The reciprocal of $\dfrac{a}{b}$ is $\dfrac{b}{a}$.

Perform the indicated operations:

49. $\begin{aligned}[(6x)(-5)] \div 2 &= [(6)(-5)x] \div 2 \\ &= -30x \div 2 \\ &= \dfrac{-30x}{2} \\ &= \dfrac{-30}{2}x = -15x \end{aligned}$

Regroup within the brackets, and multiply. Write the division as a fraction. To simplify, divide the coefficients.

53. $\dfrac{(-5z)(2)}{7} = \dfrac{-10z}{7}$ or $-\dfrac{10}{7}z$

Multiply in the numerator.

57. $\begin{aligned} &6(-2x + 4) \\ &= 6(-2x) + 6(4) \\ &= -12x + 24 \end{aligned}$

Multiply using the distributive property. Simplify.

61. $\begin{aligned} &-5(-5a + 2b - 3) \\ &= -5[-5a + 2b + (-3)] \\ &= (-5)(-5a) + (-5)(2b) + (-5)(-3) \\ &= 25a + (-10b) + (15) \\ &= 25a - 10b + 15 \end{aligned}$

Rewrite subtraction as addition. Multiply using the distributive property. Simplify. Write as subtraction.

65. $\begin{aligned} &-2a(2x + 4y - 3) \\ &= -2a[2x + 4y + (-3)] \\ &= (-2a)(2x) + (-2a)(4y) + (-2a)(-3) \\ &= -4ax + (-8ay) + 6a \\ &= -4ax - 8ay + 6a \end{aligned}$

Write the subtraction within parentheses as addition. Multiply using the distributive property. Simplify. Rewrite as subtraction.

D

69. Use the formula below to change 14°F to a Celsius measure.

$$C = \frac{5}{9}(F - 32)$$

$$C = \frac{5}{9}(14 - 32)$$

Replace F in the formula with 14.

$$C = \frac{5}{\cancel{9}}(\cancel{-18}) = -10$$

Simplify. Reduce

So 14°F is the same as -10°C.

73. If the average cash flow of the Super Duper Cookie Company is -$125.00 a day for five days, what was the cash flow for that period?

$$5(-\$125.00) = -\$625.00$$

Multiply the daily cash flow by the number of days.

The cash flow for the given period was -$625.00.

STATE YOUR UNDERSTANDING

77. What does it mean for two numbers to be reciprocals?

When two numbers are reciprocals, it means their product is one.

CHALLENGE PROBLEMS

81. Is it possible for a number to be its reciprocal? If the answer is yes, give an example.

$$1 = \frac{1}{1}$$
And the reciprocal is also $\frac{1}{1}$ or 1.

Yes, it is possible. If the number is 1, the reciprocal is 1.

EXERCISES 1.6   ORDER OF OPERATIONS

A

Perform the indicated operations:

1. $-5(-4 + 3) = -5(-1)$

Do the work indicated within parentheses. Multiply.

$$= 5$$

5. $(2 + 3)^2 = (5)^2 = 25$ Do the indicated addition
                                             within the parentheses,
                                             and then the
                                             exponentiation.

Simplify:

9. $4 \cdot 3x - 2 \cdot 5x = 12x - 10x$ Do multiplication,
                $= 2x$ then subtraction.

Evaluate the following if x = -5:

13.   x - 12   Replace x with -5 and
       $-5 - 12 = -5 + (-12) = -17$ subtract.

Perform the indicated operations:

17. $-3 \cdot 4 - 2(9 - 12) = -3 \cdot 4 - 2(-3)$ Do the subtraction
                                             within parentheses.
                $= -3 \cdot 4 + (-2)(-3)$ Simplify.
                $= -12 + 6$
                $= -6$

B

Simplify:

21. $5x + 9 - (3 - 2x)$
                $= 5x + 9 + (-3) + (2x)$ Rewrite subtraction as
                                             addition.
                $= (5x + 2x) + [9 + (-3)]$ Combine like terms.
                $= 7x + 6$ Simplify.

25. $4a - 3b - (3b + 4a)2$
       $= 4a - 3b - (6b + 8a)$ Use the distributive
                                             property.
       $= 4a - 3b + (-6b) + (-8a)$ Change subtraction to
                                             addition.
       $= -4a - 9b$ Combine like terms.

29. $3(-6x - 7) - 4(-2x + 6)$
       $= -18x - 21 + 8x - 24$ Multiply using the
                                             distributive property.
       $= -10x - 45$ Combine like terms.

Evaluate the following if a = -2 and b = 2

33. 2a + 3b - 5(a + 1)
     2(-2) + 3(2) - 5(-2 + 1)   Replace a with -2 and b
                                        with 2.
     = 2(-2) + 3(2) - 5(-1)   Perform the indicated
     = -4 + 6 + 5   operations.
     = 7

37.   4a - 3b - 2(a + b)
     = 4(-2) - 3(2) - 2(-2 + 2)   Replace a with -2 and b
     = 4(-2) - 3(2) - 2(0)   with 2. Perform the
     = 4(-2) - 3(2) - 0   indicated operations.
     = -8 - 6
     = -14

C

Perform the indicated operations:

41.   -2(8 - 12) - 3[2(8 - 4)] - 3
     = -2(-4) - 3[2(4)] - 3   Do indicated operations
                                        within parentheses first.
     = -2(-4) - 3(8) - 3   Then work within the
                                        brackets.
     = 8 - 24 - 3   Simplify.
     = -19

45.   -5 + 3(-2)^2 - [(4 - 5)^2 - 8(-2 - 4)]
     = -5 + 3(-2)^2 - [(-1)^2 - 8(-6)]   Perform operations within
                                        parentheses. Then do
     = -5 + 3(-2)^2 - [1 - 8(-6)]   exponentiation within the
                                        brackets.
     = -5 + 3(-2)^2 - [1 + 48]   Work within the brackets.
     = -5 + 3(-2)^2 - 49
     = -5 + 3(4) - 49   Do the exponentiation
     = -5 + 12 - 49   before multiplication.
     = 7 - 49   Simplify.
     = -42

Simplify:

49.   9(x - 2) - 2[3(2x - 1) - 3(-2x + 3)]
     = 9(x - 2) - [6x - 3 + 6x - 9]   Do indicated operations
     = 9(x - 2) - 2[12x - 12]   within brackets.
     = 9x - 18 - 24x + 24   Multiply using the
                                        distributive property.
     = -15x + 6   Combine like terms.

Evaluate each of the following if x = -5 and y = 6.

53.   9(x - 2y) - [3(2x - y) - 6(-3x + 2y)]
            9[(-5) - 2(6)] - [3(2·-5 - 6) - 6(-3·-5 + 2·6)]

                                     Replace each x with -5 and each y with 6.

    = 9(-5 - 12) - [3(-10 - 6) - 6(15 + 12)]   Work within each
    = 9(-17) - [3(-16) - 6(27)]   set of grouping
    = 9(-17) - (-48 - 162)   symbols.
    = -153 - (-210)
    = -153 + 210
    = 57

57.   (2x - y)(x + y)(-3x - 4y)
            (2·-5 - 6)(-5 + 6)(-3·-5 - 4·6)   Replace each x with -5
    = (-10 - 6)(1)(15 - 24)   and each y with 6. Do
    = (-16)(-9)   the indicated operations
    = 144   within each grouping
                                      symbol. Then multiply.

D

61.   Use the formula $F = \frac{9}{5}C + 32$ to convert 50°C to degrees Fahrenheit.

$$F = \frac{9}{5}C + 32 \quad\quad\quad\quad \text{Formula.}$$

$$F = \frac{9}{5}(50) + 32 \quad\quad\quad \text{Replace C with 50.}$$

$$F = \frac{9}{5}\overset{10}{\underset{1}{(50)}} + 32 \quad\quad\quad \text{Reduce and multiply.}$$

F = 90 + 32
F = 122
So 50°C is the same as 122°F.

65.   Find the total price of a TV if the down payment is $50 and there are 24 monthly payments of $29.18.

T = D + pm   Formula. T is the total price, D is the down payment, p is the payment per month, and m is the number of months of pay.

T = 50 + (29.18)(24)   Replace D with 50, p with 29.18, and m with 24.

T = 50 + 700.32   Simplify.
T = 750.32

The total price of the TV is $750.32.

69. How far does an object fall after three seconds?

$s = 16t^2$ Formula, where s is distance in feet and t represents time.

$s = 16(3)^2$ Replace t with 3.

$s = 16(9)$ Simplify.

$s = 144$

The object falls 144 ft in 3 seconds.

CHALLENGE EXERCISES

73. Is $4 - 8x = (4 - 8)x$ ever true? If the answer is yes, give an example.

$4 - 8x = (4 - 8)x$ Yes, sometimes true.

$4 - 8 \cdot 1 = (4 - 8)1$ This statement is true if

$4 - 8 \cdot 1 = (-4)1$ x is replaced by 1.

$4 - 8 = -4$ Multiplication property

$-4 = -4$ of one.

EXERCISES 1.7   SOLVING LINEAR EQUATIONS IN ONE VARIABLE

A

Solve:

1. $x + 2 = 5$

$x + 2 - 2 = 5 - 2$ Subtract 2 from both

$x = 3$ sides.

The solution set is {3}.

5. $-2x = -4$

$\dfrac{-2x}{-2} = \dfrac{-4}{-2}$ Divide both sides by -2.

$x = 2$

The solution set is {2}.

9. $-x + 1 = 3$

$-x + 1 - 1 = 3 - 1$ Subtract 1 from both

$-x = 2$ sides.

$\dfrac{-x}{-1} = \dfrac{2}{-1}$ Divide both sides by -1.

$x = -2$

The solution set is {-2}.

13.
$$-x + 15 = 36$$
$$-x + 15 - 15 = 36 - 15$$ Subtract 15 from both sides.
$$-x = 21$$
$$\frac{-x}{-1} = \frac{21}{-1}$$ Divide both sides by -1.
$$x = -21$$

The solution set is {-21}.

17.
$$x = x + 1$$
$$x - x = x - x + 1$$ Subtract x from both sides.
$$0 = 1$$ This equation is a contradiction.

The solution set is $\emptyset$.

B

Solve:

21. $2x + 3 = 3 + 2x$
$$2x + 3 = 2x + 3$$ Commutative property of addition. This equation is an identity.

The solution set is $\mathbb{R}$.

25.
$$2x + 7 = 5x + 4 - 3x$$
$$2x + 7 = 2x + 4$$ Combine like terms on the right side of the equal sign.
$$2x - 2x + 7 = 2x - 2x + 4$$ Subtract 2x from both sides.
$$7 = 4$$ This equation is a contradiction.

The solution set is $\emptyset$.

29.
$$5b + 3(b - 6) = 35$$
$$5b + 3b - 18 = 35$$ Distributive property.
$$8b - 18 = 35$$ Combine terms.
$$8b - 18 + 18 = 35 + 18$$ Add 18 to both sides.
$$8b = 53$$
$$\frac{8b}{8} = \frac{53}{8}$$ Divide both sides by 8.
$$b = \frac{53}{8}$$

The solution set is $\left\{\frac{53}{8}\right\}$.

33.
$$4x - 43 = 9x - 40$$
$$4x - 43 + 43 = 9x - 40 + 43 \qquad \text{Add 43 to both sides.}$$
$$4x = 9x + 3$$
$$4x - 9x = 9x - 9x + 3 \qquad \text{Subtract 9x from both}$$
$$-5x = 3 \qquad \text{sides.}$$
$$\frac{-5x}{-5} = \frac{3}{-5} \qquad \text{Divide both sides by -5.}$$
$$x = -\frac{3}{5}$$

The solution set is $\left\{-\frac{3}{5}\right\}$.

37.
$$5x + 6 - (10 - 2x) = (7x + 6) - (6x - 2)$$
$$5x + 6 - 10 + 2x = 7x + 6 - 6x + 2 \qquad \text{Clear parentheses,}$$
$$\qquad \qquad \qquad \qquad \qquad \qquad \qquad \qquad -(x - 1) = -x + 1.$$
$$7x - 4 = x + 8 \qquad \text{Combine like terms.}$$
$$7x - x - 4 = x - x + 8 \qquad \text{Subtract x from}$$
$$6x - 4 = 8 \qquad \text{both sides.}$$
$$6x - 4 + 4 = 8 + 4 \qquad \text{Add 4 to both sides.}$$
$$6x = 12$$
$$\frac{6x}{6} = \frac{12}{6} \qquad \text{Divide both sides}$$
$$\qquad \qquad \qquad \text{by 6.}$$
$$x = 2$$
The solution set is $\{2\}$.

C

Solve:

41.
$$2x - [3x - (8x + 2) + 3] + 6 = 0$$
$$2x - [3x - 8x - 2 + 3] + 6 = 0 \qquad \text{Remove grouping}$$
$$\qquad \qquad \qquad \qquad \qquad \qquad \qquad \text{symbols by}$$
$$\qquad \qquad \qquad \qquad \qquad \qquad \qquad \text{subtraction.}$$

$$2x - [-5x + 1] + 6 = 0$$
$$2x + 5x - 1 + 6 = 0$$
$$7x + 5 = 0 \qquad \text{Combine like terms.}$$
$$7x + 5 - 5 = 0 - 5 \qquad \text{Subtract 5 from}$$
$$7x = -5 \qquad \text{both sides.}$$
$$\frac{7x}{7} = -\frac{5}{7} \qquad \text{Divide both sides}$$
$$\qquad \qquad \qquad \text{by 7.}$$
$$x = -\frac{5}{7}$$

The solution set is $\left\{-\frac{5}{7}\right\}$.

45. $x - \{3x - [2x - (8 - x) + 2] + 3] - 6 = 0$
   $x - \{3x - [2x - 8 + x + 2] + 3\} - 6 = 0$ Remove grouping
   $x - 3x - [3x - 6] + 3\} - 6 = 0$ symbols by
   subtraction.

   $x - \{3x - 3x + 6 + 3\} - 6 = 0$
   $x - \{9\} - 6 = 0$
   $x - 9 - 6 = 0$
   $x - 15 = 0$
   $x - 15 + 15 = 0 + 15$ Add 15 to
   $x = 15$ both sides.

The solution set is $\{15\}$.

49. $6x - 8(2x + 4) = 8x - 12(6 - x)$
   $6x - 16x - 32 = 8x - 72 + 12x$ Multiply on both sides
   using the distributive
   property.
   $-10x - 32 = 20x - 72$ Combine like terms on the
   left and on the right.
   $-10x - 20x - 32 = 20x - 20x - 72$ Subtract 20x from both
   $-30x - 32 = -72$ sides.
   $-30x - 32 + 32 = -72 + 32$ Add 32 to both sides.
   $-30x = -40$
   $\dfrac{-30x}{-30} = \dfrac{-40}{-30}$ Divide both sides by -30.

   $x = \dfrac{4}{3}$ Simplify.

The solution set is $\left\{\dfrac{4}{3}\right\}$.

53. $-[3(x - 2) + 6] - [-2(x - 1) + 5] = -2[(x + 4) - 1]$
   $-[3x - 6 + 6] - [-2x + 2 + 5] = -2[x + 4 - 1]$

   Use distributive property
   within brackets on left.
   $-[3x] - [-2x + 7] = -2[x + 3]$ Combine terms within
   brackets.
   $-3x + 2x - 7 = -2[x + 3]$ Remove brackets on left by
   subtracting.
   $-3x + 2x - 7 = -2x - 6$ Distribute on the right.
   $-x - 7 = -2x - 6$ Combine on the left.
   $x - 7 = -6$ Add 2x to both sidies.
   $x = 1$ Add 7 to both sides.
The solution set is $\{1\}$.

57. $2(x - 3) - 3(2x - 7) = 4[3(x + 1) - 2]$
    $2x - 6 - 6x + 21 = 4[3x + 3 - 2]$

	Remove parentheses by using the distributive property.
$-4x + 15 = 4[3x + 1]$	Combine terms.
$-4x + 15 = 12x + 4$	Multiply on the right.
$-16x + 15 = 4$	Subtract 12x from both sides.
$-16x = -11$	Subtract 15 from both sides.
$\dfrac{-16x}{-16} = \dfrac{-11}{-16}$	Divide both sides by -16.
$x = \dfrac{11}{16}$	Simplify.

The solution set is $\left(\dfrac{11}{16}\right)$.

D

61. The cost of advertising for a minute on television is $250 more than 3 times the cost for a minute on the radio. If an advertiser spends $1000 for one minute of television time and two minutes of radio time, what is the cost per minute of each?

Simpler word form:
$$2 \binom{\text{cost per minute}}{\text{on radio}} + \binom{\text{cost per minute}}{\text{on television}} = 1000$$

Select variable:
Let x = cost of one minute of radio.
Then 3x + 250 = cost of one minute on television.

Translate to algebra:
$2x + (3x + 250) = 1000$
$\qquad 5x + 250 = 1000$
$\qquad\qquad 5x = 750$
$\qquad\qquad\ x = 150$

If x = 150, then
$3x + 250 = 3(150) + 250 = 700$

Answer: It costs $150 per minute on radio and $700 per minute on television.

65. The cost of a television set is $50 more than twice the cost of a microwave oven. If the total cost for both is $650, What is the cost of the microwave oven?

Simpler word form:

$$\left(\begin{array}{c}\text{cost of}\\\text{microwave oven}\end{array}\right) + \left(\begin{array}{c}\text{cost of}\\\text{television set}\end{array}\right) = 650$$

Select variable:
Let x = cost of microwave oven.
Then 2x + 50 = cost of television set.

Translate to algebra:

$$x + (2x + 50) = 650$$
$$3x + 50 = 650$$
$$3x = 600$$
$$x = 200$$

Answer: The cost of the microwave oven is $200.

69. One number is two more than three times another. If the sum of the two numbers is 22, find the two numbers.

Simpler word form:

$$\left(\begin{array}{c}\text{first}\\\text{number}\end{array}\right) + \left(\begin{array}{c}\text{second}\\\text{number}\end{array}\right) = 22$$

Select variable:
Let x represent the first number.
Then 3x + 2 represents the second number.

Translate to algebra:

$$x + (3x + 2) = 22$$
$$4x + 2 = 22$$
$$4x = 20$$
$$x = 5$$

If x = 5, then
3x + 2 = 3(5) + 2 = 15 + 2 = 17.

Answer: The two numbers are 5 and 17.

CHALLENGE EXERCISES

73. Given the equation 2x - 3b = 0, find a value for b so that when solving for x, the solution set is {-9}.

$2x - 3b = 0$	Original equation.
$2(-9) - 3b = 0$	Since x = -9, replace x
$-18 - 3b = 0$	with -9 and solve for b.
$-3b = 18$	
$b = -6$	

So, when x = -9, b must equal -6.

EXERCISES 1.8   LITERAL EQUATIONS AND FORMULAS

A

Solve for the indicated variable:

1. $d = rt$; t

   $d = rt$ Original formula.

   $\dfrac{d}{r} = \dfrac{rt}{r}$ Divide both sides by r.

   $\dfrac{d}{r} = t$

   or $t = \dfrac{d}{r}$ Symmetric property of equality.

5. $V = \pi r^2 h$; h

   $V = \pi r^2 h$ Original formula.

   $\dfrac{V}{\pi r^2} = \dfrac{\pi r^2 h}{\pi r^2}$ Divide both sides by πr^2.

   $\dfrac{V}{\pi r^2} = h$

   or $h = \dfrac{V}{\pi r^2}$ Symmetric property.

9. $F = \dfrac{GMm}{r^2}$; m

   $F = \dfrac{GMm}{r^2}$ Original formula.

   $r^2 (F) = r^2 \left(\dfrac{GMm}{r^2}\right)$ Multiply both sides by r^2 to eliminate the fraction.

   $Fr^2 = GMm$

   $\dfrac{Fr^2}{GM} = \dfrac{GMm}{GM}$ Divide both sides by GM to isolate m.

   $\dfrac{Fr^2}{GM} = m$

   or $m = \dfrac{Fr^2}{GM}$

13. $y = mx - 3$; x

   $y = mx - 3$ Original formula.

   $y + 3 = mx$ Add 3 to both sides.

   $\dfrac{y + 3}{m} = \dfrac{mx}{m}$ Divide both sides by m to isolate x.

   $\dfrac{y + 3}{m} = x$

   or $x = \dfrac{y + 3}{m}$

17. $2s - 3t = 8$; t

$2s - 3t = 8$ Original formula.

$\quad -3t = -2s + 8$ Subtract 2s from both sides.

$\quad\quad t = \dfrac{-2s + 8}{-3}$ Divide both sides by -3.

$\text{or } t = \dfrac{(-1)(-2s + 8)}{(-1)(-3)}$ Clear the negative sign in the denominator by multiplying both numerator and denominator by -1.

$\quad\quad = \dfrac{2s - 8}{3}$

B

21. $ax - 5y = 2$; x

$ax - 5y = 2$ Original formula.

$\quad\quad ax = 5y + 2$ Add 5y to both sides.

$\quad\quad x = \dfrac{5y + 2}{a}$ Divide both sides by a.

25. $PV = nRT$; n

$PV = nRT$ Original formula.

$\dfrac{PV}{RT} = n$ Divide both sides by RT.

29. $A = \dfrac{B - C}{n}$; B

$An = B - C$ Eliminate the fraction by multiplying both sides by n.

$An + C = B$ Add C to both sides.

33. $L = f + (n - 1)d$; d

$L = f + (n - 1)d$

$L - f = (n - 1)d$ Subtract f from both sides.

$\dfrac{L - f}{n - 1} = d$ Divide both sides by n - 1.

37. $A(P - 3t) = d$; P

$A(P - 3t) = d$

$AP - 3At = d$ Multiply by A using the distributive property.

$\quad\quad AP = 3At + d$ Add 3At to both sides.

$\quad\quad P = \dfrac{3At + d}{A}$ Divide both sides by A.

41.

$$P = 2L + 2W; \ L$$

$2L + 2W = P$	Symmetric property.
$2L = P - 2W$	Subtract 2W from both sides.
$L = \dfrac{P - 2W}{2}$	Divide both sides by 2.

45.

$$s = \tfrac{1}{2}gt^2 + vt + x; \ g$$

$\tfrac{1}{2}gt^2 + vt + x = s$	Symmetric property.
$gt^2 + 2vt + 2x = 2s$	Clear the fraction by multiplying by 2 on both sides.
$gt^2 = 2s - 2vt - 2x$	Subtract 2vt and 2x from both sides.
$g = \dfrac{2s - 2vt - 2x}{t^2}$	Divide both sides by t^2.

49.

$$y = \tfrac{4}{5}x - 5; \ x$$

$$y = \tfrac{4}{5}x - 5$$

$\tfrac{4}{5}x - 5 = y$	Symmetric property of equality.
$\tfrac{4}{5}x = y + 5$	Add 5 to both sides.
$5\left(\tfrac{4}{5}x\right) = 5(y + 5)$	Clear the fraction by multiplying by 5.
$4x = 5y + 25$	Simplify.
$x = \dfrac{5y + 25}{4}$	Divide both sides by 4.

53.

$$2A = h(b + B); \ B$$

$h(b + B) = 2A$	
$bh + Bh = 2A$	Multiply on the left.
$Bh = 2A - bh$	Subtract bh from both sides.
$B = \dfrac{2A - bh}{h}$	Divide both sides by h.

57.

$$S = \tfrac{1}{2}(a + b) - \tfrac{1}{2}(b - c); \ a$$

$2(S) = 2\left(\tfrac{1}{2}\right)(a + b) - 2\left(\tfrac{1}{2}\right)(b - c)$	Clear fractions; multiply by 2.
$2S = (a + b) - (b - c)$	
$2S = a + b - b + c$	Subtract on the right.
$2S = a + c$	Combine like terms.
$2S - c = a$	Subtract c from both sides.
or $a = 2S - c$	Symmetric property.

61.   What is a literal equation?   A literal equation is an
                                           equation that states a
                                           rule and has more than one
                                           variable.

CHALLENGE EXERCISES

65.   Using the formulas below, what Kelvin temperature
      corresponds to -40°F?

      $C = \frac{5}{9}(F - 32)$ $K = C + 273.15$

      $C = \frac{5}{9}(-40 - 32)$ Change Fahrenheit measure
                                           to Centigrade measure.

      $C = \frac{5}{9}(-72)$

      $C = -40$
      $K = C + 273.15$ Change Centigrade measure
      $K = -40 + 273.15$ to Kelvin measure.
      $K = 233.15$
      So, -40°F is equal to 233.15 K.

EXERCISES 1.9   SOLVING LINEAR INEQUALITIES IN ONE VARIABLE

A

Solve and graph the solution set:

 1. $x < -5$ The solution set contains
                                           all real numbers less than
                                           -5.
      The solution set is $\{x \mid x < -5\}$.

 5. $-y \leq 2$

      $-1(-y) \geq -1(2)$ Multiply both sides by -1.

           $y \geq -2$ The inequality sign is
                                           reversed.
      The solution set is $\{y \mid y \geq -2\}$.

9. $-x + 4 > 0$
    $-x > -4$
    $-1(-x) < -1(-4)$
    $x < 4$

Subtract 4 from both sides.
Multiply both sides by -1. The inequality sign is reversed.

The solution set is $\{x|x < 4\}$.

Solve and write the solution in interval notation:

13. $2x < 8$
     $x < 4$
     $(-\infty, 4)$

Divide both sides by 2.
Open interval notation.

Solve and write the solution in set builder notation:

17. $2x + 3 < 13$
     $2x < 10$
     $x < 5$

Subtract 3 from both sides.
Divide both sides by 2.
Set builder notation.

The solution set is $\{x|x < 5\}$.

B

21. $-3x + 6 \geq 12$
     $-3x \geq 6$
     $x \leq -2$

Subtract 6 from both sides.
Divide both sides by -3. The inequality sign is reversed.

The solution set is $\{x|x \leq -2\}$.

Solve and graph the solution set:

25. $4x + 2 \leq 3x + 2$
     $4x \leq 3x$
     $x \leq 0$

Subtract 2 from both sides.
Subtract 3x from both sides.

Solve and write the solution set in set builder notation:

29. $-5(2x + 3) \geq 4(x - 2)$
    $-10x - 15 \geq 4x - 8$ Multiply.
    $-14x - 15 \geq -8$ Subtract 4x from both sides.
    $-14x \geq 7$ Add 15 to both sides.
    $x \leq -\frac{1}{2}$ Divide both sides by -14, reversing the inequality sign.

The solution set is $\left\{ x \mid x \leq -\frac{1}{2} \right\}$.

33. $-3x + 6 < 5x - 9$
    $-3x < 5x - 15$ Subtract 6 from both sides.
    $-8x < -15$ Subtract 5x from both sides.
    $x > \frac{15}{8}$ Divide both sides by -8. Remember to reverse the inequality sign.

The solution set is $\left\{ x \mid x > \frac{15}{8} \right\}$.

Solve and write the solution in interval notation:

37. $5 - (2x + 7) \geq -1$
    $5 - 2x - 7 \geq -1$ Subtract by adding the opposite.
    $-2x - 2 \geq -1$ Combine like terms on the left.
    $-2x \geq 1$ Add 2 to both sides.
    $x \leq -\frac{1}{2}$ Divide both sides by -2. Reverse the inequality sign.
    $\left(-\infty, -\frac{1}{2} \right]$ Half-open interval notation.

c

Solve and graph the solution set:

41. $5x - (3x + 2) \geq x - 9$
    $5x - 3x - 2 \geq x - 9$ Simplify on the left.
    $2x - 2 \geq x - 9$
    $x - 2 \geq -9$ Subtract x from both sides.
    $x \geq -7$ Add 2 to both sides.

45. $4x - [5 - (2x - 3)] - 6(x + 5) \geq 10$
     $4x - [5 - 2x + 3] - 6x - 30 \geq 10$ Clear parentheses.
         $4x - 5 + 2x - 3 - 6x - 30 \geq 10$ Clear brackets and
                                                    combine like terms.
                              $-38 \geq 10$ This is a contradic-
                                                    tion, since -38 is
                                                    not to the right of
                                                    10 on the number
                                                    line.

Since the equation is a contradiction, there is no solution
to be graphed.

Solve and write the solution set in interval notation:

49. $-12 \leq 5x - 3 < 22$

       $-9 \leq 5x < 25$ Add 3 to each member.

       $-\dfrac{9}{5} \leq x < 5$ Divide each member by 5.

       $\left[-\dfrac{9}{5},\ 5\right)$ The solution set is the
                                                    set of real number greater
                                                    than or equal to $-\dfrac{9}{5}$ and
                                                    less than 5.

53. $25 \leq 7x + 3 \leq 66$

       $22 \leq 7x \leq 63$ Subtract 3 from each
                                                    member of the inequality.

       $\dfrac{22}{7} \leq x \leq 9$ Divide each member by 7.

       $\left[\dfrac{22}{7},\ 9\right]$ Closed interval notation.

Solve and write the solution in set builder notation.

57. $3x - 9 < 6$ or $2x + 1 > 13$
          $3x < 15$ or $2x > 12$ Solve each inequality
           $x < 5$ or $x > 6$ independently.
     The solution set is $\{x|x < 5\} \cup \{x|x > 6\}$.

31

61. If a bus company charges $20 plus 12 cents per mile for a ticket, could a trip of 800 miles be taken if the person had at most $120 to spend?

Simpler word form:
Cost of trip $\leq$ $120.

Select variable:
Let x represent the number of miles. The cost of the trip is then $20 + $0.12x.

Translate to algebra:

$$20 + 0.12x \leq 120$$
$$20 + 0.12(800) \leq 120 \qquad \text{Replace x with 800.}$$
$$20 + 96 \leq 120$$
$$116 \leq 120$$

Answer: Yes, since $116 is less than $120, the trip can be taken.

65. Jack can spend, at most, $55 on paint. If a gallon of paint costs $7.50 per can, what is the maximum number of cans of paint Jack can purchase?

Simpler word form:
Cost of paint $\leq$ $55.

Select variable:
Let x represent the number of cans of paint that can be bought. Then $7.50x represents the cost of the paint.

Translate to algebra:

$$7.50x \leq 55$$
$$x \leq \frac{55}{7.50} \qquad \text{Divide both sides by 7.50.}$$
$$x \leq 7.\overline{3}$$

Answer: Jack can buy at most 7 cans of paint.

69. It costs \$3.50 for the first pound and \$1.75 per pound after the first pound for a messenger to deliver a package. Will Charlie be able to get his 5-lb package delivered if he has only \$10.00?

Simpler word form:
Cost of delivery $\leq$ \$10.

Select variable:
Let x represent the number of pounds after the first pound. Then \$3.50 + \$1.75x represents the total cost of delivery.

Translate to algebra:

$$3.50 + 1.75x \leq 10$$
$$3.50 + 1.75(5) \leq 10 \qquad \text{Replace x with 5.}$$
$$3.50 + 8.75 \leq 10$$
$$12.25 \leq 10 \qquad \text{Contradiction.}$$

error
use 4

Answer: No, Charlie needs more money to get his package delivered.

CHALLENGE EXERCISES

73. If 9 < x < 25, what statement is true about -x?

Since x and -x are opposites and -x has a factor of -1, the inequality signs would be reversed and the constants would also have a factor of -1. So -9 > -x > -25, but this is the same as -25 < -x < -9.

EXERCISES 1.10   WORD PROBLEMS

1.   The length of a rectangle is 4 times its width. The
     perimeter is 78 feet. Find the length and width.

     The problem describes the   Step 1.
     relation between length and
     width of the rectangle. The
     perimeter is given. The
     question asks for the length
     and width.

     Make a drawing.   Step 2.
                ℓ

     w [rectangle]

     Select variable: Step 3.
     Let w represent the width.
     Then the length can be
     represented by 4w.

     Formula: Step 4.
      P = 2ℓ + 2w
     78 = 2(4w) + 2w   Substitute 78 for P, and
                                             4w for ℓ.

     Solve: Step 5.
      78 = 8w + 2w
      78 = 10w
     7.8 = w

     Answer: Step 6.
     The width is 7.8 ft and the
     length is 4(7.8) or 31.2 ft.

     Check: Step 7.
     78 = 2(31.2) + 2(7.8)
     78 = 62.4 + 15.6
     78 = 78
     The answer is correct. The width is 7.8 ft and the length
     is 31.2 ft.

5. Two cars leave Austin traveling in opposite directions. If their average speeds are 42 mph and 54 mph, how long will it be until they are 312 miles apart?

The problem states the speed of two cars and the distance between them. The question asks for the <u>number</u> <u>of</u> <u>hours</u> they travel.

Step 1.

Make a table.

Step 2.

	RATE (speed)	TIME	DISTANCE
Car 1	42		
Car 2	54		

Simpler *word* *form*:

$$\begin{pmatrix} \text{Distance traveled} \\ \text{by Car 1} \end{pmatrix} + \begin{pmatrix} \text{Distance traveled} \\ \text{by Car 2} \end{pmatrix} = 312 \text{ miles}$$

Select *variable*:
Let t represent the number of hours until the cars are 312 miles apart.

Step 3.

	RATE	TIME	DISTANCE
Car 1	42	t	42t
Car 2	54	t	54t

d = rt

Translate *to* *algebra*:
42t + 54t = 312

Step 4.

Solve:

$$96t = 312$$
$$t = 3.25 \text{ or } 3\frac{1}{4}$$

Step 5.

Answer:
In 3.25 hours (or 3 hours and 15 minutes), the cars will be 312 miles apart.

Step 6.

Check:

$$42t + 54t = 312$$
$$42(3.25) + 54(3.25) = 312$$
$$136.5 + 175.5 = 312$$
$$312 = 312$$

Step 7.

The answer is correct.

9. Myrna's baby girl weighed 5.75 pounds at birth. At age 7 months she weighed 16.25 pounds. What was the percent of increase in the 7 months (to the nearest percent)?

The problem gives two weights. The question asks for a comparison (in percent) between the increase and the birth weight.

Step 1.
Note that the actual increase is 16.25 - 5.75 or 10.5 pounds.

Simpler word form:
What percent of 5.75 is 10.5?

Step 2.

Select a variable:
Let x represent the percent.

Step 3.

Translate to algebra:
x(5.75) = 10.5

Step 4.

Solve:
$x = \frac{10.5}{5.75} \approx 1.83$

Step 5.

Answer:
The percent of increase in the 7 months was 183%.

Step 6.

13. The first side of a triangle is 15 inches. The third side is 3 less than 2 times the second side. If the perimeter is 51 inches, find the lengths of the second and third sides.

The problem describes the relation between the three sides of a triangle. The perimeter is given, and the length of one side. The question asks for the lengths of the other two sides.

Step 1.

Simpler word form:

first side + second side + third side = 51 in.

Step 2.
P = a + b + c

Select variable:
Let x represent the length of the second side. Then 2x - 3 represents the length of the third side.

Step 3.

Translate to algebra: Step 4.
   (15) + (x) + (2x - 3) = 51

Solve: Step 5.
                    15 + 3x - 3 = 51
                        3x + 12 = 51
                             3x = 39
                              x = 13
and 2x - 3 = 2(13) - 3 = 23

Answer: Step 6.
The second side is 13 in.,
and the third side is 23 in.

CHAPTER 2

POLYNOMIALS AND RELATED EQUATIONS

EXERCISES 2.1 POSITIVE INTEGER EXPONENTS AND MULTIPLICATION
                OF MONOMIALS

A

Multiply:

1. $r^7 \cdot r^4 \cdot r = r^7 \cdot r^4 \cdot r^1 = r^{7+4+1}$
    $= r^{12}$ Add the exponents.

5. $(a^4)^4 = a^{4(4)} = a^{16}$ Multiply the exponents.

9. $(-3x^2 y^3)^2$
    $= (-3)^2 (x^2)^2 (y^3)^2$ Raise each factor in
                                                         parentheses to the second
                                                         power.
    $= 9x^{2 \cdot 2} y^{3 \cdot 2}$ Multiply the exponents on
                                                         the variables.
    $= 9x^4 y^6$ Simplify.

13. $(x^3 y)(xy^3)(x^2 y^2)$
     $= [x^3 \cdot x \cdot x^2][y \cdot y^3 \cdot y^2]$ Group the factors to be
                                                          simplified.
     $= [x^{3+1+2}][y^{1+3+2}]$ Add the exonents.
     $= x^6 y^6$ Simplify.

17. $(4t^4)^3 = (4)^3 (t^4)^3$ Raise each factor within
                                                          parentheses to the third
                                                          power.
     $= 64t^{4(3)}$ Multiply exponents on the
                                                          variable.
     $= 64t^{12}$

B

21. $(2x^{11})(-3x^{12})(5x^{15})$
     $= [2(-3)(5)][x^{11} \cdot x^{12} \cdot x^{15}]$ Group the factors.
     $= [-30][x^{11+12+15}]$ Add the exponents on the
                                                         variables.
     $= -30x^{38}$

25. $(-14p^2 q^3)^2$
     $= (-14)^2 (p^2)^2 (q^3)^2$ Raise each factor within
                                                         parentheses to the second
                                                         power.
     $= 196p^{2(2)} q^{3(2)}$ Multiply the exponents
                                                         involved with variables.
     $= 196p^4 q^6$

29.　$(t^2)^3 (t^3)^2 = t^{2(3)} \cdot t^{3(2)}$　　　Multiply the exponents.
　　　　　　　　　　$= t^6 \cdot t^6$
　　　　　　　　　　$= t^{12}$　　　　　　　　Add the exponents.

33.　　$(-2x)^2 (-2x)^3$
　　$= (-2)^2 (x)^2 (-2)^3 (x^3)$　　　Raise each factor within
　　　　　　　　　　　　　　　each parentheses to the
　　　　　　　　　　　　　　　indicated power.

　　$= (4)(x^2)(-8)(x^3)$
　　$= (4)(-8)(x^2)(x^3)$　　　Regroup factors.
　　$= -32x^{2+3}$　　　　　　　Add the exponents.
　　$= -32x^5$

37.　　$(xy)^2 (x^2 y^2)^2$
　　$= (x)^2 (y)^2 (x^2)^2 (y^2)^2$　　　Raise each factor within
　　　　　　　　　　　　　　　parenthese to the
　　　　　　　　　　　　　　　indicated power.
　　　　　　　　　　　　　　　Multiply the exponents.

　　$= x^2 y^2 x^{2(2)} y^{2(2)}$
　　$= x^2 y^2 x^4 y^4$
　　$= (x^2)(x^4)(y^2)(y^4)$　　　Regroup the factors.
　　$= x^{2+4} y^{2+4}$　　　　　Add the exponents.
　　$= x^6 y^6$

c

41.　　$(-2xy)^3 (4x^2 y^3)$
　　$= (-2)^3 (x)^3 (y)^3 (4x^2 y^3)$　　　Raise each factor in the
　　　　　　　　　　　　　　　first parentheses to the
　　　　　　　　　　　　　　　third power.
　　$= (-8)(4)(x^3 \cdot x^2)(y^3 \cdot y^3)$　　　Regroup factors.
　　$= -32x^{3+2} y^{3+3}$　　　　Add exponents.
　　$= -32x^5 y^6$

45.　　$(2t)^2 (-3t^3)^3 (-t^2)^4$
　　$= (2)^2 (t)^2 (-3)^3 (t^3)^3 (-t^2)^4$　　　Raise factors within
　　　　　　　　　　　　　　　parentheses to the
　　　　　　　　　　　　　　　indicated power.
　　$= (4)(t^2)(-27)(t^9)(t^8)$　　　Simplify.
　　$= (4)(-27)(t^{2+9+8})$　　　Regroup.
　　$= -108t^{19}$

49.　$(x + y)^2 (x + y)^4 = (x + y)^{2+4}$　　　Add the exponents of the
　　　　　　　　　　　　　　　factor, $(x + y)$.
　　　　　　　　$= (x + y)^6$

53.　　$(2t^{4n})(t)^{2n-2}$
　　$= (2)(t^{4n})(t^{2n-2})$　　　Regroup.
　　$= 2t^{4n+(2n-2)}$　　　Add the exponents on the
　　　　　　　　　　　　　variable t.
　　$= 2t^{6n-2}$

57. $(2x)^2 (x^{m+4}) (x^{2m-3})$

= $(2)^2 (x)^2 (x^{m+4}) (x^{2m-3})$ Raise the factors in the
                                        first parenthese to the
                                        second power.

= $4[x^2 (x^{m+4}) (x^{2m-3})]$ Regroup.

= $4x^{2+(m+4)+(2m-3)}$ Add the exponents.

= $4x^{3m+3}$

D

61. If a special cylinder has a height that is the fifth power
     of the radius ($h = r^5$), express the volume of the cylinder
     in terms of the radius. Use the formula $V = \pi r^2 h$.

     $V = \pi r^2 h$
     $V = \pi r^2 (r^5)$ To express the volume
                                       replace h with r^5.

     $V = \pi r^{2+5}$
     $V = \pi r^7$ The volume expressed in
                                       terms of r.

65. The area of a circle is $A = \pi r^2$, where r is the radius of
     the circle. The radius is half the diameter, $r = \frac{1}{2}d$. Find
     a formula for the area in terms of the diameter.

     $A = \pi r^2$
     $A = \pi \left(\frac{1}{2}d\right)^2$ To find the area in terms
                                       of the diameter, d,
     $A = \pi \left(\frac{1}{2}\right)^2 (d)^2$ replace r with $\frac{1}{2}d$.

     $A = \pi \left(\frac{1}{4}\right) d^2$

     $A = \frac{1}{4}\pi d^2$ or $\frac{\pi d^2}{4}$ The area expressed in
                                       terms of d.

69. The volume of a cone is given by the formula $V = \frac{1}{3}\pi r^2 h$, where r is the radius and h is the height. If the radius is three times the cube of the height, find an expression for the volume of the cone in terms of the height.

$V = \frac{1}{3}\pi r^2 h$	Original formula.
$V = \frac{1}{3}\pi (3h^3)^2 h$	To express the volume in terms of height, h, replace r with three times the cube of h, or $3h^3$.
$V = \frac{1}{3}\pi (3)^2 (h^3)^2 h$	
$V = \frac{1}{3}\pi (9)(h^6) h$	
$V = 3\pi h^7$	

CHALLENGE EXERCISES

73. $8^x \cdot 8^y \cdot 8^z = 8^{x+y+z}$ Add the exponents.

MAINTAIN YOUR SKILLS (Sections 1.3, 1.4)

77.
$(-18.4) + (-21.7) - (13.84) - (-21.52)$
$(-18.4) + (-21.7) + (-13.84) + (21.52)$ Subtract by adding the opposite.

$= -40.1 + (-13.84) + (21.52)$ Combine terms from left to right.
$= -53.94 + (21.52)$
$= -32.42$

81.
$-\frac{4}{5}x - \frac{3}{4}y + \frac{1}{2}x - \frac{1}{2}y$

$= -\frac{4}{5}x + \frac{1}{2}x - \frac{3}{4}y - \frac{1}{2}y$ Group like terms.

$= -\frac{8}{10}x + \frac{5}{10}x - \frac{6}{8}y - \frac{4}{8}y$

$= -\frac{3}{10}x - \frac{10}{8}y$ Add.

$= -\frac{3}{10}x - \frac{5}{4}y$ Reduce.

A

Simlify, using only positive exponents:

1. $8^{-2} = \dfrac{1}{8^2}$ The negative exponent denotes a reciprocal.

        $= \dfrac{1}{64}$

5. $a^{-1}b^2 = \left(\dfrac{1}{a^1}\right)b^2$ The negative exponent on a denotes the reciprocal of a^1.

        $= \dfrac{b^2}{a}$

9. $\dfrac{a^{-3}b}{a^{-7}b^{-2}}$

    $= \dfrac{a^{-3}}{a^{-7}} \cdot \dfrac{b}{b^{-2}}$ Divide the monomials with the same base.

    $= a^{-3-(-7)} \cdot b^{1-(-2)}$ Divide the monomials by subtracting the exponents of the variables.

    $= a^4 b^3$

13. $(a^{-1}b^2)^{-2}$

    $= (a^{-1})^{-2}(b^2)^{-2}$ Raise each factor within parentheses to the indicated power, -2. Multiply the exponents.

    $= a^2 b^{-4}$

    $= \dfrac{a^2}{b^4}$ Definition of negative exponents.

17. $\dfrac{x^{-1}y^{-1}}{w^{-1}} = \dfrac{1}{xy} \cdot w$ $y^{-m} = \dfrac{1}{y^m}$ and

         $= \dfrac{w}{xy}$ $\dfrac{1}{y^{-m}} = y^m$.

B

21. $(2x^{-3}y^2)(-6x^2y^{-4})$

    $= (2)(-6)[x^{-3} \cdot x^2][y^2 \cdot y^{-4}]$ Group like factors.

    $= -12 \, x^{-3+2} \cdot y^{2+(-4)}$ Add exponents.

    $= -12x^{-1}y^{-2}$

    $= \dfrac{-12}{xy^2}$ Definition of negative exponents.

25. $(2x^2y^{-3})(-3x^{-2}y)^{-3}$

    $= (2x^2y^{-3})(-3)^{-3}(x^{-2})^{-3}(y)^{-3}$ Raise each factor within the second parenthese to the indicated power, -3. Multiply the exponents.

    $= (2x^2y^{-3})(-3)^{-3}(x^6)(y^{-3})$

    $= (2)(-3)^{-3}(x^2 \cdot x^6)(y^{-3} \cdot y^{-3})$ Regroup like factors.

    $= \dfrac{2}{-27} \cdot x^8 \cdot y^{-6}$ Apply the negative exponent to the constant, and add the exponents of the variables.

    $= -\dfrac{2x^8}{27y^6}$ Definition of negative exponents.

29. $\left(\dfrac{m^{-1}}{n}\right)^{-1} = \dfrac{(m^{-1})^{-1}}{(n)^{-1}}$ Raise monomials within parentheses to the indicated power.

    $= \dfrac{m^1}{n^{-1}}$ Multiply exponents.

    $= mn$ Definition of negative exponents.

33. $\left(\dfrac{a^5}{b^2}\right)^{-4} = \left(\dfrac{b^2}{a^5}\right)^4$ Definition of negative exponents.

    $= \dfrac{(b^2)^4}{(a^5)^4}$ Raise monomials to the indicated power.

    $= \dfrac{b^8}{a^{20}}$ Multiply the exponents.

37. $\left(\dfrac{x^2}{3y^{-3}}\right)^2 = \dfrac{(x^2)^2}{(3)^2(y^{-3})^2}$ Raise each factor within parenthese to the second power.

$= \dfrac{x^4}{9y^{-6}}$ Multiply exponents.

$= \dfrac{x^4 y^6}{9}$ Definition of negative exponents.

c

41. $(2x^2)^{-1}(2x^2)^{-2}(2x^2)^{-3}$

$= (2x^2)^{-1+(-2)+(-3)}$ Since the factor is the same in all three terms, $2x^2$, add exponents.

$= (2x^2)^{-6}$

$= \dfrac{1}{(2x^2)^6}$

$= \dfrac{1}{(2)^6(x^2)^6}$

$= \dfrac{1}{64x^{12}}$

45. $\dfrac{(5x^2 y^{-3})^2}{(10x^{-2}y^3)^{-2}} = (5x^2 y^{-3})^2(10x^{-2}y^3)^2$ Definition of negative exponents.

$= (5)^2(x^2)^2(y^{-3})^2(10)^2(x^{-2})^2(y^3)^2$ Raise each factor to the indicated power. Multiply exponents and group like variables. Add exponents.

$= (25)(100)(x^4 \cdot x^{-4})(y^{-6} \cdot y^6)$

$= 2500x^{4+(-4)}y^{(-6)+6}$

$= 2500x^0 y^0$ Simplify.

$= 2500$ x^0 and y^0 are equal to 1.

49.

$$\left(\frac{x^2 y^{-3}}{2^{-1} x^{-1}}\right)^{-1} \left(\frac{x^{-3} y^2}{3y^{-1}}\right)^{-2}$$

$$= \left(\frac{2^{-1} x^{-1}}{x^2 y^{-3}}\right)^1 \left(\frac{3y^{-1}}{x^{-3} y^2}\right)^2$$ Definition of negative exponents.

$$= \left(\frac{2^{-1} x^{-1}}{x^2 y^{-3}}\right) \left[\frac{(3)^2 y^{-2}}{(x^{-3})^2 (y^2)^2}\right]$$ Raise each factor to the indicated power.

$$= \frac{2^{-1} x^{-1}}{x^2 y^{-3}} \cdot \frac{9y^{-2}}{x^{-6} y^4}$$ Multiply exponents.

$$= \frac{y^3}{2x \cdot x^2} \cdot \frac{9x^6}{y^2 y^4}$$ Definition of negative exponents.

$$= \frac{9x^6 y^3}{2x^3 y^6}$$ Simplify. Add exponents.

$$= \frac{9x^{6-(3)} y^{3-(6)}}{2}$$ Divide by subtracting the exponents.

$$= \frac{9x^3 y^{-3}}{2}$$

$$= \frac{9x^3}{2y^3}$$ Definition of negative exponents.

53. $(x + a)^{-2} (x + a)^4 (x + a)$

     $= (x + a)^{-2+4+1}$ Add the exonents of the factor, x + a.

     $= (x + a)^3$

57. $\dfrac{(2r - 3)^5}{(2r - 3)^{-2}} = (2r - 3)^{5-(-2)}$ Divide by subtracting the exponents of the factor, 2r - 3.

               $= (2r - 3)^7$

D

61. The pH of a solution with a hydrogen ion concentration of 4.6×10^{-1} is approximately 0.3. Write the hydrogen ion concentration in decimal form.

$$4.6 \times 10^{-1} = 4.6 \times \frac{1}{10}$$ Definition of negative exponents.

$$= \frac{4.6}{10} = 0.46$$ Divide.

65. The pH of a solution with a hydrogen ion concentration of 7.3×10^{-3} is approximately 2.1. Write the hydrogen ion concentration in decimal form.

$$7.3 \times 10^{-3} = 7.3 \times \frac{1}{10^3}$$ Definition of negative exponents.

$$= \frac{7.3}{1000}$$

$$= 0.0073$$ Divide.

STATE YOUR UNDERSTANDING

69. Explain the difference between $(xy)^2$ and $(x + y)^2$.

$(xy)^2$
$= (xy)(xy)$
$= x^2 y^2$

This expression has two factors within parentheses. Raise each factor to the second power.

$(x + y)^2$
$= (x + y)(x + y)$

This expression has only one factor within parentheses, the sum $x + y$.

MAINTAIN YOUR SKILLS (Sections 1.4, 1.7, 1.9, 1.10)

Solve:

73. $3(x - 2) - 4 \geq x - (x + 3)$
$3x - 6 - 4 \geq x - x - 3$ Clear parentheses.
$3x - 10 \geq -3$ Combine like terms.
$3x \geq 7$ Add 10 to both sides.
$x \geq \frac{7}{3}$ Divide both sides by 3.

The solution set is $\left\{ x \mid x \geq \frac{7}{3} \right\}$ or $\left[\frac{7}{3}, +\infty \right)$ in half-open interval notation.

Add or subtract as indicated:

77. $5y - \{4y - 5[6 - 7(y - 8)] - 9\} - 10$
$= 5y - \{4y - 5[6 - 7y + 56] - 9\} - 10$ Clear inmost
$= 5y - \{4y - 5[-7y + 62] - 9\} - 10$ grouping symbols
$= 5y - \{4y + 35y - 310 - 9\} - 10$ first, combining
$= 5y - \{39y - 319\} - 10$ like terms when
$= 5y - 39y + 319 - 10$ possible.
$= -34y + 309$

EXERCISES 2.3   SCIENTIFIC NOTATION

A

Write in scientific notation:

1. 50,000
    5.0 is between 1 and 10   Move the decimal point
    5.0 times 10000 or 10^4 = 50,000   left four places.
    50,000 = 5.0 × 10^4

5. 430,000
    4.3 is between 1 and 10   Move the decimal point
    4.3 times 100000 or 10^5 = 430,000   left five places.
    430,000 = 4.3 × 10^5

9. 825,000
    8.25 is between 1 and 10   Move the decimal point
    825,000 = 8.25 × 10^5 left five places.

Change to place value notation:

13. 8.91 × 10^4 = 89,100   Move the decimal four
                                             place to the right.

17. 6.7 × 10^{-6} = 0.0000067   Move the decimal six
                                             places to the left.

B

Write in scientific notation:

21. 377,000
     3.77 is between 1 and 10   Move the decimal point
     377,000 = 3.77 × 10^5 left five places.

25. 611,000,000
     6.11 is between 1 and 10   Move the decimal point
     611,000,000 = 6.11 × 10^8 left eight places.

Change to place value notation:

29. 6.89 × 10^4 = 68,900   Move the decimal point
                                             four places to the right.

Perform the indicated operations using scientific notation and laws of exponents. Write the result in scientific notation:

33. $(5.4 \times 10^{-3})(2 \times 10^{-4})$
    $= (5.4 \times 2)(10^{-3} \times 10^{-4})$ Group the decimals and powers of 10.

    $= 10.8 \times 10^{-7}$
    $= (1.08 \times 10^{1}) \times 10^{-7}$ Write 10.8 in scientific notation.

    $= 1.08 \times 10^{-6}$ Add the exponents on the factor 10.

37. $\dfrac{8.1 \times 10^{-5}}{3.0 \times 10^{2}} = \dfrac{8.1}{3.0} \times \dfrac{10^{-5}}{10^{2}}$ Regroup.

    $= 2.7 \times 10^{-7}$ Perform the operations for each fraction.

c

Write in scientific notation:

41.   3784
    3.784 is between 1 and 10   Move the decimal point left three places.

    $3784 = 3.784 \times 10^{3}$

45.   0.000484
    4.84 is between 1 and 10   Move the decimal point right four places.

    $0.000484 = 4.84 \times 10^{-4}$

Change to place value notation:

49. $2.36 \times 10^{-9} = 0.00000000236$ Move the decimal point nine places to the left.

Perform the indicated operations using scientific notation and the laws of exponents. Write the result in both scientific notation and place value notation:

53. $\dfrac{(7.2 \times 10^2)(3.6 \times 10^{-5})}{(4 \times 10^{-1})(5.4 \times 10^3)}$

$= \dfrac{(7.2)(3.6)}{(4)(5.4)} \times \dfrac{10^{2+(-5)}}{10^{-1+(3)}}$ Regroup, separating the division of powers of 10 and the other numbers. Add exponents. Perform the operations for each fraction.

$= 1.2 \times \dfrac{10^{-3}}{10^2}$

$= 1.2 \times 10^{-5}$ Subtract the exponents. Scientific notation.

$= 0.000012$ Place value notation.

57. $\dfrac{16,000,000}{(0.00815)(80,000)}$

$= \dfrac{1.6 \times 10^7}{(8.15 \times 10^{-3})(8 \times 10^4)}$ Express each number in scientific notation.

$= \dfrac{1.6 \times 10^7}{(8.15)(8) \times (10^{-3})(10^4)}$ Perform the multiplication in the denominator.

$= \dfrac{1.6 \times 10^7}{65.2 \times 10}$

$= \dfrac{1.6 \times 10^7}{6.52 \times 10^2}$ Write the denominator in scientific notation.

$= \dfrac{1.6}{6.52} \times \dfrac{10^7}{10^2}$ Regroup.

$= 0.25 \times 10^5$ Perform the indicated operations and round to the nearest hundredth.

$= (2.5 \times 10^{-1}) \times 10^5$

$= 2.5 \times 10^4$ Scientific notation.

$= 25,000$ Place value notation.

D

Write in scientific notation:

61. In one year, light travels approximately 5,870,000,000,000 miles.

    5.87 is between 1 and 10   Move the decimal point
                                             twelve places to the left.

    $5{,}870{,}000{,}000{,}000 = 5.87 \times 10^{12}$

65. The length of a long x-ray is approximately 0.000001 centimeter.

    1.0 is between 1 and 10.   Move the decimal point six
                                              places to the right.

    $0.000001 = 1.0 \times 10^{-6}$

Write in place value notation:

69. The distance from Earth to the nearest star is approximately 2.55×10^{13} miles.

    $2.55 \times 10^{13} = 25{,}500{,}000{,}000{,}000$ Move the decimal point
                                                        thirteen places to the
                                                        right.

STATE YOUR UNDERSTANDING

73. Why is 18.25 not written in scientific notation?

    Scientific notation is a way of writing very large or very small numbers so as to require less space. The number 18.25 requires less space written the way it is, in place value notation. Therefore, it is generally not written in scientific notation.

CHALLENGE EXERCISES

77. One Joule = 0.23901 calorie. Write the number of calories in scientific notation.

    0.23901

    2.3901 is between 1 and 10   Move the decimal point one
                                               place to the right.

    $0.23901 = 2.3901 \times 10^{-1}$

MAINTAIN YOUR SKILLS

81. $3.7p - 2.6pq + 1 - 2.9p + 3.8pq + q + 7.3pq + 2.1q - 3.4$

    $= (3.7p - 2.9p) + (-2.6pq + 3.8pq + 7.3pq) + (q + 2.1q)$
    $\quad + (1 - 3.4)$

                                                       Group like terms.

    $= 0.8p + 8.5pq + 3.1q - 2.4$ Combine like terms.

85. Phil's weight dropped from 227 pounds to 183 pounds in six
months. What was the percent of weight loss, to the nearest
percent?

Simpler word form:
What percent of 227 is 44?   Note that the weight loss
                                     is 227 - 183 or 44.

Select a variable:
Let x represent the percent.

Translate to algebra:
x(227) = 44

Solve:
$x = \frac{44}{227} \approx 0.19$

Answer:
The percent of weight loss was 19%.

EXERCISES 2.4 POLYNOMIALS (PROPERTIES AND CLASSIFICATIONS)

A

Add or subtract as indicated:

1.   (4x + 3y) + (6x + 12y)
   = (4x + 6x) + (3y + 12y)   Group like terms.
   = 10x + 15y

5.   (19x - 20) - (11x + 16)
   = (19x - 20) + (-11x - 16)   Definition of subtraction.
   = (19x - 11x) + (-20 - 16)   Group like terms.
   = 8x - 36   Combine like terms.

Subtract:

9. $16x^2 + 5x + 20$
      $5x^2 + 2x + 6$
     $11x^2 + 3x + 14$ Subtract. Add the
                                     opposite of the term being
                                     subtracted.

Add:

13.   (2.7p - 2.4q + 8.4) + (-3.8p + 3.1q - 2.7)
    = (2.7p - 3.8p) + (-2.4q + 3.1q) + (8.4 - 2.7)   Group like
                                                        terms.
    = -1.1p + 0.7q + 5.7

52

Add or subtract as indicated:

17. $(4r - 2s + 3) - (-3r + 2s - 4) + (8r - 5s - 1)$

    $= (4r - 2s + 3) + (3r - 2s + 4) + (8r - 5s - 1)$ Definition of sub-traction.

    $= (4r + 3r + 8r) + (-2s - 2s - 5s) + (3 + 4 - 1)$ Group like terms.

    $= 15r - 9s + 6$ Combine like terms.

B

Add or subtract as indicated:

21. $(5x^3 + 4x^2 - 5x) + (3x^2 + 2x + 1) - (8x - 5)$

    $= (5x^3 + 4x^2 - 5x) + (3x^2 + 2x + 1) + (-8x + 5)$

                Rewrite subtraction as addition.

    $= 5x^3 + (4x^2 + 3x^2) + (-5x + 2x - 8x) + (1 + 5)$

                Group like terms.

    $= 5x^3 + 7x^2 - 11x + 6$ Combine.

25.   Add: $4a^2 - 3ab + b^2$

                    $12ab - 7b^2 + 2$

        $3a^2 - 6ab + 7b^2 - 9$

         $\underline{5a^2 + 7b^2 + 10}$ Arrange the polynomials vertically.

      $12a^2 + 3ab + 8b^2 + 3$

29.   Subtract: $4a^4 - a^3 - a^2$ from $7a^4 - 3a^3 + a$

    $7a^4 - 3a^3 + a$ Arrange the polynomials vertically.

    $\underline{4a^4 - a^3 - a^2}$

    $3a^4 - 2a^3 + a^2 + a$ Subtract. Add the opposites of the terms being subtracted.

33.   Subtract: $2x + 3$ and $8 - x$ from $4x - 2$

      $(4x - 2) - (2x + 3) - (8 - x)$ Write horizontally.

  $= (4x - 2) + (-2x - 3) + (-8 + x)$ Rewrite subtractions.

  $= (4x - 2x + x) + (-2 - 3 - 8)$ Group like terms.

  $= 3x - 13$ Combine like terms.

Solve:

37. $(5x + 2) - (4x + 1) - 3x = (4x - 5) - (-3x + 2)$
$= (5x + 2) + (-4x - 1) - 3x = (4x - 5) + (3x - 2)$

Rewrite both sides of the equal sign using the definition of subtraction.

$= (5x - 4x - 3x) + (2 - 1) = (4x + 3x) + (-5 - 2)$

Group and combine like terms on both sides.

$-2x + 1 = 7x - 7$

$-2x = 7x - 8$ Subtract 1 from both sides.

$-9x = -8$ Subtract 7x from both sides.

$x = \dfrac{8}{9}$ Divide both sides by -9.

The solution set is $\left\{\dfrac{8}{9}\right\}$.

C

Add or subtract as indicated:

41. $\left(\dfrac{1}{5}p + 2q - \dfrac{1}{2}\right) + \left(\dfrac{3}{2}p - \dfrac{3}{4}q - 1\right) - \left(-\dfrac{1}{2}p - \dfrac{1}{2}q - \dfrac{1}{2}\right)$

$= \left(\dfrac{1}{5}p + 2q - \dfrac{1}{2}\right) + \left(\dfrac{3}{2}p - \dfrac{3}{4}q - 1\right) + \left(\dfrac{1}{2}p + \dfrac{1}{2}q + \dfrac{1}{2}\right)$

Rewrite, group, and combine, using like denominators within groups.

$= \left(\dfrac{1}{5}p + \dfrac{3}{2}p + \dfrac{1}{2}p\right) + \left(2q - \dfrac{3}{4}q + \dfrac{1}{2}q\right) + \left(-\dfrac{1}{2} - 1 + \dfrac{1}{2}\right)$

$= \left(\dfrac{2}{10}p + \dfrac{15}{10}p + \dfrac{5}{10}p\right) + \left(\dfrac{8}{4}q - \dfrac{3}{4}q + \dfrac{2}{4}q\right) + \left(-\dfrac{1}{2} - \dfrac{2}{2} + \dfrac{1}{2}\right)$

$= \dfrac{22}{10}p + \dfrac{7}{4}q - 1$

$= \dfrac{11}{5}p + \dfrac{7}{4}q - 1$

45. $x^2 - [2x^2 - (5x^2 + 6x) - x + 8] + 3$
$= x^2 - [2x^2 - 5x^2 - 6x - x + 8] + 3$ Rewrite all subtractions as additions of opposites beginning with innermost grouping symbols.

$= x^2 - [-3x^2 - 7x + 8] + 3$ Combine like terms.

$= x^2 + 3x^2 + 7x - 8 + 3$

$= 4x^2 + 7x - 5$

49. Add $7x + 2y$ to the difference of $6x - y$ and $2x + 3y$.

$(6x - y) - (2x + 3y) + (7x + 2y)$ Rewrite horizontally.
$= (6x - y) + (-2x - 3y) + (7x + 2y)$ Definition of subtraction.
$= (6x - 2x + 7x) + (-y - 3y + 2y)$ Regroup and combine like terms.
$= 11x - 2y$

53. Subtract $(x - y)$ from the difference of $(2x + 3y)$ and $(7x - 8y)$.

$(2x + 3y) - (7x - 8y) - (x - y)$ Rewrite.
$= (2x + 3y) + (-7x + 8y) + (-x + y)$ Definition of subtraction.
$= (2x - 7x - x) + (3y + 8y + y)$ Group like terms combine.
$= -6x + 12y$

Solve:

57. $2x^2 - [-3x^2 + (5x^2 + 2x + 1)] = -3x + 5$
$2x^2 - [-3x^2 + 5x^2 + 2x + 1] = -3x + 5$ Clear parentheses, then brackets.

$2x^2 - [2x^2 + 2x + 1] = -3x + 5$
$2x^2 - 2x^2 - 2x - 1 = -3x + 5$
$-2x - 1 = -3x + 5$
$x - 1 = 5$ Add 3x to both sides.
$x = 6$ Add 1 to both sides.

The solution set is {6}.

D

61. The cost of manufacturing n brushes is given by $C = \frac{1}{5}n^2 - 14n + 20$. The cost of marketing the same items is given by $C = 5n^2 - 30n + 50$. Find the formula for the total cost of manufacturing and marketing the brushes. Find the total cost when n = 50.

$\left(\frac{1}{5}n^2 - 14n + 20\right) + (5n^2 - 30n + 50)$ Add to find the total cost.

$= \left(\frac{1}{5}n^2 + 5n^2\right) + (-14n - 30n) + (20 + 50)$

$= \frac{26}{5}n^2 - 44n + 70$

$\frac{26}{5}(50)^2 - 44(50) + 70$ Find the total cost when n = 50. Replace n with 50.

$= \frac{26}{5}(2500) - 2200 + 70$

$= 13000 - 2200 + 70$

$= 10870$

The total cost when n = 50 is $10,870.

65. The volume of one container is given by the formula $V_1 = 2x^3 - 3x^2 + x + 40$ (where x is the length of one side), and the volume of a second container is $V_2 = x^3 + 3x^2 - 5x + 10$. Find a formula for the combined volume, V, of the two containers. Find the combined volume if x = 2.

$\begin{array}{l} 2x^3 - 3x^2 + x + 40 \\ \underline{x^3 + 3x^2 - 5x + 10} \\ 3x^3 - 4x + 50 \end{array}$ Add to find the combined volume.

$3(2)^3 - 4(2) + 50$ Replace x with 2.

$= 3(8) - 8 + 50$

$= 24 - 8 + 50$

$= 66$

The combined volume if x = 2 is 66.

69. Girl scout troop #52 can sell $\frac{3}{2}t^2 - 2t + 20$ boxes of cookies in t days. Troop #63 can sell $2t^2 - 4t + 10$ boxes of cookies in t days. Find a formula for the total number of boxes of cookies, T, both troops can sell in t days. How many boxes will both troops have sold in 2 days?

$$\left(\frac{3}{2}t^2 - 2t + 20\right) + (2t^2 - 4t + 10) \qquad \text{Add to find the total number.}$$

$$= \left(\frac{3}{2}t^2 + 2t^2\right) + (-2t - 4t) + (20 + 10)$$

$$= \frac{7}{2}t^2 - 6t + 30$$

$$\frac{7}{2}(2)^2 - 6(2) + 30 \qquad \text{Replace to with 2.}$$

$$= \frac{7}{2}(4) - 12 + 30$$

$$= 14 - 12 + 30$$

$$= 32$$

In 2 days, both troops together will sell 32 boxes.

CHALLENGE EXERCISES

Add or subtract as indicated; assume all variables represent positive integers:

73. $(4x^{2n} + 5x^n - 12) - (2x^{2n} - 5x^n - 4)$

$= (4x^{2n} + 5x^n - 12) + (-2x^{2n} + 5x^n + 4)$ Write the subtraction as addition of opposites.

$= (4x^{2n} - 2x^{2n}) + (5x^n + 5x^n) + (-12 + 4)$ Group like terms.

$= (4 - 2)x^{2n} + (5 + 5)x^n + (-8)$

$= 2x^{2n} + 10x^n - 8$

MAINTAIN YOUR SKILLS

Perform the indicated operations:

77. $(128) + (43) + (-219) + (-71)$

$= 171 + (-219) + (-71)$

$= -48 + (-71)$

$= -119$

81. $-[415 + (-329) + (-38) + (18)]$
     $= -[66]$
     $= -66$

EXERCISES 2.6 MULTIPLICATION OF POLYNOMIALS

A

Multiply:

1. $2(x + 3) = 2(x) + 2(3)$ Distributive property.
         $= 2x + 6$ Multiply.

5. $2(3x + 7) = 2(3x) + 2(7)$ Distributive property.
         $= 6x + 14$ Multiply.

9. $y(y^2 + 3y - 8)$
     $= y(y^2) + y(3y) + y(-8)$ Distributive property.
     $= y^3 + 3y^2 - 8y$ Multiply.

13. $3xy(x^2 - 6xy + y^2)$
      $= 3xy(x^2) + 3xy(-6xy) + 3xy(y^2)$ Distributive property.
      $= 3x^3 y - 18x^2 y^2 + 3xy^3$ Multiply.

17. $-a^3(-2a^4 + 5a^3 - 3a - 1)$
      $= (-a^3)(-2a^4) + (-a^3(5a^3) + (-a^3)(-3a) + (-a^3)(-1)$
                                       Distributive property.
      $= 2a^7 - 5a^6 + 3a^4 + a^3$ Multiply.

B

21. $(x - 3)(x + 4) = (x - 3)(x) + (x - 3)(4)$ Distribute the
                 $= x(x) - 3(x) + x(4) - 3(4)$ binomial $(x-3)$.
                 $= x^2 - 3x + 4x - 12$
                 $= x^2 + x - 12$

25. $(3m - 7t)(t - 4m)$
     $= (3m - 7t)t + (3m - 7t)(-4m)$ Distribute the binomial,
                                        $(3m - 7t)$.
     $= 3m(t) - 7t(t) + 3m(-4m) - 7t(-4m)$ Distribute again using
                                             the monomials, (t) and
                                             $(-4m)$.
     $= 3mt - 7t^2 - 12m^2 + 28mt$
     $= -12m^2 + 31mt - 7t^2$

29. $(x^2 + 1)(2x^2 - 7x + 5)$

$= (x^2 + 1)(2x^2) + (x^2 + 1)(-7x) + (x^2 + 1)(5)$

                                      Distribute $(x^2 + 1)$.

$= x^2(2x^2) + 1(2x^2) + x^2(-7x) + 1(-7x) + x^2(5) + 1(5)$

                                      Distribute the factors $2x^2$, $-7x$, and 5.

$= 2x^4 + 2x^2 - 7x^3 - 7x + 5x^2 + 5$ Multiply.

$= 2x^4 - 7x^3 + 7x^2 - 7x + 5$ Combine like terms.

Solve:

33. $3(x + 2) - 2(2x + 9) = 4(x + 2)$

        $3x + 6 - 4x - 18 = 4x + 8$ Multiply.

              $-x - 12 = 4x + 8$ Combine like terms.

                $-5x - 12 = 8$ Subtract $4x$ from both sides.

                      $-5x = 20$ Add 12 to both sides.

                       $x = -4$ Divide both sides by -5.

The solution set is $\{-4\}$.

37. $2x^2 + 18 = (2x + 3)(x + 5)$

     $2x^2 + 18 = (2x + 3)(x) + (2x + 3)5$ Distributive property.

     $2x^2 + 18 = 2x^2 + 3x + 10x + 15$ Multiply.

     $2x^2 + 18 = 2x^2 + 13x + 15$

            $18 = 13x + 15$ Subtract $2x^2$ from both sides.

               $3 = 13x$ Subtract 15 from both sides.

             $\dfrac{3}{13} = x$ Divide both sides by 13.

The solution set is $\left\{\dfrac{3}{13}\right\}$.

C

Multiply:

41. $(a + b - x)(a - b + x)$

$= (a + b - x)(a) + (a + b - x)(-b) + (a + b - x)(x)$

$= a(a) + b(a) - x(a) + a(-b) + b(-b) - x(-b) + a(x) + b(x)$
      $- x(x)$

$= a^2 + ab - ax - ab - b^2 + bx + ax + bx - x^2$

$= a^2 - b^2 + 2bx - x^2$

45. $(2x + 3y)(x^2 - 3xy + y^2 + 6)$

$= (2x + 3y)(x^2) + (2x + 3y)(-3xy) + (2x + 3y)(y^2) +$
      $(2x + 3y)(6)$

$= 2x(x^2) + 3y(x^2) + 2x(-3xy) + 3y(-3xy) + 2x(y^2) + 3y(y^2)$
      $+ 2x(6) + 3y(6)$

$= 2x^3 + 3x^2y - 6x^2y - 9xy^2 + 2xy^2 + 3y^3 + 12x + 18y$

$= 2x^3 - 3x^2y - 7xy^2 + 12x + 3y^3 + 18y$

49.

$$(x + 1)(x - 2)(x + 3)$$
$$= [(x + 1)(x) + (x + 1)(-2)](x + 3)$$
$$= [x(x) + 1(x) + x(-2) + 1(-2)](x + 3)$$
$$= [x^2 + x - 2x - 2](x + 3)$$

Multiply the first two binomials using the distributive property.

$$= (x^2 - x - 2)(x + 3)$$

Distribute $(x^2 - x - 2)$.

$$= (x^2 - x - 2)(x) + (x^2 - x - 2)(3)$$
$$= x^2(x) - x(x) - 2(x) + x^2(3) - x(3) - 2(3)$$
$$= x^3 - x^2 - 2x + 3x^2 - 3x - 6$$
$$= x^3 + 2x^2 - 5x - 6$$

Solve:

53.

$$x(x + 3) + (x + 2)(x - 7) = 2x(x + 3) - 12$$
$$(x)x + (x)3 + (x + 2)(x) + (x + 2)(-7) = 2x(x) + 2x(3)$$

Multiply, using the distributive property.

$$x^2 + 3x + x(x) + 2(x) + x(-7) + 2(-7) = 2x^2 + 6x - 12$$
$$x^2 + 3x + x^2 + 2x - 7x - 14 = 2x^2 + 6x - 12$$
$$2x^2 - 2x - 14 = 2x^2 + 6x - 12$$

Combine like terms.

$$-2x - 14 = 6x - 12$$

Subtract $2x^2$ from both sides.

$$-8x - 14 = -12$$

Subtract $6x$ from both sides.

$$-8x = 2$$

Add 14 to both sides.

$$x = -\frac{1}{4}$$

Divide both sides by -8 and reduce.

57.

$$3x(4x - 3) - (2x - 1)(3x + 2) = 2x(3x - 4)$$
$$12x^2 - 9x - [6x^2 - 3x + 4x - 2] = 6x^2 - 8x$$

Distribute and multiply.

$$12x^2 - 9x - [6x^2 + x - 2] = 6x^2 - 8x$$

Combine like terms and clear parentheses.

$$12x^2 - 9x - 6x^2 - x + 2 = 6x^2 - 8x$$
$$6x^2 - 10x + 2 = 6x^2 - 8x$$
$$-10x + 2 = -8x$$

Subtract $6x^2$ from both sides.

$$-10x = -8x - 2$$

Subtract 2 from both sides.

$$-2x = -2$$

Add $8x$ to both sides.

$$x = 1$$

Divide both sides by -2.

61. Mr. Ewing averaged 50 mph on his recent trip to Houston. On the return trip, he took a scenic route part of the way and took 2 hours longer and averaged only 42 mph. If the return trip was 30 miles longer, how long did it take Mr. Ewing to drive each direction?

Simpler word form:

$$\left(\begin{array}{c}\text{Distance} \\ \text{traveled from} \\ \text{Houston}\end{array}\right) - \left(\begin{array}{c}\text{Distance} \\ \text{traveled to} \\ \text{Houston}\end{array}\right) = 30$$

Select a variable:

Let t represent time used driving to Houston. Then t + 2 represents the time used to make the return trip.

	RATE ·	TIME =	DISTANCE
To Houston	50	t	50t
From Houston	42	t + 2	42(t + 2)

Translate to algebra:

42(t + 2) − 50t = 30

Solve:

$$42t + 84 - 50t = 30$$
$$-8t + 84 = 30$$
$$-8t = 54$$
$$t = 6\frac{3}{4}$$
$$\text{and } t + 2 = 8\frac{3}{4}$$

Answer:

It took Mr. Ewing $6\frac{3}{4}$ hours to travel to Houston, and $8\frac{3}{4}$ hours to return.

65. A 15-pound candy mixture selling for $1.90 per pound is made
     up of caramels costing $2.00 per pound and hard candy
     costing $1.75 per pound. How many pounds of each kind of
     candy make up the mixture?

     Simpler word form:
     Cost of   cost of
     caramels   +   hard candy   = total cost
     in mixture   in mixture

     Select variable:
     Let x represent the number of pounds of caramels; so 15 - x
     represents the number of pounds of hard candy.

     Translate to algebra:
     $2.00x + 1.75(15 - x) = 1.90(15)$

     Solve:
     $$2x + 26.25 - 1.75x = 28.50$$
     $$0.25x + 26.25 = 28.50$$
     $$0.25x = 2.25$$
     $$x = 9$$
     $$\text{and} \quad 15 - x = 15 - 9 = 6$$

     Answer:
     There were 9 lbs of caramels and 6 lbs of hard candy in the
     mixture.

69. Mari rode her bicycle to work at a rate of 6 miles/hr and
     walked home at a rate of 2 miles/hr. If it took her 2 hours
     longer to walk home, how long did it take her to ride her
     bicycle to work?

     Simpler word form:
     Distance   Distance
       to   =   from
      work   work

     Select variable:
     Let t represent the time it took for Mari to bicycle to
     work. Then t + 2 represents the time it took to walk home.

     Translate to algebra:
     $6t = 2(t + 2)$ Rate times time equals
                              distance.

     Solve:
     $$6t = 2t + 4$$
     $$4t = 4$$
     $$t = 1$$

     Answer:
     It took Mari one hour to bicycle to work.

CHALLENGE EXERCISES

Multiply; assume that all variables name positive integers:

73. $(x^n + y^n)^2$

   $= (x^n + y^n)(x^n + y^n)$

   $= (x^n)(x^n) + (x^n)y^n + x^n(y^n) + y^n(y^n)$ Distributive property.

   $= x^{2n} + 2x^n y^n + y^{2n}$ Add exponents.

MAINTAIN YOUR SKILLS

Multiply or divide as indicated:

77. $(-256) \div (-50) = 5.12$ or $5\frac{3}{25}$ The quotient of negatives is positive.

81. $-17(-2x + 8w) = 34x - 136w$ Distributive property.

EXERCISES 2.7   SPECIAL PRODUCTS

A

Multiply:

EXAMPLE 1

Multiply: $(x + 3)(x - 6)$

$$\overset{\quad\ \text{F}\qquad\ \text{O}\qquad\ \text{I}\qquad\ \text{L}}{(x + 3)(x - 6) = x^2 - 6x + 3x - 18}$$
$$= x^2 - 3x - 18$$

1. $\overset{\qquad\qquad\qquad\ \ \text{F}\qquad\ \text{O}\qquad\ \text{I}\qquad\ \text{L}}{(x + 3)(x + 4) = x^2 + 4x + 3x + 12}$
   $= x^2 + 7x + 12$ Simplify.

5. $\overset{\qquad\qquad\qquad\ \ \text{F}\qquad\ \text{O}\quad\ \text{I}\quad\ \text{L}}{(b - 1)(b - 4) = b^2 - 4b - b + 4}$
   $= b^2 - 5b + 4$ Simplify.

9. $\overset{\qquad\qquad\qquad\ \ \text{F}\qquad\ \text{O}\qquad\ \text{I}\qquad\ \text{L}}{(x + 6)(x - 7) = x^2 - 7x + 6x - 42}$
   $= x^2 - x - 42$ Simplify.

13. $(x - 9)^2 = x^2 - 18x + 81$ Square the first term. Double the product $(-9x)$ of the first and last terms. Square the last term.

17. $(2x + 11)(2x - 11) = 4x^2 - 121$ Multiply the first terms. Multiply the last terms.

B

21. $$(3x - 5)(x - 6) \overset{F \quad\;\; O \quad\;\; I \quad\;\; L}{= 3x^2 - 18x - 5x + 30}$$
$$= 3x^2 - 23x + 30 \qquad \text{Simplify.}$$

25. $$(3x + 7)(5x - 6) \overset{F \quad\;\; O \quad\;\; I \quad\;\; L}{= 15x^2 - 18x + 35x - 42}$$
$$= 15x^2 + 17x - 42$$

29. $(2a + 5)^2 = 4a^2 + 20a + 25$ Square first term. Double the product $(10a)$ of the first and last terms. Square the last term.

33. $(8w - 13)^2 = 64w^2 - 208w + 169$ Square the binomial.

Solve:

37. $(3t - 4)(3t + 4) = (3t + 2)^2 + 4$
$$9t^2 - 16 = 9t^2 + 12t + 4 + 4 \qquad \text{Multiply on both sides to remove the parenthese.}$$
$$-16 = 12t + 8 \qquad \text{Simplify.}$$
$$-24 = 12t$$
$$-2 = t$$
The solution set is $\{-2\}$.

C

Multiply:

41. $$(5x - 11)(3x + 9) = 15x^2 + 45x - 33x - 99 \qquad \text{FOIL}$$
$$= 15x^2 + 12x - 99$$

45. $$\left(\frac{1}{2}x + \frac{1}{3}\right)\left(\frac{1}{3}x + \frac{1}{2}\right) \overset{F \qquad\; O \qquad\; I \qquad\; L}{= \frac{1}{6}x^2 + \frac{1}{4}x + \frac{1}{9}x + \frac{1}{6}}$$
$$= \frac{1}{6}x^2 + \frac{13}{36}x + \frac{1}{6} \qquad \text{Simplify.}$$

49. $[(2x + y) - 5]^2$ Square the "binomial"
                                               with terms (2x + y) and
                                               -5.

    $= (2x + y)^2 + 2[-5(2x + y)] + (-5)^2$ Square the first "term"
                                               (2x + y). Double the
                                               product [-5(2x + y)] of
                                               the first and last
                                               terms. Square the last
                                               term.

    $= 4x^2 + 4xy + y^2 - 20x - 10y + 25$ Simplify.

53. $[(a + b) + (c + d)][(a + b) - (c + d)]$
        $= (a + b)^2 - (c + d)^2$ Multiply. Conjugate
                                               "binomials" with terms
                                               (a + b) and (c + d).

        $= a^2 + 2ab + b^2 - (c^2 + 2cd + d^2)$
        $= a^2 + 2ab + b^2 - c^2 - 2cd - d^2$

Solve:

57. $(2x - 3)(3x + 4) + (3x - 1)(x + 1) = (3x + 5)^2 + 20$
            $6x^2 - x - 12 + 3x^2 + 2x - 1 = 9x^2 + 30x + 25 + 20$
                        $9x^2 + x - 13 = 9x^2 + 30x + 45$
                            $x - 13 = 30x + 45$
                              $-29x = 58$
                                $x = -2$

The solution set is {-2}.

D

61. A square building lot had to be increased by 5 feet on each
     side to meet city code. If this increase added 525 square
     feet to the area, what was the length of a side of the
     original lot?

     Simpler word form:
     Original area   New area
        of lot   + 525 =   of lot

     Select variable:
     Let s represent a side of the original square lot. Then
     s + 5 represents a side of the enlarged square lot.

     Translate to algebra:
     $s^2 + 525 = (s + 5)^2$

Solve:
$$s^2 + 525 = s^2 + 10s + 25$$
$$525 = 10s + 25$$
$$500 = 10s$$
$$50 = s$$

Answer:
The length of each side of the original lot was 50 ft.

65. The length of a vegetable garden plot is to be increased by 2 ft while the width is increased by 4 ft. These changes will increase the area by 100 sq ft. If thelength was originally 5 ft more than the width, find the new dimensions of the garden plot.

Simpler word form:

Original garden area	+	100	=	New garden area

Select variable:
Let x represent the original width. Then x + 5 represents the original length, x + 7 represents new length, and x + 4 represents new width.

Translate to algebra:
$$x(x + 5) + 100 = (x + 4)(x + 7)$$

Solve:
$$x^2 + 5x + 100 = x^2 + 11x + 28$$
$$5x + 100 = 11x + 2$$
$$72 = 6x$$
$$12 = x \qquad \text{Original width.}$$

New width: x + 4 = 12 + 4 = 16
New length: x + 7 = 12 + 7 = 19

Answer:
The dimensions of the new garden are 19 ft (length) by 16 ft (width).

69. Johnny is six years older than his brother, Jason. Two years ago, the product of their ages was 20 less than the product of their present ages. How old are the boys now?

Simpler word form:

$$\left(\begin{array}{c} \text{The product of} \\ \text{their ages} \\ \text{two years ago} \end{array} \right) = \left(\begin{array}{c} \text{product of} \\ \text{their} \\ \text{present ages} \end{array} \right) - 20$$

Select variable:
Let x represent Jason's age now. Then x + 6 represents Johnny's age now, x - 2 represents Jason's age two years ago, and x + 4 represents Johnny's age two years ago.

	Age Now	Age Two Years Ago
Jason	x	x - 2
Johnny	x + 6	x + 6 - 2 or x + 4

Translate to algebra:
$(x - 2)(x + 4) = x(x + 6) - 20$

Solve:
$$x^2 + 2x - 8 = x^2 + 6x - 20$$
$$2x - 8 = 6x - 20$$
$$12 = 4x$$
$$3 = x$$ Jason's age now.
and $x + 6 = 3 + 6 = 9$ Johnny's age now.

Answer:
Jason's age now is 3, while Johnny's age now is 9.

CHALLENGE EXERCISES

73. Give an example to show that
$$(x - y)^2 \neq x^2 - y^2$$
$$(4 - 2)^2 \neq (4)^2 - (2)^2 \qquad \text{Replace x with 4 and}$$
$$(2)^2 \neq 16 - 4 \qquad \text{y with 2.}$$
$$4 \neq 12$$

77. Multiply: $(a - b)(a^2 + ab + b^2)$
$$= (a)a^2 + (a)ab + (a)b^2 + (-b)a^2 + (-b)ab + (-b)b^2$$
$$= a^3 + a^2b + ab^2 - a^2b - ab^2 - b^3$$
$$= a^3 - b^3$$

MAINTAIN YOUR SKILLS

81. Evaluate: $-5a - 3(2a - 3b)$ if $a = -12$ and $b = 15$.

   $-5a - 6a + 9b$ Simplify the expression
= $-11a + 9b$ by clearing the parentheses and combining like terms.

   $-11(-12) + 9(15)$ Replace a with -12 and b with 15.

= $132 + 135$
= 267 Simplify.

67

85. Find the total price of a set of collector's plates if the down payment is \$12.75 and there are fifteen monthly payments of \$18.75.

$$12.75 + 15(18.75) = 12.75 + 281.25$$
$$= 294$$

The total price is \$294.00

EXERCISES 2.8 COMMON FACTOR AND FACTORING BY GROUPING

A

Factor:

1. $12m - 12n = 12(m - n)$

    The number 12 is a common factor.

5. $6ab - 12ac + 18ad$
    $= 6a(b - 2c + 3d)$

    Use 6a as the common monomial factor.

9. $3xy + 6y + xz + 2z$
    $= (3xy + 6y) + (xz + 2z)$

    The first two terms have common factor 3y, and the last two terms have common factor z.

    $= 3y(x + 2) + z(x + 2)$ Factor each group.
    $= (3y + z)(x + 2)$ The terms $3y(x + 2)$ and $z(x + 2)$ have common factor $(x + 2)$.

13. $ax + bx + cx + ay + by + cy$
     $= (ax + bx + cx) + (ay + by + cy)$ Group the terms of the polynomial.

     $= x(a + b + c) + y(a + b + c)$ Factor out the greatest common factor from each group.

     $= (x + y)(a + b + c)$ Factor out the common trinomial $(a + b + c)$ from each term.

Solve:

17. $(2x - 1)(4x - 1) = 0$
     $2x - 1 = 0$ or $4x - 1 = 0$ Zero-product property.
     $2x = 1 4x = 1$
     $x = \frac{1}{2}$ or $x = \frac{1}{4}$

     The solution set is $\left\{\frac{1}{2}, \frac{1}{4}\right\}$.

B
Factor:

21. $18x^2 y^2 - 30xy^2 = 6xy^2 (3x - 5)$

The greatest common factor is $6xy^2$.

25. $\begin{aligned} &\quad 20x^2 - 15x + 8xy - 6y \\ &= (20x^2 - 15x) + (8xy - 6y) \end{aligned}$

Group the terms of the polynomial.

$= 5x(4x - 3) + 2y(4x - 3)$

Factor out the greatest common factor from each pair of terms.

$= (5x + 2y)(4x - 3)$

Factor out the common binomial, $4x - 3$.

29. $\begin{aligned} &\quad 2x^2 y - 6x^2 + y - 3 \\ &= (2x^2 y - 6x^2) + (y - 3) \end{aligned}$

Group the polynomial in pairs.

$= 2x^2 (y - 3) + 1(y - 3)$

Factor out the common factor from each group.

$= (2x^2 + 1)(y - 3)$

Factor out the common binomial, $y - 3$.

Solve:

33. $3x^2 + 18x = 0$
$3x(x + 6) = 0$
$3x = 0$ or $x + 6 = 0$
$\quad x = 0$ or $\quad\quad x = -6$
The solution set is $\{0, -6\}$.

Factor the left side.
Zero-product property.

37. $14x^2 - 35x + 6x - 15 = 0$
$(14x^2 - 35x) + (6x - 15) = 0$

Group the terms on the left.

$7x(2x - 5) + 3(2x - 5) = 0$
$(7x + 3)(2x - 5) = 0$

Factor the left side.

$7x + 3 = 0$ or $2x - 5 = 0$

Zero-product property.

$x = -\frac{3}{7}$ or $\quad x = \frac{5}{2}$

The solution set is $\left\{-\frac{3}{7}, \frac{5}{2}\right\}$.

C

Factor:

41. $\begin{aligned} &\quad 6x^3 y^2 z - 9x^2 y^3 z - 24x^2 y^2 z^2 \\ &= 3x^2 y^2 z(2x - 3y - 8z) \end{aligned}$

Greatest common factor is $3x^2 y^2 z$.

45. $42x^2 + 35xy - 18xy - 15y^2$
    $= (42x^2 + 35xy) + (-18xy - 15y^2)$ Group the terms.
    $= 7x(6x + 5y) - 3y(6x + 5y)$ Factor out the greatest common monomial from each group.

    $= (7x - 3y)(6x + 5y)$ Factor out the common binomial factor, $(6x + 5y)$.

49. $2a^2x + 2bx + 2cx - 3a^2 - 3b - 3c$
    $= (2a^2x + 2bx + 2cx) + (-3a^2 - 3b - 3c)$ Group the polynomial.

    $= 2x(a^2 + b + c) - 3(a^2 + b + c)$ Factor out the greatest common monomial.

    $= (2x - 3)(a^2 + b + c)$ Factor out the common trinomial.

Solve:

53. $x^2 - 5x - 11x + 55 = 0$
    $(x^2 - 5x) + (-11x + 55) = 0$ Factor the left side.
    $x(x - 5) - 11(x - 5) = 0$
    $(x - 11)(x - 5) = 0$
    $x - 11 = 0$ or $x - 5 = 0$ Zero-product property.
    $x = 11$ or $x = 5$
    The solution set is {5, 11}.

57. $x^2 - 2ax + 3ax - 6a^2 = 0$
    $(x^2 - 2ax) + (3ax - 6a^2) = 0$ Factor the left side by grouping.
    $x(x - 2a) + 3a(x - 2a) = 0$
    $(x + 3a)(x - 2a) = 0$
    $x + 3a = 0$ or $x - 2a = 0$ Zero-product property.
    $x = -3a$ or $x = 2a$
    The solution set is {-3a, 2a}.

D

61. If the average profit (p) per set of dishes sold is given by $p = \frac{1}{4}q^2 - 36q + 42$, where q is the number of sets sold, how many sets must be sold to have an average profit of $42?

Formula:
$p = \frac{1}{4}q^2 - 36q + 42$

Substitute:
$42 = \frac{1}{4}q^2 - 36q + 42$ Substitute 42 for p.

Solve:

$$0 = \frac{1}{4}q^2 - 36q$$ Subtract 42 from both sides.

$$0 = q\left(\frac{1}{4}q - 36\right)$$ Factor the right side.

$q = 0$ $\frac{1}{4}q - 36 = 0$

$q = 0$ $q = 144$ Reject $q = 0$, since there is no average cost if no sets are sold.

Answer:
The number of sets that must be sold to have a profit of $42 is 144.

65. Given the equation $s = 16t - 4t^2$, where s is the height of a falling object and t is the time the object falls, how long does it take the object to hit the ground? (s = 0)

Formula:
$s = 16t - 4t^2$

Substitute:
$0 = 16t - 4t^2$ Substitute 0 for s.

Solve:
$0 = 4t(4 - t)$ Factor on the right.
$4t = 0$ or $4 - t = 0$
 $t = 0$ or $4 = t$ Reject $t = 0$. The object won't hit the ground until it starts falling.

Answer:
It takes the object 4 seconds to hit the ground.

69. The cost of manufacturing x toy soldiers is $C = 2x^2 - 1600x + 5000$. How many soldiers can be produced at a cost of $5000?

Formula:
$C = 2x^2 - 1600x + 5000$

Substitute:
$5000 = 2x^2 - 1600x + 5000$ Replace C with 5000.

Solve:
$0 = 2x^2 - 1600x$
$0 = 2x(x - 800)$
$2x = 0$ or $x - 800 = 0$
 $x = 0$ or $x = 800$ Reject $x = 0$.

Answer:
800 toy soldiers can be produced at a cost of $5000.

CHALLENGE EXERCISES

Factor; assume all variables represent positive integers:

73. $x^{3n} - x^{2n} = x^{2n}(x^n) - x^{2n}(1)$
    $= x^{2n}(x^n - 1)$

The greatest common factor is x^{2n} since the first term is $x^{2n} \cdot x^n$ or x^{3n} and the second term is $x^{2n} \cdot 1$.

MAINTAIN YOUR SKILLS

Perform the indicated operations:

77. $6 \cdot 3 + 4(8 \cdot 6 - 2) + 3 \cdot 2$
    $= 18 + 4(48 - 2) + 6$
    $= 18 + 4(46) + 6$
    $= 18 + 184 + 6$
    $= 202 + 6$
    $= 208$

Solve:

81. $x - 2(5x + 3) = 4 - 7x$
    $x - 10 - 6 = 4 - 7x$
    $-9x - 6 = 4 - 7x$
    $-2x - 6 = 4$
    $-2x = 10$
    $x = -5$
    The solution set is $\{-5\}$.

EXERCISES 2.9   FACTORING TRINOMIALS

A

Factor:

1. $x^2 + 12x + 35$
    $(x + m)(x + n) = x^2 + 12x + 35$

$mn = 35$	$m + n = 12$
$1 \cdot 35$	36
$5 \cdot 7$	12

    $x^2 + 12x + 35 = (x + 5)(x + 7)$

Look for two integers, m and n, such that mn = 35 and m + n = 12.

The numbers needed are 5 and 7.

72

5. $a^2 - 7a - 18$
   $mn = -18 \quad m + n = -7$
      $1 \cdot -18 \qquad -17$
      $2 \cdot -9 \qquad\quad -7$

   The numbers m and n are 2 and -9.

   $a^2 - 7a - 18 = (a + 2)(a - 9)$

9. $a^2 - 4a - 10$
   $mn = -10 \quad m + n = -4$
      $1 \cdot -10 \qquad -9$
      $2 \cdot -5 \qquad\quad -3$

   Prime polynomial.

   There are no integers whose product is -10 and whose sum is -4.

Solve:

13. $\qquad x^2 - 3x + 2 = 0$
    $(x - 1)(x - 2) = 0$
    $x - 1 = 0 \text{ or } x - 2 = 0$
    $\qquad x = 1 \text{ or } \qquad x = 2$

    Factor the left side. Zero-product property.

    The solution set is {1, 2}.

17. $\quad x^2 + 16x + 63 = 0$
    $(x + 7)(x + 9) = 0$
    $x + 7 = 0 \text{ or } x + 9 = 0$
    $\qquad x = -7 \text{ or } \qquad x = -9$

    Factor the left side. Zero-product property.

    The solution set is {-9, -7}.

B

Factor:

21. $x^2 + x - 110$
    $mn = -110 \quad m + n = 1$
       $-1 \cdot 110 \qquad 109$
       $-2 \cdot 55 \qquad\quad 53$
       $-5 \cdot 22 \qquad\quad 17$
       $-10 \cdot 11 \qquad\quad 1$

    Since the middle term is positive, consider only factors whose sum is positive.

    $x^2 + x - 110 = (x - 10)(x + 11)$

73

25. $2a^2 + 9a - 18$

$mn = 2 \cdot -18 = -36 \qquad m + n = 9$

$\qquad -1 \cdot 36 \qquad\qquad\qquad\qquad 35$

$\qquad -2 \cdot 18 \qquad\qquad\qquad\qquad 16$

$\qquad -3 \cdot 12 \qquad\qquad\qquad\qquad 9$

List possible combinations of m and n.

$2a^2 - 3a + 12a - 18$

$(2a^2 - 3a) + (12a - 18)$

$a(2a - 3) + 6(2a - 3)$

$(a + 6)(2a - 3)$

Rewrite the middle term:
$9a = -3a + 12a.$
Factor by grouping.

29. $2t^2 - 9t - 35$

$mn = 2 \cdot -35 = -70 \qquad m + n = -9$

$\qquad 1 \cdot -70 \qquad\qquad\qquad\qquad -69$

$\qquad 2 \cdot -35 \qquad\qquad\qquad\qquad -33$

$\qquad 5 \cdot -14 \qquad\qquad\qquad\qquad -9$

$2t^2 + 5t - 14t - 35$

Rewrite the middle term:
$-9t = 5t - 14t$
Factor by grouping.

$t(2t + 5) - 7(2t + 5)$

$(t - 7)(2t + 5)$

Solve:

33. $9x^2 - 5x - 4 = 0$

$(9x + 4)(x - 1) = 0$

$9x + 4 = 0 \text{ or } x - 1 = 0$

$\qquad x = -\dfrac{4}{9} \text{ or } \quad x = 1$

Factor the left side.
Zero-product property.

The solution set is $\left\{ -\dfrac{4}{9}, \ 1 \right\}$.

37. $12x^2 + 7x - 12 = 0$

$(3x + 4)(4x - 3) = 0$

$3x + 4 = 0 \text{ or } 4x - 3 = 0$

$\qquad x = -\dfrac{4}{3} \text{ or } \quad x = \dfrac{3}{4}$

Factor the left side.
Zero-product property.

The solution set is $\left\{ -\dfrac{4}{3}, \ \dfrac{3}{4} \right\}$.

C

Factor:

41. $x^2 - 6x - 216$
     Try $(x + 2)(x - 108)$.
     The product is $x^2 - 106x - 216$.

     Factor by trial and error.
     The middle term, $-106x$, is
     not close to $-6x$, so try
     factors of -216 that are
     closer together than 2 and
     -108.

     Try $(x + 12)(x - 18)$.

     The product is $x^2 - 6x - 216$,
     so the factors are $(x + 12)(x - 18)$.

45. $6y^2 + 38y + 55$
     $mn = 6 \cdot 55 = 330 \qquad m + n = 38$
     $\qquad 1 \cdot 330 \qquad\qquad 331$
     $\qquad 2 \cdot 165 \qquad\qquad 167$
     $\qquad 3 \cdot 110 \qquad\qquad 113$
     $\qquad 5 \cdot 66 \qquad\qquad\; 71$
     $\qquad 6 \cdot 55 \qquad\qquad\; 61$
     $\qquad 10 \cdot 33 \qquad\qquad 43$
     $\qquad 11 \cdot 30 \qquad\qquad 41$
     $\qquad 15 \cdot 33 \qquad\qquad 37$

     There are no other factors
     to try.

     Prime polynomial.

49. $(x + 2y)^2 - (x + 2y) - 42$
     Let $x + 2y = p$

     Since the polynomial has
     two expressions in
     $(x + 2y)$, the factors will
     be easier to see using
     substitution.

     $p^2 - p - 42$

     Factor by trial and error.
     The middle term, $-1p$, is
     small so the factors of
     -42 should be close.

     Try $(p - 7)(p + 6)$.
     The product is $p^2 - p - 42$, so
     the factors are $(p - 7)(p + 6)$.
     $(x + 2y - 7)(x + 2y + 6)$

     Replace p with $x + 2y$.

75

53. $(6t - 1)^2 - 17(6t - 1) - 38$
    Let $6t - 1 = p$

Since the polynomial has two expressions in $(6t - 1)$, the factors will be easier to see using substitution.

$p^2 - 17p - 38$
$mn = -38 \qquad m + n = -17$
$\quad 1 \cdot -38 \qquad\qquad -37$
$\quad 2 \cdot -19 \qquad\qquad -17$

List possible combinations of m and n.

$(p + 2)(p - 19)$
$(6t - 1 + 2)(6t - 1 - 19)$
$(6t + 1)(6t - 20)$

Replace p with 6t - 1. Combine like terms within parentheses.

$(6t + 1)(2)(3t - 10)$
or $2(6t + 1)(3t - 10)$

Notice that the terms of the second binomial have a common factor, 2.

Solve:

57. $10x^2 - 31x - 63 = 0$
    $(2x - 9)(5x + 7) = 0$
    $2x - 9 = 0$ or $5x + 7 = 0$
    $\qquad x = \dfrac{9}{2}$ or $\qquad x = -\dfrac{7}{5}$

Factor the left side.

The solution set is $\left\{-\dfrac{7}{5}, \dfrac{9}{2}\right\}$.

D

61. The perimeter of a rectangular frame is 36 inches. If the area enclosed by the frame is 72 square inches, find the dimensions of the frame.

    Formula:
    $\ell w = A$

Formula for the area of a rectangle.

    Select variable:
    Let x represent the width of the frame: Since half the perimeter or 18 inches is the length plus the width, 18 - x represents the length.

    Substitute:
    $(18 - x)x = 72$

Replace ℓ with 18 - x, and w with x. Replace A with 72.

Solve:
$$18x - x^2 = 72$$
$$0 = x^2 - 18x + 72$$
$$0 = (x - 6)(x - 12) \qquad \text{Factor.}$$
$$x - 6 = 0 \text{ or } x - 12 = 0 \qquad \text{Zero-product property.}$$
$$x = 6 \text{ or } \qquad x = 12$$

Answer:
The dimensions of the frame are 12 in. by 6 in.

65. A strip of metal 30 inches wide is to be bent into a rectangular conduit that is enclosed on all four sides. If the area of the cross section is 50 sq in., what are the dimensions of the cross section?

Simpler word form:

$$\begin{pmatrix} \text{Area of} \\ \text{cross} \\ \text{section} \end{pmatrix} = \begin{pmatrix} \text{Width of} \\ \text{cross} \\ \text{section} \end{pmatrix} \begin{pmatrix} \text{Length of} \\ \text{cross} \\ \text{section} \end{pmatrix}$$

Select variable:
Let x represent the width of the cross section. Since half the perimeter, 15, is the sum of the width and the length, then 15 - x represents the length.

Translate to Algebra:
$$50 = x(15 - x)$$

Solve:
$$50 = 15x - x^2$$
$$x^2 - 15x + 50 = 0$$
$$(x - 10)(x - 5) = 0$$
$$x - 10 = 0 \text{ or } x - 5 = 0$$
$$x = 10 \text{ or } \qquad x = 5$$

If x = 10, then the length, 15 - x = 15 - 10 = 5.
If x = 5, then the length, 15 - x = 15 - 5 = 10.

Answer:
The dimensions of the cross section are 5 in. by 10 in.

CHALLENGE EXERCISES

Factor the following; assuming all variables represent positive integers:

69. $x^4 + 8x^2 + 12$

   $(x^2 + m)(x^2 + n)$ — Factors of the first term are $(x)(x^3)$ and $(x^2)(x^2)$. Use $(x^2)(x^2)$.

   $(x^2 + 6)(x^2 + 2)$ — Find m and n such that $mn = 12$ and $m + n = 8$.

77

Solve for x:

73. $7x - 6x = 2x + 4$
           $x = 2x + 4$
          $-x = 4$
           $x = -4$

77. $t = \frac{3}{5}x - 9$

        $t + 9 = \frac{3}{5}x$

     $\frac{5}{3}(t + 9) = x$

             $x = \frac{5(t + 9)}{3} = \frac{5t + 45}{3}$

     or $\frac{5}{3}t + \frac{5}{3}(9) = \frac{5}{3}t + 15$

EXERCISES 2.10   FACTORING SPECIAL CASES

A

Factor, if possible:

1. $a^2 - 1$
    $(a)^2 - (1)$

    $(a + 1)(a - 1)$

    The difference of the squares of a and 1.
    The factors are conjugate binomials.

5. $y^2 + 16$
    $(y)^2 + (4)^2$

    $y^2 + 16$ is a prime polynomial.

    The sum of two squares cannot be factored by the methods of this section.

9. $c^2 + 6c + 9$
    $(c)^2 + 2(3)(c) + (3)^2$

    $(c + 3)^2$

    The polynomial begins and ends with squares, c^2 and 9. The middle term is twice the product of the expressions that are squared. The middle term is positive, so the binomial is a sum.

13. $x^3 - 1$

    $(x)^3 - (1)^3$

The difference of two cubes.
The factors are a binomial and a trinomial. The binomial is the difference of x and 1. The trinomial is the square of the first term, x, plus the product of the two terms, x·1, plus the square of the last term, 1.

    $(x - 1)(x^2 + x + 1)$

Solve:

17. $\qquad x^2 - 49 = 0$
$\quad (x - 7)(x + 7) = 0$
$\quad x - 7 = 0$ or $x + 7 = 0$
$\qquad x = 7$ or $\qquad x = -7$

Factor the difference of squares on the left.
Zero-product law.

The solution set is $\{-7, 7\}$.

B

Factor, if possible:

21. $16a^2 + 49$

Sum of two squares.

    Prime polynomial.

The sum of two squares that have no common factor is a prime polynomial.

25. $\quad 16y^2 + 24y + 9$
$= (4y)^2 + 24y + (3)^2$

$= (4y + 3)^2$

A perfect square trinomial. The middle term is twice the product of the squared expressions:
$2(4y)(3) = 24y$

29. $\quad a^3 b^3 - c^3$
$= (ab)^3 - (c)^3$
$= (ab - c)(a^2 b^2 + abc + c^2)$

Difference of two cubes. The factors are a binomial, (ab - c) and a trinomial, which is the square of ab plus (ab)(c) plus the square of c.

Solve:

33.
$$4x^2 - 25 = 0$$
$$(2x - 5)(2x + 5) = 0$$

Factor the difference of squares on the left.
Zero-product law.

$$2x - 5 = 0 \text{ or } 2x + 5 = 0$$
$$x = \frac{5}{2} \text{ or } \qquad x = -\frac{5}{2}$$

The solution set is $\left\{-\frac{5}{2}, \frac{5}{2}\right\}$.

37.
$$4x^2 - 36x = -81$$
$$4x^2 - 36x + 81 = 0$$
$$(2x - 9)^2 = 0$$

Add 81 to both sides.
Factor the perfect square trinomial on the left.
Zero-product law.

$$2x - 9 = 0 \text{ or } 2x - 9 = 0$$
$$x = \frac{9}{2} \text{ or } \qquad x = \frac{9}{2}$$

The solution set is $\left\{\frac{9}{2}\right\}$.

C

Factor, if possible.

41.
$$81x^4 - 16y^4$$
$$= (9x^2)^2 - (4y^2)^2$$
$$= (9x^2 - 4y^2)(9x^2 + 4y^2)$$
$$= (3x - 2y)(3x + 2y)(9x^2 + 4y^2)$$

Difference of squares.
Difference of squares in the first binomial.

45.
$$36a^2b^2 - 132abc + 121c^2$$
$$= (6ab)^2 - 2(6ab)(11c) + (11c)^2$$
$$= (6ab - 11c)^2$$

Perfect square trinomial.

49.
$$64a^3 - 27b^3$$
$$= (4a)^3 - (3b)^3$$
$$= (4a - 3b)(16a^2 + 12ab + 9b^2)$$

Difference of cubes.

53.
$$(3x + y)^2 - (2x + 5)^2$$
$$= [3x + y + 2x + 5][3x + y - (2x + 5)]$$
$$= [5x + y + 5[3x + y - 2x - 5]$$
$$= (5x + y + 5)(x + y - 5)$$

Difference of squares

57.
$$(x^2 - 10x + 25) - y^2$$
$$= (x - 5)^2 - (y)^2$$
$$= (x - 5 - y)(x - 5 + y)$$

Difference of squares.

Solve:

61.
$$4x^2 - 169 = 0$$
$$(2x - 13)(2x + 13) = 0 \qquad \text{Factor the left side.}$$
$$2x - 13 = 0 \text{ or } 2x + 13 = 0 \qquad \text{Zero-product property.}$$
$$x = \frac{13}{2} \text{ or } \qquad x = -\frac{13}{2}$$

The solution set is $\left\{-\frac{13}{2}, \frac{13}{2}\right\}$.

D

65.   A variable electrical current is given by $i = t^2 - 16t + 75$. If t is in seconds, in how many seconds will the current (i) equal 11 amperes?

Formula:
$i = t^2 - 16t + 75$.

Substitute:

$11 = t^2 - 16t + 75$ $i = 11$.

$0 = t^2 - 16t + 64$ Rewrite the equation with 0 on one side.

$0 = (t - 8)^2$ Factor.

$t - 8 = 0$ or $t - 8 = 0$

$t = 8$ or $\quad t = 8$ A double root.

Answer:
The current will be 11 amperes in 8 seconds.

69.   During a 12-hour shift at the Clear Picture television assembly plant, the number of sets assembled in a given hour is shown by $N = 140 + 18t - t^2$ $t > 1$, where N is the number of sets and t is the hour of the shift. During what hour are 221 sets assembled?

Formula:
$N = 140 + 18t - t^2$

Substitute:

$221 = 140 + 18t - t^2$ Replace N with 221.

$t^2 - 18t + 81 = 0$ Rewrite the equation with zero on one side.

$(t - 9)^2 = 0$ Factor.

$t - 9 = 0$ or $t - 9 = 0$ Zero-product property.

$t = 9$ or $\quad t = 9$

Answer:
During the 9th hour, 221 sets are assembled.

73. How do we recognize when a polynomial is a perfect square trinomial?

    A polynomial is a perfect square trinomial when the first term and the last term of the trinomial are squares, and the middle term is twice the product of the squared expressions from the first and last terms.

CHALLENGE EXERCISES

Factor; assume that all variables represent positive integers:

77. $x^{3n} - 1$

    $= (x^n)^3 - (1)^3$ Rewrite as the difference of cubes.

    $= (x^n - 1)(x^{2n} + x^n + 1)$ The middle term is $(x^n)(1)$.

EXERCISES 2.11 FACTORING: A REVIEW

A

Factor, if possible:

1. $9c + 6b - 15$
   $3(3c + 2b - 5)$ Look for a greatest common factor first. The GCF is 3. The resulting trinomial cannot be further factored.

5. $x^2 - 8x + 12$
   $mn = 12$ $m + n = -8$
   $(-1)(-12)$ -13
   $(-2)(-6)$ -8 List possible combinations of m and n.
                          No more combinations are needed.
   $(x - 2)(x - 6)$

9. $a^2 + 81$
   $(a)^2 + (9)^2$ The sum of two squares cannot be factored by methods in this section.

   $a^2 + 81$ is a prime polynomial.

Solve:

13. $4x^2 - 8x = 0$
      $4x(x - 2) = 0$
      $4x = 0$ or $x - 2 = 0$
       $x = 0$ or $x = 2$

The GCF is $4x$.
Zero-product law.

      The solution set is $\{0, 2\}$.

17. $x^2 - 8x + 15 = 0$

There are only two combinations of factors for 15: $1 \cdot 15$ and $3 \cdot 5$.
Trial and error.
Zero-product law.

      $(x - 3)(x - 5) = 0$
      $x - 3 = 0$ or $x - 5 = 0$
         $x = 3$ or $x = 5$

      The solution set is $\{3, 5\}$.

B
Factor, if possible:

21. $x^2 - 16x + 55$

The two combinations of factors of 55 are: $1 \cdot 55$ and $5 \cdot 11$.

      $(x - 11)(x - 5)$

25. $3x^2 + 9x - 30$
      $3(x^2 + 3x - 10)$
      $3(x + 5)(x - 2)$

The GCF is 3.

29. $25x^2 + 1$
      $(5x)^2 + (1)^2$

The sum of squares cannot be factored by the methods of this section.

      $25x^2 + 1$ is a prime polynomial.

33. $1000c^3 - d^3$
      $(10c)^3 - (d)^3$
      $(10c - d)(100\,c^2 + 10cd + d^2)$

Difference of cubes.

37. $2x^2 + 13x + 21$
      $(2x + 1)(x + 21)$
      $(2x + 21)(x + 1)$
      $(2x + 3)(x + 7)$
      $(2x + 7)(x + 3)$

There are two combinations of factors of 21: $1 \cdot 21$ and $3 \cdot 7$.
Trial and error.

      $2x^2 + 13x + 21 = (2x + 7)(x + 3)$

Solve:

41. $x^2 - 11x + 24 = 0$
     $(x - 3)(x - 8) = 0$

     $x - 3 = 0$ or $x - 8 = 0$
     $x = 3$ or $x = 8$

     The solution set is $\{3, 8\}$.

There is no common factor. For 24, there are four combinations of factors: $1 \cdot 24$, $2 \cdot 12$, $3 \cdot 8$, and $4 \cdot 6$. Zero-product law.

45. $6x^2 + 6x - 12 = 0$
     $6(x^2 + x - 2) = 0$
     $6(x + 2)(x - 1) = 0$
     $(x + 2)(x - 1) = 0$
     $x + 2 = 0$ or $x - 1 = 0$
     $x = -2$ $x = 1$

     The solution set is $\{-2, 1\}$.

The GCF is 6.
Trial and error.
Divide both sides by 6.
Zero-product law.

C

Factor completely:

49. $36w^2 + 60w + 25$

$mn = 900$	$m + n = 60$
$10 \cdot 90$	100
$12 \cdot 75$	87
$20 \cdot 45$	64
$30 \cdot 30$	60

$36w^2 + 30w + 30w + 25$
$6w(6w + 5) + 5(6w + 5)$
$(6w + 5)(6w + 5)$ or $(6w + 5)^2$

There is no common factor. Use an m, n chart. Use only positive factors since both the sum, 60, and product, 900, are are positive.

Rewrite $60w$ as $30w + 30w$.

53. $250x^2 - 16$
     $2(125x^3 - 8)$
     $2[(5x)^3 - (2)^3]$
     $2(5x - 2)(25x^2 + 10x + 4)$

The GCF is 2.
Difference of cubes.

57. $28a^3 + 58a^2 - 30a$
     $2a(14a^2 + 29a - 15)$
     $2a(2a + 5)(7a - 3)$

The GCF is 2a.
Factor by trial and error.

61. $x^4 - x^2 - 12$
     $(x^2 - 4)(x^2 + 3)$

Trial and error. Note that the factors of x^4 are $x^2 \cdot x^2$. Also the first binomial is the difference of squares.

     $(x + 2)(x - 2)(x^2 + 3)$

65. $x^6 - 1$

    $(x^3 + 1)(x^3 - 1)$ Difference of squares.

    $(x + 1)(x^2 - x + 1)(x - 1)(x^2 + x + 1)$ Sum of cubes, and difference of cubes.

Solve:

69. $6x^2 - 41x + 70 = 0$

    $m \cdot n = 420 \quad m + n = -41$ Use an m, n chart.

    $-42 \cdot -10 \quad\quad -52$

    $-21 \cdot -20 \quad\quad -41$

       $6x^2 - 21x - 20x + 70 = 0$ Rewrite $-41x$ as $-21x - 20x$.

    $3x(2x - 7) - 10(2x - 7) = 0$ Factor by grouping.

         $(3x - 10)(2x - 7) = 0$

    $3x - 10 = 0$ or $2x - 7 = 0$ Zero-product law.

        $x = \dfrac{10}{3}$ or $\quad x = \dfrac{7}{2}$

The solution set is $\left\{ \dfrac{10}{3}, \dfrac{7}{2} \right\}$.

73. $96w^2 + 1 = 28w$

    $96w^2 - 28w + 1 = 0$ Subtract $28w$ from both sides.

    $(4w - 1)(24w - 1) = 0$ Factor.

    $4w - 1 = 0$ or $24w - 1 = 0$ Zero-product law.

       $x = \dfrac{1}{4}$ or $\quad w = \dfrac{1}{24}$

The solution set is $\left\{ \dfrac{1}{24}, \dfrac{1}{4} \right\}$.

STATE YOUR UNDERSTANDING

77. What is meant by "factor completely."

First, factor out the greatest common factor. If the polynomial is a binomial, see if it is the difference of two squares or if it is the sum or difference of two cubes. If the polynomial is a trinomial, factor by any appropriate method. If the polynomial contains four or more terms, try factoring by grouping.

CHALLENGE EXERCISES

81.　　$x^2 + 2x + 1 - y^2$

　　$= (x^2 + 2x + 1) - y^2$　　　　Group the first three
　　　　　　　　　　　　　　　　terms and factor them.
　　$= (x + 1)^2 - (y)^2$　　　　　Difference of squares.
　　$= (x + 1 - y)(x + 1 + y)$

MAINTAIN YOUR SKILLS

Solve:

85.　$6a - [4a - (1 - a) + 6] \leq$

　　　$6a - [4a - 1 + a + 6] \leq 12$　　Remove grouping symbols

　　　　　$6a - [5a + 5] \leq 12$　　and combine terms when

　　　　　　$6a - 5a - 5 \leq 12$　　possible.

　　　　　　　　$a - 5 \leq 12$

　　　　　　　　　$a \leq 17$

86

CHAPTER 3

RATIONAL EXPRESSIONS AND RELATED EQUATIONS

EXERCISES 3.1 PROPERTIES OF RATIONAL EXPRESSIONS

A

For what value of the variable is each of the following undefined?

1. $\dfrac{5x}{x - 3}$

   $x - 3 = 0$ Set $x - 3 = 0$ and solve.

       $x = 3$

   The expression is not defined when $x = 3$.

5. $\dfrac{3x + 1}{(x - 2)(x + 5)}$

   $(x - 2)(x + 5) = 0$ Set the denominator equal
   $x - 2 = 0$ or $x + 5 = 0$ to zero and solve.
       $x = 2$ or $x = -5$

   The expression is not defined when $x = 2$ or $x = -5$.

9. $\dfrac{x + 8}{x^2 - 8x - 20}$

       $x^2 - 8x - 20 = 0$ Set $x^2 - 8x - 20 = 0$.
   $(x - 10)(x + 2) = 0$ Solve.
       $x = 10$ or $x = -2$

   The expression is not defined when $x = 10$ or $x = -2$.

Build these rational expressions by finding the missing numerator. You do not need to state the variable restrictions:

13. $\dfrac{m}{x} = \dfrac{?}{x^2}$

    $\dfrac{m}{x} = \dfrac{m(x)}{x(x)} = \dfrac{mx}{x^2}$

    To find the missing numerator find the factor that was introduced into the denominator (x). This can be done by division ($x^2 \div x$). Now introduce this factor (x) into the numerator.

    The missing numerator is mx.

87

17. $\dfrac{10}{2w + 7} = \dfrac{?}{2w^2 + 7w}$

To find the missing numerator find the factor that was introduced into the denominator by factoring $2w^2 + 7w$.
The factors of $2w^2 + 7w$ are w and $2w + 7$.
Since w was introduced as a factor in the denominator, the Basic Principle of Fractions calls for it to be introduced in the numerator as well.

$\dfrac{10}{2w + 7} = \dfrac{10(w)}{(2w + 7)(w)}$

$= \dfrac{10w}{2w^2 + 7w}$

The missing numerator is $10w$.

B

Reduce. You do not need to state the variable restrictions.

21. $\dfrac{3abc}{18abx} = \dfrac{(3ab)c}{(3ab)6x}$

Identify the common factors, $3ab$, in the numerator and denominator.

$= \dfrac{(\cancel{3ab})c}{(\cancel{3ab})6x} = \dfrac{c}{6x}$

Reduce by dividing out the common factor.

25. $\dfrac{5(y - 2)}{6(y - 2)} = \dfrac{5(\cancel{y - 2})}{6(\cancel{y - 2})}$

Reduce by dividing out the common factor, $y - 2$.

$= \dfrac{5}{6}$

Build these rational expressions by finding the missing numerator. You do not need to state the variable restrictions:

29. $\dfrac{-3}{6mn} = \dfrac{?}{-12m^2 n}$

Divide $-12m^2 n$ by $6mn$ to find the factors that were introduced into the denominator. $-12m^2 n \div 6mn = -2m$
Introduce $-2m$ as a factor into the numerator.

$\dfrac{-3}{6mn} = \dfrac{-3(-2m)}{6mn(-2m)}$

$= \dfrac{6m}{-12m^2 n}$

The missing numerator is $6m$.

33. $\dfrac{3}{x-1} = \dfrac{?}{x^2-1}$

Find the factor that was introduced into the denominator by factoring x^2-1. The factors of x^2-1 are $x+1$ and $x-1$.

Since $(x+1)$ was introduced into the denominator it must be introduced into the numerator as well.

$\dfrac{3}{x-1} = \dfrac{3(x+1)}{(x-1)(x+1)}$

$= \dfrac{3x+3}{x^2-1}$

The missing numerator is $3x + 3$.

Reduce. State the restrictions on the variables so that the denominator is not zero:

37. $\dfrac{54a^2 b^2 c^3}{72a^4 bc^2}$

$a \neq 0,\ b \neq 0,\ c \neq 0$

$72a^4 bc^2 = 0$ when a, b, or c = 0.

$= \dfrac{\overset{1}{\cancel{18a^2 bc^2}}(3bc)}{\underset{1}{\cancel{18a^2 bc^2}}(4a^2)} = \dfrac{3bc}{4a^2},\ abc \neq 0$

Reduce by dividing out the common factors, $18a^2 bc^2$, in the numerator and denominator.

C

41. $\dfrac{x^2-1}{4x+4}$

$x \neq -1$

$4x + 4 = 4(x+1) \neq 0$.

$\dfrac{x^2-1}{4x+4} = \dfrac{(\cancel{x+1})(x-1)}{4(\cancel{x+1})}$

Factor both numerator and denominator.

$= \dfrac{x-1}{4},\ x \neq -1$

Reduce by dividing out the common factor $(x+1)$.

Build these rational expressions. State the restrictions on the variables so that the denoinator is not zero.

45. $\dfrac{m + 2}{2m + 7} = \dfrac{?}{4m^2 - 49}$

$m \neq \pm \dfrac{7}{2}$

The expression $4m^2 - 49 = (2m + 7)(2m - 7)$ and will equal zero if either factor is zero.

$\dfrac{m + 2}{2m + 7} = \dfrac{(m + 2)(2m - 7)}{(2m + 7)(2m - 7)}, \; m \neq \pm \dfrac{7}{2}$

Since $2m - 7$ was introduced as a factor in the denominator, introduce it as a factor in the numerator as well.

$= \dfrac{2m^2 - 3m - 14}{4m^2 - 19}, \; m \neq \pm \dfrac{7}{2}$ Multiply.

The missing numerator is $2m^2 - 3m - 14$, $m \neq \pm \dfrac{7}{2}$.

49. $\dfrac{7x + 2}{2x - 1} = \dfrac{?}{6x^2 - 5x + 1}$

$x \neq \dfrac{1}{2}, \; x \neq \dfrac{1}{3}$

The expression $6x^2 - 5x + 1 = (2x - 1)(3x - 1)$ and will equal zero if either factor is zero.

$\dfrac{7x + 2}{2x - 1} = \dfrac{(7x + 2)(3x - 1)}{(2x - 1)(3x - 1)}, \; x \neq \dfrac{1}{2}, \dfrac{1}{3}$ Introduce $(3x - 1)$ into the numerator.

$= \dfrac{21x^2 - x - 2}{6x^2 - 5x + 1}, \; x \neq \dfrac{1}{2}, \dfrac{1}{3}$ Multiply.

The missing numerator is $21x^2 - x - 2$, $x \neq \dfrac{1}{2}, \dfrac{1}{3}$.

Reduce. You need not state the variable restrictions:

53. $\dfrac{x^2 - 16}{x^2 - x - 12} = \dfrac{(x + 4)(x - 4)}{(x + 3)(x - 4)}$ Identify the common factor $(x - 4)$ in the numerator and denominator.

$= \dfrac{x + 4}{x + 3}$ Reduce by dividing out the common factor.

57. $\dfrac{2x^2 - x - 15}{2x^2 + 7x + 15} = \dfrac{(2x + 5)(x - 3)}{(2x + 5)(x + 1)}$ Identify the common factor $(2x + 5)$ in the numerator and denominator.

$\qquad\qquad = \dfrac{x - 3}{x + 1}$ Reduce by dividing out the common factor.

61. $\dfrac{x^2 - 2x - 3}{3x^3 + 3} = \dfrac{\overset{1}{\cancel{(x + 1)}}(x - 3)}{3\underset{1}{\cancel{(x + 1)}}(x^2 - x + 1)}$ Factor. Divide out the common factor, $(x + 1)$.

$\qquad\qquad = \dfrac{x - 3}{3(x^2 - x + 1)}$

STATE YOUR UNDERSTANDING

65. Explain why the following rational expression will not reduce in the manner indicated.

$\dfrac{\cancel{x} - 3}{\cancel{x} + 1} = -3$ The x's have been eliminated incorrectly. The x's are not __factors__. Therefore they cannot be eliminated by reducing.

$\dfrac{-3\cancel{x}}{\cancel{x}} = -3$ These x's have been eliminated correctly.

69. Using the formula of Exercise 68, write a rational expression that will represent the tensil stress on a bar if a force of $x^2 - 2x - 8$ Newtons is applied and the cross-sectional area is $x^2 + 1$.

Formula:
$S = \dfrac{F}{A}$

Substitute:
$S = \dfrac{x^2 - 2x - 8}{x^2 + 1}$ Replace F with $x^2 - 2x - 8$ and A with $x^2 + 1$.

MAINTAIN YOUR SKILLS

Perform the indicated operations:

73. $\left(-\dfrac{\overset{3}{\cancel{39}}}{\underset{5}{\cancel{45}}}\right) \cdot \left(\dfrac{\overset{1}{\cancel{9}}}{\underset{2}{\cancel{26}}}\right) = -\dfrac{3}{10}$

77. $-\dfrac{46}{39} \div \dfrac{23}{39} = -\dfrac{\overset{2}{\cancel{46}}}{\underset{3}{\cancel{39}}} \cdot \dfrac{\overset{1}{\cancel{39}}}{\underset{1}{\cancel{23}}} = -\dfrac{2}{3}$

EXERCISES 3.2   MULTIPLICATION AND DIVISION OF RATIONAL EXPRESSIONS

Multiply or divide and simplify. You need not state variable restrictions:

A

1. $\dfrac{y}{8} \cdot \dfrac{16}{w} = \dfrac{16y}{8w}$ Multiply the numerators and denominators.

   $= \dfrac{(8)(2y)}{(8)(w)} = \dfrac{2y}{w}$ Reduce.

   or

   $\dfrac{y}{8} \cdot \dfrac{16}{w} = \dfrac{y}{\underset{1}{\cancel{8}}} \cdot \dfrac{\overset{2}{\cancel{16}}}{w} = \dfrac{2y}{w}$ Cancel common factors before multiplying.

5. $\dfrac{a}{\underset{2}{\cancel{12}}} \cdot \dfrac{\overset{-1}{\cancel{-6}}}{a - 3} = \dfrac{-a}{2a - 6}$ Cancel the common factor 6, and multiply.

9. $\dfrac{2a - b}{\underset{1}{\cancel{4}}} \cdot \dfrac{\overset{3}{\cancel{12}}}{2a - b} = 3$ Cancel common factors.

Divide and simplify. State the restrictions on the variables:

13. $\dfrac{7y}{3} \div \dfrac{3z}{4} = \dfrac{7y}{3} \cdot \dfrac{4}{3z},\ z \neq 0$ Change division to multiplication by multiplying by the reciprocal of the divisor. Restrict the variable z.

   $= \dfrac{28y}{9z},\ z \neq 0$

17. $\dfrac{t - 4}{6} \div 5 = \dfrac{t - 4}{6} \cdot \dfrac{1}{5}$ Change division to multiplication.

   $= \dfrac{t - 4}{30}$ Multiply. There are no variables in the denominators so there are no restrictions needed.

B

Multiply or divide and simplify. You do not need to state the restrictions:

21. $\dfrac{6ax}{5by} \cdot \dfrac{10y}{18x} = \dfrac{60axy}{90bxy}$ Multiply the numerators and denominators.

$\qquad\qquad = \dfrac{(30xy)(2a)}{(30xy)(3b)} = \dfrac{2a}{3b}$ Reduce.

25. $\dfrac{\overset{1}{\cancel{x}}}{3\cancel{(x+2)}} \cdot \dfrac{\cancel{x+2}}{\underset{2}{\cancel{14}}} = \dfrac{1}{6}$ Reduce and multiply.

29. $(x^2 - 8x - 9)\left[\dfrac{12}{3x - 27}\right]$

$\qquad = (x \cancel{- 9})(x + 1)\left[\dfrac{\overset{4}{\cancel{12}}}{\underset{1}{\cancel{3}(x \cancel{- 9})}}\right]$ Factor. Eliminate the common factors.

$\qquad = 4(x + 1)$ or $4x + 4$

Multiply or divide and simplify. State the restrictions on the variables:

33. $\dfrac{5x - 3}{y^2 - y} \cdot \dfrac{8y - 8}{3 - 5x}$

$\qquad = \dfrac{\cancel{5x - 3}}{y(\cancel{y - 1})} \cdot \dfrac{8(\cancel{y - 1})}{-1(\cancel{5x - 3})}, \ y \neq 0, 1; \ x \neq \dfrac{3}{5}$ Factor. Note that $5x - 3$ and $3 - 5x$ are opposites. State the restrictions.

$\qquad = -\dfrac{8}{y}; \ y \neq 0, 1$ and $x \neq \dfrac{3}{5}$

37. $\dfrac{2x + 6}{x - 5} \div \dfrac{x^2 - 9}{x^2 - 10x + 25}$

$\qquad = \dfrac{2(\cancel{x + 3})}{\cancel{x - 5}} \cdot \dfrac{(x - 5)(\cancel{x - 5})}{(x - 3)(\cancel{x + 3})}, \ x \neq 3, 5, -3$ Rewrite division as multiplication. Restrict the variables.

$\qquad = \dfrac{2(x - 5)}{x - 3}, \ x \neq \pm 3, 5$ Eliminate common factors.

C

Multiply or divide and simplify. You need not state the variable restrictions.

41. $\dfrac{5y + 10}{y^2 - 4} \cdot \dfrac{y^2 - 7y + 10}{8y - 40}$

$= \dfrac{5(\cancel{y + 2})}{(\cancel{y + 2})(y - 2)} \cdot \dfrac{(\cancel{y - 5})(y - 2)}{8(\cancel{y - 5})} = \dfrac{5}{8}$ Factor and reduce.

45. $\dfrac{x^2 + 9x + 18}{x^2 + 2x - 15} \div \dfrac{2x + 12}{3x + 15}$

$= \dfrac{x^2 + 9x + 18}{x^2 + 2x - 15} \cdot \dfrac{3x + 15}{2x + 12}$ Invert and multiply.

$= \dfrac{(\cancel{x + 6})(x + 3)}{(\cancel{x + 5})(x - 3)} \cdot \dfrac{3(\cancel{x + 5})}{2(\cancel{x + 6})}$ Factor and reduce.

$= \dfrac{3(x + 3)}{2(x - 3)} = \dfrac{3x + 9}{2x - 6}$ Multiply

49. $\dfrac{6x^2 + 39x + 63}{10x^2 - 17x + 3} \div \dfrac{6x^2 + 33x + 45}{10x^2 + 33x - 7}$

$= \dfrac{6x^2 + 39x + 63}{10x^2 - 17x + 3} \cdot \dfrac{10x^2 + 33x - 7}{6x^2 + 33x + 45}$ Invert and multiply.

$= \dfrac{(\cancel{3x + 9})(2x + 7)}{(\cancel{5x - 1})(2x - 3)} \cdot \dfrac{(\cancel{5x - 1})(2x + 7)}{(\cancel{3x + 9})(2x + 5)}$ Factor and reduce.

$= \dfrac{(2x + 7)(2x + 7)}{(2x - 3)(2x + 5)} = \dfrac{4x^2 + 28x + 49}{4x^2 + 4x - 15}$ Multiply.

Multiply or divide and simplify. State the restrictions on the variables.

53. $\dfrac{3y^2 + 6y + 12}{9y - y^3} \cdot \dfrac{y^3 + 9y}{y^3 - 8}$

$= \dfrac{3(\cancel{y^2 + 2y + 4})}{y(3 - y)(3 + y)} \cdot \dfrac{y(y^2 + 9)}{(y - 2)(\cancel{y^2 + 2y + 4})}$ Factor. Divide out common factors.

$= \dfrac{3(y^2 + 9)}{(y - 2)(3 - y)(3 + y)}, \; y \neq -3, \, 0, \, 2, \, 3$

94

Multiply or divide and simplify.

57. $\dfrac{ac + ac + bc + bd}{ac - ad + bc - bd} \cdot \dfrac{c^2 - cd + bc - bd}{ac + ad - bc - bd}$

$= \dfrac{a(c + d) + b(c + d)}{a(c - d) + b(c - d)} \cdot \dfrac{c(c - d) + b(c - d)}{a(c + d) - b(c + d)}$ Factor by grouping.

$= \dfrac{(a + b)(c + d)}{(a + b)(c - d)} \cdot \dfrac{(c + b)(c - d)}{(a - b)(c + d)}$

$= \dfrac{b + c}{a - b}$ Eliminate common factors.

61. $\dfrac{6x^2 + 11x - 10}{4x^2 - 7x - 2} \div \dfrac{x^2 - 3x - 40}{x^2 + 3x - 10} \div \dfrac{2x^2 - 3x - 20}{4x^2 - 31x - 8}$

$= \dfrac{(3x - 2)(2x + 5)}{(4x + 1)(x - 2)} \cdot \dfrac{(x + 5)(x - 2)}{(x - 8)(x + 5)} \cdot \dfrac{(4x + 1)(x - 8)}{(2x + 5)(x - 4)}$

$= \dfrac{3x - 2}{x - 4}$

STATE YOUR UNDERSTANDING

65. Explain the difference in the processes for multiplication and division of rational expressions.

The only difference is that in division, we must rewrite the divisor as its reciprocal and <u>then</u> multiply.

CHALLENGE EXERCISES

Perform the indicated operations, reduce to lowest terms. You do not need to state the restrictions on the variables. Assume that all exponents are positive integers.

69. $\dfrac{x^{3n} - 8}{x^{2n} - 16} \div \dfrac{x^{2n} + 2x^n + 4}{x^n + 4}$

$= \dfrac{(x^n)^3 - (2)^3}{(x^n)^2 - (4)^2} \cdot \dfrac{x^n + 4}{x^{2n} + 2x^n + 4}$ Rewrite the division as multiplication.

$= \dfrac{(x^n - 2)(x^{2n} + 2x^n + 4)}{(x^n - 4)(x^n + 4)} \cdot \dfrac{x^n + 4}{x^{2n} + 2x^n + 4}$ Factor.

$= \dfrac{x^n - 2}{x^n - 4}$ Eliminate common factors.

Add or subtract as indicated:

73. $\dfrac{11}{9} - \dfrac{8}{7} = \dfrac{77}{63} - \dfrac{72}{63} = \dfrac{5}{63}$ Rewrite the fractions with common denominators and subtract.

77. $\dfrac{5}{7} - \dfrac{32}{21} = \dfrac{15}{21} - \dfrac{32}{21} = -\dfrac{17}{21}$

EXERCISES 3.3   ADDITION AND SUBTRACTION OF RATIONAL EXPRESSIONS

A

Find the LCD of the fractions:

1. $\dfrac{1}{5ab} + \dfrac{1}{5ac}$

     $5ab = 5 \cdot a \cdot b$ Factor each denominator.
     $5ac = 5 \cdot a \cdot c$ The different factors are 5, a, b, and c.

     $LCD = 5abc$ The LCD is the product of the highest power of each factor.

Add or subtract. Reduce, if possible. You need not state the variable restrictions:

5. $\dfrac{3}{x} + \dfrac{6}{x} - \dfrac{5}{x}$ The fractions have a common denominator. Combine the numerators and retain the common denominator.

     $= \dfrac{3 + 6 - 5}{x} = \dfrac{4}{x}$

9. $\dfrac{a + 1}{3} - \dfrac{a + 2}{2}$

     $LCD = 6$
     $= \dfrac{(a + 1)(2)}{3(2)} - \dfrac{(a + 2)(3)}{2(3)}$ Build the fractions so that each has the LCD for the denominator.

     $= \dfrac{2a + 2}{6} - \dfrac{3a + 6}{6}$ Multiply.

     $= \dfrac{(2a + 2) - (3a + 6)}{6}$ Subtract the numerators, and retain the common denominator.

     $= \dfrac{2a + 2 - 3a - 6}{6}$

     $= \dfrac{-a - 4}{6}$

13. $\dfrac{1}{5ab} + \dfrac{1}{5ac}$

LCD $= 5abc$

$\dfrac{1}{5ab} + \dfrac{1}{5ac} = \dfrac{1}{5ab} \cdot \dfrac{c}{c} + \dfrac{1}{5ac} \cdot \dfrac{b}{b}$

Build the fractions so that they have the LCD as the denominator.

$= \dfrac{c}{5abc} + \dfrac{b}{5abc}$

Multiply.

$= \dfrac{b + c}{5abc}$

Add the numerators and retain the common denominator.

17. $\dfrac{5}{2y - 2} + \dfrac{4}{3y - 3}$

LCD $= 6(y - 1)$

Find the LCD of the fractions.

$\dfrac{5}{2(y - 1)} + \dfrac{4}{3(y - 1)} = \dfrac{5}{2(y - 3)} \cdot \dfrac{3}{3} + \dfrac{4}{3(y - 1)} \cdot \dfrac{2}{2}$

Build the fractions to have the LCD as denominators.

$= \dfrac{15}{6(y - 1)} + \dfrac{8}{6(y - 1)}$

Multiply.

$= \dfrac{23}{6(y - 1)}$

B

Find the LCD of the following fractions:

21. $\dfrac{5x}{x - 2} + \dfrac{6}{2 - x}$

LCD $= x - 2$ or $2 - x$

The denominators are opposites, so either can be the LCD.

97

Add or subtract. Simplify, if possible. You need not state the variable restrictions:

25. $\dfrac{7}{w - 4} + \dfrac{1}{4 - w}$

LCD = w - 4

$\dfrac{7}{w - 4} + \dfrac{1}{4 - w} = \dfrac{7}{w - 4} + \dfrac{1}{4 - w} \cdot \dfrac{-1}{-1}$ Build each fraction using the LCD as denominator.

$= \dfrac{7}{w - 4} + \dfrac{-1}{w - 4}$ Multiply.

$= \dfrac{6}{w - 4}$ Add.

<u>or</u>
LCD = 4 - w

$\dfrac{7}{w - 4} + \dfrac{1}{4 - w} = \dfrac{7}{w - 4} \cdot \dfrac{-1}{-1} + \dfrac{1}{4 - w}$

$= \dfrac{-7}{4 - w} + \dfrac{1}{4 - w} = \dfrac{-6}{4 - w}$

29. $\dfrac{3t + 4}{5} + t - 1$

LCD = 5

$\dfrac{3t + 4}{5} + t - 1 = \dfrac{3t + 4}{5} + t \cdot \dfrac{5}{5} - 1 \cdot \dfrac{5}{5}$ Use the LCD to build each term.

$= \dfrac{3t + 4}{5} + \dfrac{5t}{5} - \dfrac{5}{5}$ Multiply.

$= \dfrac{3t + 4 + 5t - 5}{5}$

$= \dfrac{8t - 1}{5}$

33. $\dfrac{3}{x + 6} + \dfrac{5}{2x - 3}$

LCD = (x + 6)(2x - 3)

$\dfrac{3}{x + 6} + \dfrac{5}{2x - 3} = \dfrac{3}{x + 6} \cdot \dfrac{2x - 3}{2x - 3} + \dfrac{5}{2x - 3} \cdot \dfrac{x + 6}{x + 6}$

Building using the LCD as the denominator.

$= \dfrac{6x - 9}{(x + 6)(2x - 3)} + \dfrac{5x + 30}{(x + 6)(2x - 3)}$

Multiply.

$= \dfrac{11x + 21}{(x + 6)(2x - 3)}$ Add.

Add or subtract. Reduce, if possible. State the restrictions on the variables:

37. $\dfrac{1}{2x - 3} - \dfrac{x + 6}{x + 3}$

$x \neq \dfrac{3}{2}, -3;$ LCD = $(2x - 3)(x + 3)$ Restrict the variable.

$$\dfrac{1}{2x - 3} - \dfrac{x + 6}{x + 3} = \dfrac{1}{2x - 3} \cdot \dfrac{x + 3}{x + 3} - \dfrac{x + 6}{x + 3} \cdot \dfrac{2x - 3}{2x - 3}$$

$$= \dfrac{x + 3}{(2x - 3)(x + 3)} + \dfrac{-1(2x^2 + 9x - 18)}{(x + 3)(2x - 3)}$$

$$= \dfrac{-2x^2 - 8x + 21}{(2x - 3)(x + 3)}, \quad x \neq \dfrac{3}{2}, -3$$

c

Find the LCD of the following fractions:

41. $\dfrac{x + 1}{x^2 + 2x - 15} + \dfrac{2x + 1}{x^2 - 4x + 3}$

$x^2 + 2x - 15 = (x + 5)(x - 3)$ Factor each denominator.
$x^2 - 4x + 3 = (x - 3)(x - 1)$
LCD = $(x - 1)(x - 3)(x + 5)$ The LCD is the product of the highest power of each factor.

Add or subtract. Reduce, if possible. State the restrictions on the variables:

45. $\dfrac{2}{x + 3} - \dfrac{3}{x - 5} + \dfrac{x}{x^2 - 2x - 15}$

$\qquad = \dfrac{2}{x + 3} - \dfrac{3}{x - 5} + \dfrac{x}{(x - 5)(x + 3)}$ Factor the third denominator.

$x \neq -3, 5; \text{ LCD} = (x + 3)(x - 5)$ Restrict the variable.

$\dfrac{2}{x + 3} - \dfrac{3}{x - 5} + \dfrac{x}{(x + 3)(x - 5)}$

$\qquad = \dfrac{2}{x + 3} \cdot \dfrac{x - 5}{x - 5} - \dfrac{3}{x - 5} \cdot \dfrac{x + 3}{x + 3} + \dfrac{x}{(x + 3)(x - 5)}$

$\qquad = \dfrac{2(x - 5)}{(x + 3)(x - 5)} + \dfrac{-3(x + 3)}{(x + 3)(x - 5)} + \dfrac{x}{(x + 3)(x - 5)}$

$\qquad = \dfrac{2x - 10}{(x + 3)(x - 5)} + \dfrac{-3x - 9}{(x + 3)(x - 5)} + \dfrac{x}{(x + 3)(x - 5)}$

$\qquad = \dfrac{-19}{(x + 3)(x - 5)}, \; x \neq -3, 5$

49. $\dfrac{6}{x^2 + x - 6} - \dfrac{5}{x^2 + 2x - 8} + \dfrac{4}{x^2 + 7x + 12}$

$\qquad \begin{aligned} x^2 + x - 6 &= (x + 3)(x - 2) \\ x^2 + 2x - 8 &= (x + 4)(x - 2) \\ x^2 + 7x + 12 &= (x + 4)(x + 3) \\ x &\neq -3, 2, -4 \end{aligned}$ Factor each denominator.

Restrict the variable.

$\text{LCD} = (x + 3)(x - 2)(x + 4)$

$\dfrac{6}{(x + 3)(x - 2)} \cdot \dfrac{x + 4}{x + 4} - \dfrac{-5}{(x + 4)(x - 2)} \cdot \dfrac{x + 3}{x + 3}$
$\qquad + \dfrac{4}{(x + 4)(x + 3)} \cdot \dfrac{x - 2}{x - 2}$

$\qquad = \dfrac{6(x + 4)}{(x + 3)(x - 2)(x + 4)} + \dfrac{-5(x + 3)}{(x + 4)(x - 2)(x + 3)}$
$\qquad + \dfrac{4(x - 2)}{(x + 4)(x + 3)(x - 2)}$

$\qquad = \dfrac{6x + 24}{(x + 3)(x - 2)(x + 4)} + \dfrac{-5x - 15}{(x + 3)(x - 2)(x + 4)}$
$\qquad + \dfrac{4x - 8}{(x + 3)(x - 2)(x + 4)}$

$\qquad = \dfrac{5x + 1}{(x + 3)(x - 2)(x + 4)}, \; x \neq -3, 2, -4$

53. $\dfrac{1}{x} - \dfrac{2}{x+2} - \dfrac{2}{x^2+3x+2}$

$x = 0,\ -2,\ -1;\ \text{LCM} = x(x+2)(x+1)$

$\dfrac{1(x+2)(x+1)}{x(x+2)(x+1)} - \dfrac{2(x)(x+1)}{(x+2)(x)(x+1)} - \dfrac{2(x)}{(x+2)(x+1)(x)}$

$= \dfrac{x^2+3x+2}{x(x+2)(x+1)} + \dfrac{-(2x^2+2x)}{x(x+2)(x+1)} + \dfrac{-(2x)}{x(x+2)(x+1)}$

$= \dfrac{x^2+3x+2-2x^2-2x-2x}{x(x+1)(x+2)}$

$= \dfrac{-x^2-x+2}{x(x+1)(x+2)} = \dfrac{-1(x^2+x-2)}{x(x+1)(x+2)}$ Factor and reduce.

$= \dfrac{-1(x-1)(x+2)}{x(x+2)(x+1)}$

$= \dfrac{-x+1}{x(x+1)},\ x = 0,\ -1$

57. $\dfrac{p}{q^{-1}} - \dfrac{p^{-1}}{q}$

$\dfrac{pq}{1} - \dfrac{1}{pq}$ Rewrite in fraction form. Recall that $x^{-m} = \dfrac{1}{x^m}$.

$p \neq 0,\ q \neq 0;\ \text{LCM} = pq$

$\dfrac{(pq)(pq)}{1(pq)} - \dfrac{1}{pq}$ Build each fraction.

$= \dfrac{p^2q^2}{pq} - \dfrac{1}{pq} = \dfrac{p^2q^2-1}{pq},\ p \neq 0,\ q \neq 0$

Perform the indicated operations. You do not need to state the variable restrictions.

61. $x^{-1} + (x + 1)^{-1} - x^{-2}$

$$= \frac{1}{x} + \frac{1}{x + 1} - \frac{1}{x^2}$$ Rewrite with positive exponents.

$$= \frac{1}{x} \cdot \frac{x(x + 1)}{x(x + 1)} + \frac{1}{x + 1} \cdot \frac{x^2}{x^2} - \frac{1}{x^2} \cdot \frac{x + 1}{x + 1}$$ The LCD is $x^2(x + 1)$.

$$= \frac{x(x + 1)}{x^2(x + 1)} + \frac{x^2}{x^2(x + 1)} - \frac{x + 1}{x^2(x + 1)}$$

$$= \frac{x^2 + x + x^2 - (x + 1)}{x^2(x + 1)}$$

$$= \frac{x^2 + x + x^2 - x - 1}{x^2(x + 1)}$$

$$= \frac{2x^2 - 1}{x^2(x + 1)}$$

65. $(x^2 + x - 6)^{-1} - 2(x^2 - 4)^{-1} + 3(x^2 + 5x + 6)^{-1}$

$$= \frac{1}{x^2 + x - 6} - \frac{2}{x^2 - 4} + \frac{3}{x^2 + 5x + 6}$$
Rewrite with positive exponents.

$$= \frac{1}{(x + 3)(x - 2)} - \frac{2}{(x - 2)(x + 2)} + \frac{3}{(x + 2)(x + 3)}$$
Factor the denominators.

$$= \frac{x + 2}{(x + 2)(x + 3)(x - 2)} - \frac{2(x + 3)}{(x + 2)(x + 3)(x - 2)} + \frac{3(x - 2)}{(x + 2)(x + 3)(x - 2)}$$

LCD: $(x+2)(x+3)(x-2)$

$$= \frac{x + 2 - 2(x + 3) + 3(x - 2)}{(x + 2)(x + 3)(x - 2)}$$

$$= \frac{x + 2 - 2x - 6 + 3x - 6}{(x + 2)(x + 3)(x - 2)}$$

$$= \frac{2x - 10}{(x + 2)(x + 3)(x - 2)}$$

69. Define least common denominator (LCD).

The least common denominator is the polynomial with the least number of factors that is a multiple of each of the denominators in the expression being considered.

MAINTAIN YOUR SKILLS

Solve the following equations:

73. $6x + 7 = 7x - 1$
$-x + 7 = -1$
$-x = -8$
$x = 8$
The solution set is $\{8\}$.

77. $(2x + 4)(x + 4) = (x - 5)(2x + 5)$
$2x^2 + 12x + 16 = 2x^2 - 5x - 25$
$12x + 16 = -5x - 25$
$17x + 16 = -25$
$17x = -41$
$x = -\dfrac{41}{17}$
The solution set is $\left\{-\dfrac{41}{17}\right\}$.

EXERCISES 3.4   COMPLEX FRACTIONS

Simplify. You need not state variable restrictions:

A

1. $\dfrac{\frac{3}{x}}{\frac{6}{y}} = \dfrac{3}{x} \div \dfrac{6}{y}$

Rewrite as a division problem.

$= \dfrac{3}{x} \cdot \dfrac{y}{6} = \dfrac{y}{2x}$

Multiply by the reciprocal of the divisor and simplify.

5. $\dfrac{\frac{1}{x + 1}}{\frac{1}{x - 1}} = \dfrac{(x + 1)(x - 1)\left(\dfrac{1}{x + 1}\right)}{(x + 1)(x - 1)\left(\dfrac{1}{x - 1}\right)}$

Multiply the numerator and denominator by the LCD of $(x+1)$ and $(x-1)$ which is $(x+1)(x-1)$.

$= \dfrac{x - 1}{x + 1}$

Simplify.

9. $\dfrac{q - \frac{1}{3}}{p + \frac{1}{3}} = \dfrac{3\left(q - \frac{1}{3}\right)}{3\left(p + \frac{1}{3}\right)}$ Multiply by the LCM, 3.

$= \dfrac{3q - 1}{3q + 1}$ Simplify.

Simplify. State the restrictions on the variables:

13. $\dfrac{2 - \frac{1}{a}}{3 + \frac{1}{a}} = \dfrac{a\left(2 - \frac{1}{1}\right)}{a\left(3 + \frac{1}{a}\right)}$, $a \neq 0$ Multiply by the LCM, a.

$= \dfrac{2a - 1}{3a + 1}$, $a \neq 0, -\frac{1}{3}$ Simplify.

B

Simplify. You need not state variable restrictions:

17. $\dfrac{\frac{x}{3} + 2}{\frac{x + 1}{3}} = \left(\dfrac{x}{3} + 2\right) \div \dfrac{x + 1}{3}$ Rewrite as a division problem.

$= \dfrac{x + 6}{3} \div \dfrac{x + 1}{3}$ Express the dividend, $\frac{x}{3} + 2$, using 3 as the LCD.

$= \dfrac{x + 6}{3} \cdot \dfrac{3}{x + 1}$ Multiply by the reciprocal of the divisor.

$= \dfrac{x + 6}{x + 1}$ Simplify.

21.

$$\frac{t - \dfrac{t^2}{w}}{t + \dfrac{t}{w^2}} = \frac{w^2\left(t - \dfrac{t^2}{w}\right)}{w^2\left(t + \dfrac{t}{w^2}\right)}$$

Multiply by the LCM, w^2.

$$= \frac{tw^2 - t^2w}{tw^2 + t}$$

Simplify.

$$= \frac{tw(w - t)}{t(w^2 + 1)}$$

Factor both numerator and denominator to identify any possible common factors.
Divide out the common factor, t.

$$= \frac{w(w - t)}{w^2 + 1}$$

$$= \frac{w^2 - tw}{w^2 + 1}$$

Multiply.

25.

$$\frac{\dfrac{5y - 8}{3}}{\dfrac{y + 1}{4} - \dfrac{y - 1}{6}}$$

$$= \frac{5y - 8}{3} \div \left(\frac{y + 1}{4} - \frac{y - 1}{6}\right)$$

Rewrite the divisor using 12 as the LCD. Remember that the fraction bar is a grouping symbol.

$$= \frac{5y - 8}{3} \div \left(\frac{y + 1}{4}\cdot\frac{3}{3} - \frac{y - 1}{6}\cdot\frac{2}{2}\right)$$

$$= \frac{5y - 8}{3} \div \left(\frac{3y + 3}{12} + \frac{-2y + 2}{12}\right)$$

$$= \frac{5y - 8}{3} \div \frac{y + 5}{12}$$

Simplify the divisor by adding.

$$= \frac{5y - 8}{\cancel{3}_{1}} \cdot \frac{\cancel{12}^{4}}{y + 5}$$

Multiply by the reciprocal of the divisor.

$$= \frac{20y - 32}{y + 5}$$

Simplify.

Simplify.   State the restrictions on the variables:

29. $\dfrac{2a^{-1} - 3b^{-1}}{b - a}$

$= \dfrac{\dfrac{2}{a} - \dfrac{3}{b}}{b - a}$

Rewrite the negative exponents.

$= \dfrac{ab\left(\dfrac{2}{a} - \dfrac{3}{b}\right)}{ab(b - a)}$

Multiply by the LCD, ab.

$= \dfrac{2b - 3a}{ab(b - a)}$, $a \neq 0$, $b \neq 0$, $a \neq b$

Multiply.   Restrict the variables.

C

Simplify.   You need not state the variable restrictions:

33. $\dfrac{\dfrac{1}{a + b} - \dfrac{1}{a - b}}{\dfrac{1}{a + b} + \dfrac{1}{a - b}} = \dfrac{(a + b)(a - b)\left(\dfrac{1}{a+b} - \dfrac{1}{a-b}\right)}{(a + b)(a - b)\left(\dfrac{1}{a+b} + \dfrac{1}{a-b}\right)}$

Multiply by the LCD, $(a + b)(a - b)$.

$= \dfrac{(a - b) - (a + b)}{(a - b) + (a + b)}$ Simplify.

$= \dfrac{a - b - a - b}{a - b + a + b} = \dfrac{-2b}{2a} = -\dfrac{b}{a}$

37. $$\dfrac{\dfrac{2}{a-1} + \dfrac{1}{a+1}}{\dfrac{6}{a^2-1}}$$

$$\dfrac{(a+1)(a-1)\left[\dfrac{2}{a-1} + \dfrac{1}{a+1}\right]}{(a+1)(a-1)\left[\dfrac{6}{a^2-1}\right]}$$ Multiply both numerator and denominator by the LCD which is $(a+1)(a-1)$.

$$= \dfrac{2(a+1) + 1(a-1)}{6}$$ Simplify.

$$= \dfrac{2a + 2 + a - 1}{6} = \dfrac{3a+1}{6}$$

41. $$((x+2)^{-1} - (x-2)^{-1})^{-1}$$

$$\left(\dfrac{1}{x+2} - \dfrac{1}{x-2}\right)^{-1}$$ Rewrite the negative exponents of the inner expressions using fractions.

$$= \left(\dfrac{1}{x+2}\cdot\dfrac{x-2}{x-2} - \dfrac{1}{x-2}\cdot\dfrac{x+2}{x+2}\right)^{-1}$$ Build each fraction to have the LCD, $(x+2)(x-2)$, as its denominator.

$$= \left(\dfrac{x-2}{(x+2)(x-2)} - \dfrac{x+2}{(x-2)(x+2)}\right)^{-1}$$

$$= \left(\dfrac{x-2-x-2}{(x+2)(x-2)}\right)^{-1} = \left(\dfrac{-4}{(x+2)(x-2)}\right)^{-1}$$

$$= \dfrac{(x+2)(x-2)}{-4}$$ Rewrite the negative exponent as a reciprocal.

$$= \dfrac{x^2-4}{-4}$$ Simplify.

or $$\dfrac{-1(x^2-4)}{-1(-4)} = \dfrac{4-x^2}{4}$$

Simplify:

45.
$$\dfrac{2a + \dfrac{1}{6 - \dfrac{1}{a}}}{3a + \dfrac{2}{6 + \dfrac{1}{a}}}$$

$$\dfrac{2a + \dfrac{1}{6 - \dfrac{1}{a}} \cdot \dfrac{a}{a}}{3a + \dfrac{2}{6 + \dfrac{1}{a}} \cdot \dfrac{a}{a}}$$

To eliminate the complex fraction in the numerator and denominator, multiply by $\dfrac{a}{a}$.

$$= \dfrac{2a + \dfrac{a}{6a - 1}}{3a + \dfrac{2a}{6a + 1}}$$

Simplify.

$$= \dfrac{(6a - 1)(6a + 1)\left[2a + \dfrac{1}{6a - 1}\right]}{(6a - 1)(6a + 1)\left[3a + \dfrac{2a}{6a + 1}\right]}$$

Multiply the numerator and denominator by $(6a - 1)(6a + 1)$, the LCD.

$$= \dfrac{2a(36a^2 - 1) + a(6a + 1)}{3a(36a^2 - 1) + 2a(6a - 1)}$$

Distributive law.

$$= \dfrac{72a^3 - 2a + 6a^2 + a}{108a^3 - 3a + 12a^2 - 2a}$$

Multiply.

$$= \dfrac{72a^3 + 6a^2 - a}{108a^3 + 12a^2 - 5a}$$

Combine like terms.

$$= \dfrac{a(72a^2 + 6a - 1)}{a(108a^2 + 12a - 5)}$$

Factor out the common monomial, a.

$$= \dfrac{a(12a - 1)(6a + 1)}{a(6a - 1)(18a + 5)}$$

$$= \dfrac{(12a - 1)(6a + 1)}{(6a - 1)(18a + 5)}$$

49.

$$\dfrac{1 + \dfrac{1}{1 + 1}}{1 + \dfrac{1}{1 + \dfrac{1}{1 + 1}}} = \dfrac{1 + \dfrac{1}{2}}{1 + \dfrac{1}{1 + \dfrac{1}{2}}}$$ Add in numerator and denominator.

$$= \dfrac{\dfrac{3}{2}}{1 + \dfrac{1}{\dfrac{3}{2}}}$$ Write mixed numbers as improper fractions.

$$= \dfrac{\dfrac{3}{2}}{1 + \dfrac{2}{3}}$$ Divide in denominator.

$$= \dfrac{\dfrac{3}{2}}{\dfrac{5}{3}}$$ Add in denominator.

$$= \dfrac{9}{10}$$ Divide.

D

53. Show that $\dfrac{x^{-1} + 3^{-1}}{x^{-1} - 3^{-1}} \neq \dfrac{(x + 3)^{-1}}{(x - 3)^{-1}}$

$$\dfrac{2^{-1} + 3^{-1}}{2^{-1} - 3^{-1}} \neq \dfrac{(2 + 3)^{-1}}{(2 - 3)^{-1}}$$ Replace x with 2.

$$\dfrac{\dfrac{1}{2} + \dfrac{1}{3}}{\dfrac{1}{2} - \dfrac{1}{3}} \neq \dfrac{5^{-1}}{(-1)^{-1}}$$

$$\dfrac{\dfrac{5}{6}}{\dfrac{1}{6}} \neq \dfrac{-1}{5}$$

$$5 \neq -\dfrac{1}{5}$$

Therefore, the original expressions are not equivalent.

Solve:

57. $\frac{3}{4}x + \frac{3}{4} = \frac{1}{2}x - 4$

$3x + 3 = 2x - 16$ Multiply both sides by

$x + 3 = -16$ the LCD, 4.

$x = -19$

The solution set is $\{-19\}$.

61. $\frac{7}{9}y - 1 + \frac{2}{3}y = 12$

$7y - 9 + 6y = 108$ Multiply both sides by

$13y - 9 = 108$ the LCD, 9.

$13y = 117$

$y = 9$

The solution set is $\{9\}$.

EXERCISES 3.5   EQUATIONS CONTAINING RATIONAL EXPRESSIONS

A

Solve:

1. $\frac{x}{4} + \frac{1}{3} = \frac{x}{12}$ There are no restrictions
   on the variable.

   $12\left(\frac{x}{4} + \frac{1}{3}\right) = 12\left(\frac{x}{12}\right)$ Multiply each side of the
   equation by the LCD of 4,
   3, and 12.   LCD = 12.

   $3x + 4 = x$ Simplify.

   $2x = -4$

   $x = -2$

   The solution set is $\{-2\}$.

5. $\frac{5}{t - 2} = 4$

   $(t - 2)\left[\frac{5}{t - 2}\right] = (t - 2)4$ Multiply each side of the
   equation by the LCD,
   t - 2.

   $5 = 4t - 8$ Simplify.

   $13 = 4t$

   $\frac{13}{4} = t$

   The solution set is $\left\{\frac{13}{4}\right\}$.

9.
$$\frac{4}{x + 3} = \frac{5}{2x + 7}$$

$$(x + 3)(2x + 7)\left(\frac{4}{x + 3}\right) = (x + 3)(2x + 7)\left(\frac{5}{2x + 7}\right)$$

Multiply each side by the LCD: $(x+3)(2x+7)$.

$$4(2x + 7) = 5(x + 3)$$
$$8x + 28 = 5x + 15$$
$$3x = -13$$
$$x = -\frac{13}{3}$$

The solution set is $\left\{-\frac{13}{3}\right\}$.

13.
$$\frac{4}{x - 3} + \frac{3}{x - 3} = 5$$

$$x - 3\left(\frac{4}{x - 3} + \frac{3}{x - 3}\right) = (x - 3)5$$
$$4 + 3 = 5x - 15$$
$$22 = 5x$$
$$\frac{22}{5} = x$$

The solution set is $\left\{\frac{22}{5}\right\}$.

17.
$$\frac{2}{x + 4} + \frac{3}{x + 4} = 10$$

$$x + 4\left[\frac{2}{x + 4} + \frac{3}{x + 4}\right] = (x + 4)10$$
$$2 + 3 = 10x + 40$$
$$5 = 10x + 40$$
$$-35 = 10x$$
$$-\frac{35}{10} = x$$
$$-\frac{7}{2} = x$$

The solution set is $\left\{-\frac{7}{2}\right\}$.

21.
$$\frac{1}{x + 2} - \frac{3}{4} = \frac{7}{x + 2}$$

$$4(x + 2)\left(\frac{1}{x + 2} - \frac{3}{4}\right) = 4(x + 2)\left(\frac{7}{x + 2}\right)$$

Multiply each side by the LCD of 4 and x + 2. LCD = 4(x + 2).

$$4 - 3(x + 2) = 4(7)$$
$$4 - 3x - 6 = 28$$
$$-3x = 30$$
$$x = -10$$

The solution set is {-10}.

25.
$$\frac{3}{y + 3} - \frac{9}{y^2 - 9} = \frac{2}{y - 3}$$
$$3(y - 3) - 9 = 2(y + 3)$$

Multiply each side by (y - 3)(y + 3).

$$3y - 9 - 9 = 2y + 6$$
$$3y - 18 = 2y + 6$$
$$y - 18 = 6$$
$$y = 24$$

The solution set is {24}.

29.
$$\frac{4}{x - 3} - \frac{8}{x^2 - 9} = \frac{3}{3 - x}$$

$$\frac{4}{x - 3} - \frac{8}{x^2 - 9} = \frac{-1(3)}{-1(3 - x)}$$

Multiply the numerator and denominator by -1.

$$\frac{4}{x - 3} - \frac{8}{x^2 - 9} = \frac{-3}{x - 3}$$

$$(x + 3)(x - 3)\left(\frac{4}{x - 3} - \frac{8}{x^2 - 9}\right) = (x + 3)(x - 3)\left(\frac{-3}{x - 3}\right)$$

Now multiply both sides by the LCM: $x^2 - 9$ or (x+3)(x-3). Simplify.

$$4(x - 3) - 8 = -3(x + 3)$$
$$4x + 12 - 8 = -3x - 9$$
$$7x = -13$$
$$x = -\frac{13}{7}$$

The solution set is $\left\{-\frac{13}{7}\right\}$.

33. Solve for a: $\dfrac{3}{b+a} - \dfrac{2}{b} = 5$

$$b(b+1)\left(\dfrac{3}{b+a} - \dfrac{2}{b}\right) = b(b+a)(5)$$

Multiply both sides by the LCM, $b(b+a)$.

$$3b - 2(b+a) = 5(b^2 + ab)$$
$$3b - 2b - 2a = 5b^2 + 5ab$$
$$-2a - 5ab = 5b^2 - b$$
$$a(-2 - 5b) = 5b^2 - b$$

On the left, factor out an a.

$$a = \dfrac{5b^2 - b}{-2 - 5b} \cdot \dfrac{-1}{-1}$$
$$a = \dfrac{b - 5b^2}{5b + 2}$$

Solve:

37.
$$\dfrac{7}{x+2} - \dfrac{3}{x+4} = \dfrac{2}{x^2 + 6x + 8}$$
$$7(x+4) - 3(x+2) = 2$$
$$7x + 28 - 3x - 6 = 2$$
$$4x + 22 = 2$$
$$4x = -20$$
$$x = -5$$

Multiply both sides by the LCD: $(x+2)(x+4)$.

The solution set is $\{-5\}$.

c

41.
$$\dfrac{1}{x^2 + 4x + 3} - \dfrac{1}{x^2 - 5x - 6} = \dfrac{1}{x^2 - 3x - 18}$$

$$\dfrac{1}{(x+3)(x+1)} - \dfrac{1}{(x-6)(x+1)} = \dfrac{1}{(x-6)(x+3)}$$

$$(x-6) - (x+3) = x + 1$$
$$x - 6 - x - 3 = x + 1$$
$$-9 = x + 1$$
$$-10 = x$$

Multiply both sides by the LCD, $(x+3)(x+1)(x-6)$, and simplify.

The solution set is $\{-10\}$.

45.

$$\frac{x + 3}{x^2 - x - 2} + \frac{x + 6}{x^2 + 3x + 2} = \frac{2x - 1}{x^2 - 4}$$

$$\frac{x + 3}{(x - 2)(x + 1)} + \frac{x + 6}{(x + 2)(x + 1)} = \frac{2x - 1}{(x - 2)(x + 2)}$$

The LCD is
$(x-2)(x+1)(x+2)$.

$$(x + 2)(x + 3) + (x + 6)(x - 2) = (2x - 1)(x + 1)$$

Multiply by the LCD and
simplify.

$$x^2 + 5x + 6 + x^2 + 4x - 12 = 2x^2 + x - 1$$
$$9x - 6 = x - 1$$
$$8x - 6 = -1$$
$$8x = 5$$
$$x = \frac{5}{8}$$

The solution set is $\left\{\frac{5}{8}\right\}$.

49.

$$\frac{4 + x}{x^2 - 2x - 8} - \frac{3 + 2x}{2x^2 + 5x + 2} = \frac{7}{2x^2 - 7x - 4}$$

$$= \frac{4 + x}{(x - 4)(x + 2)} - \frac{3 + 2x}{(2x + 1)(x + 2)} = \frac{7}{(2x + 1)(x - 4)}$$

Factor.

$$= (4 + x)(2x + 1) - (3 + 2x)(x - 4) = 7(x + 2)$$

Multiply both sides by the
LCD: $(x-4)(x+2)(2x+1)$.

$$= 2x^2 + 9x + 4 - [2x^2 - 5x - 12] = 7x + 14$$
$$= 2x^2 + 9x + 4 - 2x^2 + 5x + 12 = 7x + 14$$
$$14x + 16 = 7x + 14$$
$$7x = -2$$
$$x = -\frac{2}{7}$$

The solution set is $\left\{-\frac{2}{7}\right\}$.

53. $x^{-1} + 5^{-1} = 3x^{-1} + 1$

$$\frac{1}{x} + \frac{1}{5} = \frac{3}{x} + 1$$

Rewrite with positive
exponents.

$$5 + x = 15 + 5x$$
$$-4x = 10$$
$$x = -\frac{5}{2}$$

Multiply both sides by
the LCD: 5x.

The solution set is $\left\{-\frac{5}{2}\right\}$.

114

57. $\dfrac{\dfrac{5}{x} + \dfrac{2}{3}}{\dfrac{3}{x}} = 1$

Simplify the complex fraction.

$\dfrac{3x\left[\dfrac{5}{x} + \dfrac{2}{3}\right]}{3x\left[\dfrac{3}{x}\right]} = 1$

Multiply the left side by $\dfrac{3x}{3x}$.

$\dfrac{15 + 2x}{9} = 1$

$15 + 2x = 9$ Multiply both sides by 9.

$2x = -6$

$x = -3$

The solution set is $\{-3\}$.

D

61. Frank and Freda can pick a quart of blueberries in 12 minutes. Frank alone takes 18 minutes longer than Freda. How long does it take each of them to pick a quart?

Simpler word form:

$\begin{pmatrix}\text{Fraction of work} \\ \text{done by Frank} \\ \text{in one minute}\end{pmatrix} + \begin{pmatrix}\text{Fraction of work} \\ \text{done by Freda} \\ \text{in one minute}\end{pmatrix} = \begin{pmatrix}\text{Fraction of work} \\ \text{done by both} \\ \text{in one minute}\end{pmatrix}$

Select variable:

Let t represent the time it takes Freda to pick a quart of berries.

	Time alone	Fraction done in one minute
Freda	t	$\dfrac{1}{t}$
Frank	t + 18	$\dfrac{1}{t + 18}$
Both	12	$\dfrac{1}{12}$

$$\frac{1}{t + 18} + \frac{1}{t} = \frac{1}{12}$$

$12t + 12(t + 18) = t(t + 18)$ Multiply each side by the

$12t + 12t + 216 = t^2 + 18t$ LCD: $12t(t + 18)$.

$t^2 - 6t - 216 = 0$

$(t - 18)(t + 12) = 0$

$t - 18 = 0$ or $t + 12 = 0$

$t = 18$ or $t = -12$ Reject $t = -12$ as t represents a positive number.

Frank takes 18 minutes longer than Freda, therefore, 18 (Freda's time) + 18 = 36.
It takes Freda 18 minutes to pick a quart of blueberries, and it takes Frank 36 minutes.

65.   Charlie and Jim decide to drive to Seattle. They both leave from the same place at the same time, but Charlie drives 5 miles/hr faster than Jim. If Charlie has traveled 65 miles in the same time that Jim has traveled 60 miles, find how fast each was driving.

Simpler word form:

$$\left(\begin{array}{c}\text{Time it takes Charlie}\\ \text{to travel 65 miles}\end{array}\right) = \left(\begin{array}{c}\text{Time it takes Jim}\\ \text{to travel 60 miles}\end{array}\right)$$

Select a variable:
Let r represent the rate of speed that Jim travels. Then r + 5 represents the rate of speed that Charlie travels.

	Distance traveled	Rate of Speed	Time
Jim	60	r	$\frac{60}{r}$
Charlie	65	r + 5	$\frac{65}{r + 5}$

Translate to algebra:

$$\frac{60}{r} = \frac{65}{r + 5}$$

$60(r + 5) = 65r$ Cross multiply.

$60r + 300 = 65r$ Simplify.

$300 = 5r$

$r = 60$ Jim's rate of speed.

Charlie travels 5 miles faster
than Jim so 60 + 5 = 65.   Charlie's rate of speed.

Charlie was traveling at 65 mph, and Jim was traveling at 60 mph.

69. Ariel saved $900 for vacation. She figures that by spending $15 a day less, she can stay away 3 days longer than originally planned. How many days did she originally plan to be gone?

(Hint: $\dfrac{\text{total cost}}{\text{number of days}}$ = cost per day)

Simpler word form:

$$\left(\dfrac{\text{total cost}}{\substack{\text{number of days} \\ \text{originally planned}}}\right) - 15 = \left(\dfrac{\text{total cost}}{\substack{\text{number of days} \\ \text{in new plan}}}\right)$$

Select a variable:
Let x represent the number of days Ariel originally planned to be gone. Then x + 3 represents the number of days she will be away in her new plan.

Translate to algebra:
$$\dfrac{900}{x} - 15 = \dfrac{900}{x + 3}$$

Solve:

$x(x + 3)\left(\dfrac{900}{x} - 15\right) = x(x + 3)\left(\dfrac{900}{x + 3}\right)$ Multiply each side by the LCM: $x(x + 3)$.

$900(x + 3) - 15(x)(x + 3) = 900x$	Simplify.
$900 + 2700 - 15x^2 - 45x = 900x$	
$-15x^2 - 45x + 2700 = 0$	
$15x^2 + 45x - 2700 = 0$	Multiply each side by -1.
$15(x^2 + 3x - 180) = 0$	Factor.
$(x + 15)(x - 12) = 0$	Divide each side by 15 and factor.
$x + 15 = 0$ or $x - 12 = 0$	Zero-product law.
$x = -15$ $x = 12$	Reject -15 because it cannot represent the number of vacation days.

So Ariel orginally planned to be away for 12 days.

STATE YOUR UNDERSTANDING

73. Explain the procedure for solving an equation containing rational expressions.

   Determine what the lowest common denominator is. Use it to multiply both sides of the equation. The result will be an equation that contains no rational expressions, and can be solved by the usual methods.

77.
$$(x + 5)^{-1} - (x + 2)^{-1} + (x - 2)^{-1} = x(x^2 - 4)^{-1}$$

$$\frac{1}{x + 5} - \frac{1}{x + 2} + \frac{1}{x - 2} = \frac{x}{x^2 - 4}$$

$$(x^2 - 4) - (x + 5)(x - 2) + (x + 5)(x + 2) = x(x + 5)$$

The LCD is $(x+5)(x^2-4)$. Multiply both sides by the LCD.

$$x^2 - 4 - [x^2 + 3x - 10] + [x^2 + 7x + 10] = x^2 + 5x$$
$$x^2 - 4 - x^2 - 3x + 10 + x^2 + 7x + 10 = x^2 + 5x$$
$$x^2 + 4x + 16 = x^2 + 5x$$
$$4x + 16 = 5x$$
$$-x + 16 = 0$$
$$-x = -16$$
$$x = 16$$

The solution set is {16}.

MAINTAIN YOUR SKILLS

Solve for w:

81. $1 - 8w < w + 16$

$1 - 9w < 16$

$-9w < 15$

$\dfrac{-9w}{-2} > \dfrac{15}{-9}$

$w > -\dfrac{15}{9}$

$w > -\dfrac{5}{3}$

The solution in interval notation is $\left(-\dfrac{5}{3}, +\infty\right)$.

Multiply:

85. $(b - 5)(2b^2 - 4b + 5)$

$= (b - 5)(2b^2) + (b - 5)(-4b) + (b - 5)(5)$

$= 2b^3 - 10b^2 - 4b^2 + 20b + 5b - 25$

$= 2b^3 - 14b^2 + 25b - 25$

A

Divide:

1. $$\frac{5x^3 + 15x^2 + 10x}{5x}$$

$$= \frac{5x^3}{5x} + \frac{15x^2}{5x} + \frac{10x}{5x}$$ Rewrite as the sum of fractions.
$$= x^2 + 3x + 2$$ Reduce.

5. $$\frac{14x^3 - 28x^2 - 6x}{7x}$$

$$= \frac{14x^3}{7x} - \frac{28x^2}{7x} - \frac{6x}{7x}$$ Rewrite as the sum of fractions.
$$= 2x^2 - 4x - \frac{6}{7}$$ Reduce.

9. $$\frac{10x^2y^2 - 15xy^3 - 20y}{5xy}$$

$$= \frac{10x^2y^2}{5xy} - \frac{15xy^3}{5xy} - \frac{20y}{5xy}$$
$$= 2xy - 3y^2 - \frac{4}{x}$$

Divide:

13. $$a^2 - 3 \overline{)\begin{array}{l} a \\ a^2 - 3a \\ \underline{a^2 - 3a} \\ 0 \end{array}}$$

a times $a-3$ equals a^2-3a. Place the a in the answer above $-3a$ and multiply. Subtract. The remainder is 0.

The quotient is a.

17. $$6x + 11 \overline{)\begin{array}{l} 3x \\ 18x^2 + 33x \\ \underline{18x^2 + 33x} \\ 0 \end{array}}$$

$3x$ times $6x - 11$ equals $18x^2 - 33x$. Place the $3x$ in the answer above $33x$. Multiply. Subtract.

The quotient is $3x$.

B

21.

$$
\begin{array}{r}
x + 3 \\
x + 2\overline{)x^2 + 5x + 6} \\
\underline{x^2 + 2x} \\
3x + 6 \\
\underline{3x + 6} \\
0
\end{array}
$$

x times x equals x^2. Place the x in the answer above 5x. Multiply. Subtract. 3 times x equals 3x. Place the 3 in the answer above 6, and multiply. Subtract.

The quotient is x + 3.

25. $(x^2 + 7x - 18) \div (x + 9)$

$$
\begin{array}{r}
x - 2 \\
x + 9\overline{)x^2 + 7x - 18} \\
\underline{x^2 + 9x} \\
-2x - 18 \\
\underline{-2x - 18} \\
0
\end{array}
$$

Rewrite as a long division problem. $x \cdot x = x^2$. $-2 \cdot x = -2x$.

The quotient is x - 2.

29.

$$
\begin{array}{r}
2x + 7 \\
x - 3\overline{)2x^2 + x - 21} \\
\underline{2x^2 - 6x} \\
7x - 21 \\
\underline{7x - 21} \\
0
\end{array}
$$

$2x \cdot x = 2x^2$.

$7 \cdot x = 7x$.

The quotient is 2x + 7

33.

$$
\begin{array}{r}
3x + 7 \\
2x + 3\overline{)6x^2 + 23x + 21} \\
\underline{6x^2 + 9x} \\
14x + 21 \\
\underline{14x + 21} \\
0
\end{array}
$$

3x times 2x equals $6x^2$.

7 times 2x equals 14x.

The qotient is 3x + 7.

120

37. $(2x^2 - 5x + 16) \div (2x + 1)$

$$
\begin{array}{r}
x - 3 \\
2x + 1 \overline{)\, 2x^2 - 5x + 16} \\
\underline{2x^2 + x } \\
-6x + 16 \\
\underline{-6x - 3} \\
19
\end{array}
$$

The remainder is 19.

So $(2x^2 - 5x + 16) \div (2x + 1) = x - 3 + \dfrac{19}{2x + 1}$.

C

41. $(x^3 + 3x^2 + 3x + 1) \div (x + 1)$

$$
\begin{array}{r}
x^2 + 2x + 1 \\
x + 1 \overline{)\, x^3 + 3x^2 + 3x + 1} \\
\underline{x^3 + x^2 } \\
2x^2 + 3x + 1 \\
\underline{2x^2 + 2x } \\
x + 1 \\
\underline{x + 1} \\
0
\end{array}
$$

The quotient is $x^2 + 2x + 1$.

45.

$$
\begin{array}{r}
x^2 + 3 \\
x - 2 \overline{)\, x^3 - 2x^2 + 3x - 6} \\
\underline{x^3 - 2x^2 } \\
3x - 6 \\
\underline{3x - 6} \\
0
\end{array}
$$

The quotient is $x^2 + 3$.

49.

$$
\begin{array}{r}
2x^2 + x - 1 \\
3x + 1 \overline{)\, 6x^3 + 5x^2 - 2x + 1} \\
\underline{6x^3 + 2x^2 } \\
3x^2 - 2x + 1 \\
\underline{3x + x } \\
-3x + 1 \\
\underline{-3x - 1} \\
2
\end{array}
$$

The quotient is $2x^2 + x - 1 + \dfrac{2}{3x + 1}$.

53.

$$
\begin{array}{r}
x^5 - x^4 + x^3 - x^2 + x - 1 \\
x + 1{\overline{)}}\,x^6 + 0x^5 + 0x^4 + 0x^3 + 0x^2 + 0x + 1 \\
\underline{x^6 + x^5} \\
-x^5 + 0x^4 + 0x^3 + 0x^2 + 0x + 1 \\
\underline{-x^5 - x^4} \\
x^4 + 0x^3 + 0x^2 + 0x + 1 \\
\underline{x^4 + x^3} \\
-\,x^3 + 0x^2 + 0x + 1 \\
\underline{-\,x^3 - x^2} \\
x^2 + 0x + 1 \\
\underline{x^2 + x} \\
-x + 1 \\
\underline{-x - 1} \\
2
\end{array}
$$

Insert zeros as place holders for the terms that are missing.

The quotient is $x^5 - x^4 + x^3 - x^2 + x - 1 + \dfrac{2}{x + 1}$.

D

57.   Is $x + 5$ a factor of $x^3 + 10x^2 + 36x + 55$? If so, write the factor.

$$
\begin{array}{r}
x^2 + 5x + 11 \\
x + 5{\overline{)}}\,x^3 + 10x^2 + 36x + 55 \\
\underline{x^3 + 5x^2} \\
5x^2 + 36x + 55 \\
\underline{5x^2 + 25x} \\
11x + 55 \\
\underline{11x + 55} \\
0
\end{array}
$$

Yes, $x + 5$ is a factor of $x^3 + 10x^2 + 36x + 55$.
The factors are $x + 5$ and $x^2 + 5x + 11$ or
$(x + 5)(x^2 + 5x + 11)$.

61.   Factor $x^3 + 9x^2 + 26x + 24$ completely if $x + 4$ is one of its factors.

$$
\begin{array}{r}
x^2 + 5x + 6 \\
x + 4{\overline{)}}\,x^3 + 9x^2 + 26x + 24 \\
\underline{x^3 + 4x^2} \\
5x^2 + 26x \\
\underline{5x^2 + 20x} \\
6x + 24 \\
\underline{6x + 24}
\end{array}
$$

Divide $x^3 + 9x^2 + 26x + 24$ by $x + 4$.

$x^2 + 5x + 6 = (x + 2)(x + 3)$ Now factor the quotient.

So $x^3 + 9x^2 + 26x + 24 = (x + 4)(x^2 + 5x + 6)$
$ = (x + 4)(x + 3)(x + 2)$

65. Factor $x^5 + 3x^4 - 16x - 48$ completely if $x + 3$ is one of its factors.

$$
\begin{array}{r}
x^4 \qquad\qquad\qquad -16 \\
x + 3 \overline{)x^5 + 3x^4 + 0x^3 + 0x^2 - 16x - 48} \\
\underline{x^5 + 3x^4 \qquad\qquad\qquad\qquad} \\
-16x - 48 \\
\underline{-16x - 48}
\end{array}
$$

Divide $x^5 + 3x^4 - 16x - 48$ by $x + 3$.

$$
\begin{aligned}
x^4 - 16 &= (x^2 - 4)(x^2 + 4) \\
&= (x - 2)(x + 2)(x + 4)
\end{aligned}
$$

Now factor the quotient which is the difference of squares.

$$
\begin{aligned}
\text{So } x^5 + 3x^4 - 16x - 48 &= (x + 3)(x^4 - 16) \\
&= (x + 3)(x - 2)(x + 2)(x^2 + 4)
\end{aligned}
$$

CHALLENGE EXERCISES

Divide; assume all exponents are positive integers:

69. $(x^{2n} + 4x^n + 3) \div (x^n + 1)$

$$
\begin{array}{r}
x^n + 3 \\
x^n + 1 \overline{)x^{2n} + 4x^n + 3} \\
\underline{x^{2n} + x^n \qquad\quad} \\
3x^n + 3 \\
\underline{3x^n + 3}
\end{array}
$$

So $(x^{2n} + 4x^n + 3) \div (x^n + 1) = (x^n + 3)$

73. Find a value for k so that if $x^3 + 6x^2 - x - k$ is divided by $x - 2$, the remainder is 1.

$$
\begin{array}{r}
x^2 + 8x + 15 \\
x - 2 \overline{)x^3 + 6x^2 - x - k} \\
\underline{x^3 - 2x^2 \qquad\qquad\quad} \\
8x^2 - x \\
\underline{8x^2 - 16x} \\
15x - k \\
\underline{15x - 30} \\
1
\end{array}
$$

So, $-k - (-30) = 1$
$-k + 30 = 1$
$-k = -29$
$k = 29$

In the expression, $x^3 + 6x^2 - x - k$, if k = 29 then the remainder is 1.

Multiply:

77. $(2y - 1)(2y + 1)(4y^2 + 1)$
   $= (4y^2 - 1)(4y^2 + 1)$
   $= 16y^4 - 1$

81. $(3y + 2)^3 = (3y + 2)(3y + 2)(3y + 2)$
   $= (9y^2 + 12y + 4)(3y + 2)$
   $= 9y^2(3y + 2) + 12y(3y + 2) + 4(3y + 2)$
   $= 27y^3 + 18y^2 + 36y^2 + 24y + 12y + 8$
   $= 27y^3 + 54y^2 + 36y + 8$

EXERCISES 3.7   SYNTHETIC DIVISION

A

Divide.   Use synthetic division.

1. $(x^2 - 2x - 15) \div x - 5$

$$\begin{array}{r|rrr} 5 & 1 & -2 & -15 \\ & & +5 & +15 \\ \hline & 1 & +3 & 0 \end{array}$$

Divide by 5, the opposite of -5.

The quotient is $x + 3$.

5. $(4x^2 - 31x + 55) \div x - 5$

$$\begin{array}{r|rrr} 5 & 4 & -32 & 55 \\ & & 20 & -55 \\ \hline & 4 & -11 & 0 \end{array}$$

Divide by 5, the opposite of -5.

The quotient is $4x - 11$.

9. $(x^3 + 2x^2 - 8x + 5) \div x - 1$

$$\begin{array}{r|rrrr} 1 & 1 & 2 & -8 & 5 \\ & & 1 & 3 & -5 \\ \hline & 1 & 3 & -5 & 0 \end{array}$$

Divide by 1.

The quotient is $x^2 + 3x - 5$.

B

Divide. Use synthetic division.

13. $(x^2 - 3x + 7) \div (x + 2)$

$$-2 \begin{array}{|cccc} 1 & - & 3 & + & 7 \\ & & - & 2 & + & 10 \\ \hline 1 & & - & 5 & + & 17 \end{array}$$

The remainder is 17.

The quotient is $x - 5 + \dfrac{17}{x + 2}$.

17. $(x^3 - 5x^2 + 12x - 15) \div (x - 3)$

$$3 \begin{array}{|cccc} 1 & -5 & 12 & -15 \\ & 3 & -6 & 18 \\ \hline 1 & -2 & 6 & 3 \end{array}$$

The quotient is $x^2 - 2x + 6 + \dfrac{3}{x - 3}$.

21. $(2x^3 - 5x^2 + 4x + 3) \div (x - 1)$

$$1 \begin{array}{|cccc} 2 & -5 & 4 & 3 \\ & 2 & -3 & 1 \\ \hline 2 & -3 & 1 & 4 \end{array}$$

The quotient is $2x^2 - 3x + 1 + \dfrac{4}{x - 1}$.

C

Divide. Use synthetic division.

25. $(x^3 - 3x - 7) \div (x + 2)$

$$-2 \begin{array}{|cccc} 1 & + 0 & - 3 & - 7 \\ & - 2 & + 4 & - 2 \\ \hline 1 & - 2 & + 1 & - 9 \end{array}$$

Insert zero for the missing x term.

The missing quotient is $x^2 - 2x + 1 - \dfrac{9}{x + 2}$.

29. $(2x^4 - 13x^3 + 17x^2 + 18x - 24) \div (x - 4)$

$$4 \begin{array}{|ccccc} 2 & - 13 & + 17 & + 18 & - 24 \\ & + 8 & - 20 & - 12 & + 24 \\ \hline 2 & - 5 & - 3 & + 6 & + 0 \end{array}$$

The quotient is $2x^3 - 5x^2 - 3x + 6$

33. $(x^4 - 16) \div (x + 2)$

$$
\begin{array}{r|rrrrr}
-2 & 1 & +\,0 & +\,0 & +\,0 & -\,16 \\
   &   & -\,2 & +\,4 & -\,8 & +\,16 \\
\hline
   & 1 & -\,2 & +\,4 & -\,8 & +\,0
\end{array}
$$

The quotient is $x^3 - 2x^2 + 4x - 8$.

D

37. Is $x - 5$ a factor of $x^5 - 19x^3 + 20x^2 - 19x + 5$?

$$
\begin{array}{r|rrrrrr}
5 & 1 & -\,0 & -\,19 & +\,20 & -\,19 & +\,5 \\
  &   & +\,5 & +\,25 & +\,30 & +\,250 & +\,1150 \\
\hline
  & 1 & +\,5 & +\,6 & +\,50 & +\,231 & +\,1155
\end{array}
$$

Divide. Use synthetic division. Insert zero for the missing x^2 term. The remainder is 1155.

Since there is a remainder, $x - 5$ is not a factor of the given polynomial.

41. Is $x + 2$ a factor of $2x^2 - 5x - 3$? If so, write the factors.

$$
\begin{array}{r}
2x - 9 \\
x + 2 \overline{)\, 2x^2 - 5x - 3} \\
\underline{2x^2 + 4x} \\
-9x - 3 \\
\underline{-9x - 18} \\
15
\end{array}
$$

Since there is a remainder, $x + 2$ is not a factor of the given polynomial.

45. Is 1 a solution of $4x^2 - x - 3 = 0$?

$$
\begin{array}{r|rrr}
1 & 4 & -1 & -3 \\
  &   & 4 & 3 \\
\hline
  & 4 & 3 & 0
\end{array}
$$

Divide the polynomial by 1. The remainder is zero.

Yes, 1 is a solution.

49. Is -5 a solution of $x^3 + 10x^2 + 36x + 55 = 0$?

$$
\begin{array}{r|rrrr}
-5 & 1 & 10 & 36 & 55 \\
   &   & -5 & -25 & -55 \\
\hline
   & 1 & 5 & 11 & 0
\end{array}
$$

Divide the polynomial by -5. The remainder is zero.

Yes, -5 is a solution.

Use synthetic division to divide the following. Assume all exponents are positive integers:

53. $(x^{4n} - 5x^{3n} - 10x^{2n} - x^n + 15) \div x^n - 1$

$$
\begin{array}{r|rrrrr}
1 & 1 & -5 & -10 & -1 & 15 \\
  &   & 1 & -4 & -14 & -15 \\
\hline
  & 1 & -4 & -14 & -15 & 0
\end{array}
$$

So $x^{4n} - 5x^{3n} - 10x^{2n} - x^n + 15 \div x^n - 1 =$

$$x^{3n} - 4x^{2n} - 14x^n - 15$$

MAINTAIN YOUR SKILLS

Factor completely:

57. $x^4 - x^2 - 12 = (x^2 - 4)(x^2 + 3)$
$$= (x - 2)(x + 2)(x^2 + 3)$$

61. $mx^2 + nx^2 - 9m - 9n = (mx^2 + nx^2) + (-9m - 9n)$
$$= x^2(m + n) - 9(m + n)$$
$$= (x^2 - 9)(m + n)$$
$$= (x - 3)(x + 3)(m + n)$$

CHAPTER 4

ROOTS, RADICALS, AND COMPLEX NUMBERS

EXERCISES 4.1   RATIONAL EXPONENTS

B

Simplify, using only positive exponents.

1. $144^{1/2} = 12$

The exponent, 1/2, denotes the positive square root.

5. $5^{3/5} \cdot 5^{1/5} = 5^{4/5}$

First law of exponents. When multiplying powers with the same base, add the exponents.

9. $\dfrac{x^{5/9}}{x^{2/9}} = x^{5/9 - 2/9} = x^{3/9}$

Second law of exponents. When dividing powers with the same base, subtract the exponents.

$= x^{1/3}$

Reduce the fractional exponent.

13. $(10^{1/2})^3 = 10^{3/2}$

Third law of exponents. When raising a power to another power, multiply the exponents.

17. $(x^{3/2})^4 = x^6$

Multiply the exponents.

B

Simplify, using only positive exponents:

21. $3x^{1/2} \cdot 5x^{1/4} = (3 \cdot 5)(x^{1/2} \cdot x^{1/4})$

Commutative and associative laws of multiplication.

$= 15x^{1/2 + 1/4}$

First law of exponents.

$= 15x^{3/4}$

25. $(4^{1/3}x^{3/7})(4^{2/3}x^{3/7}) = (4^{1/3} \cdot 4^{2/3})(x^{3/7} \cdot x^{3/7})$ Commutative and associative laws of multiplication.

$$= 4^{3/3} \cdot x^{6/7}$$

First law of exponents.

$$= 4x^{6/7}$$

Simplify.

29. $\dfrac{-72d^{1/3}}{9d^{-1/6}} = -8d^{1/3-(-1/6)}$ Divide the coefficients.

Second law of exponents. When dividing powers with the same base, subtract the exponents.

$$= -8d^{1/2}$$

33. $(27x^{3/4})^{2/3} = 27^{2/3} \cdot x^{(3/4)(2/3)}$ Fourth law of exponents.

$$= 9x^{1/2} \qquad\qquad 27^{2/3} = (27^{1/3})^2 = (3)^2$$

37. $(-4c^{1/4}d^{1/2})^3 = (-4)^3 \cdot (c)^{3/4} \cdot (d)^{3/2}$ Fourth law of exponents.

$$= -64c^{3/4}d^{3/2}$$

c

Simplify, using only positive exponents:

41. $(x^{1/2}y^2)(x^{2/3}y^{3/5}) = (x^{1/2} \cdot x^{2/3})(y^2 \cdot y^{3/5})$ Commutative and associative laws of multiplication.

$$= x^{7/6}y^{13/5}$$

45. $(3a^{3/5}b^{7/8})^2 = 3^2 \cdot a^{6/5} \cdot b^{14/8}$ Fourth law of exponents.

$$= 9a^{6/5}b^{7/4}$$

49. $(x^{-1/2}y^{2/3})^{-2/3} = (x^{-1/2})^{-2/3}(y^{2/3})^{-2/3}$ Fourth law of exponents.

$$= x^{1/3}y^{-4/9}$$ Third law of exponents.

$$= \frac{x^{1/3}}{y^{4/9}}$$ $y^{-4/9} = \dfrac{1}{y^{4/9}}$

53. $\left[\dfrac{x^{-5/6}}{z^{-2/3}}\right]^{-1/2} = \dfrac{(x^{-5/6})^{-1/2}}{(z^{-2/3})^{-1/2}}$ Fifth law of exponents.

$$= \frac{x^{5/12}}{z^{1/3}}$$

57. $(a^{2/3} + 3b^{1/3})(a^{1/3} - 2b^{2/3})$

$= \underset{F}{(a^{2/3})(a^{1/3})} + \underset{0}{(a^{2/3})(-2b^{2/3})} + \underset{I}{(3b^{1/3})(a^{1/3})} +$

$\quad + \underset{L}{(3b^{1/3})(-2b^{2/3})}$

$= a^{3/3} - 2a^{2/3}b^{2/3} + 3a^{1/3}b^{1/3} - 6b^{3/3}$

$= a \quad\;\; - 2a^{2/3}b^{2/3} + 3a^{1/3}b^{1/3} - 6b$

D

61. Two lines of a computer program read

LINE NUMBER	COMMAND
310	y = 16
320	x = y ** 1.5

What is the numerical value of x?

(The symbol "**" is used to tell the computer that the next number is an exponent.)

$x = y ** 1.5$ Computer statement 320.
$x = 16** 1.5$ Substitution, y = 16, from computer statement 310.

So $x = 16^{1.5} = 16^{3/2}$ Algebra statement.

$x = (16^{1/2})^3$ Third law of exponents. Find the root first, then raise to the power, 3.

$x = (4)^3$ Since $16^{1/2} = 4$.
$x = 64$

65. Two lines of a computer program read

LINE NUMBER	COMMAND
530	w = 16
540	y = (w + 84) ** 1.5

What is the numerical value of y?
(Remember that the number following "**" is an exponent.)

$y = (w + 84)**1.5$	Computer statement 540.
$y = (16 + 84)**1.5$	Substitution, w = 16, from computer statement 530.

So $y = 100^{1.5} = 100^{3/2}$	Algebra statement.
$y = (100^{1/2})^3$	Third law of exponents.
$y = 10^3$	Since $100^{1/2} = 10$.
$y = 1000$	

CHALLENGE EXERCISES

Simplify; assume all variable exponents represent positive integers:

69. $(x^3)^{n/3} \cdot (x^{2n})^{1/2} = x^n \cdot x^n$	Multiply exponents.
$= x^{2n}$	Add exponents.

Factor as indicated:

73. $x^{3/2} + x^{1/2} = x^{1/2}(? + ?)$	
$= x^{1/2}(x^1 + ?)$	Since 3/2 = 1/2 + 1, replace the first ? with x^1.
$= x^{1/2}(x^1 + x^0)$	Since 1/2 = 1/2 + 0, replace the second ? with x^0.
$= x^{1/2}(x + 1)$	Simplify.

MAINTAIN YOUR SKILLS

Calculate. Write answer in scientific notation:

77. $(8.46 \times 10^2)(5 \times 10^{-3})(0.2 \times 10^4)$

$= (8.46 \times 5 \times 0.2)(10^2 \times 10^{-3} \times 10^4)$

$= 8.46 \times 10^3$

132

Factor completely:

81. $(x - 4)^2 - (y - 3)^2$
    $= (x - 4 + y - 3)[(x - 4) - (y - 3)]$ Difference of
                                                    squares.
    $= (x + y - 7)(x - 4 - y + 3)$
    $= (x + y - 7)(x - y - 1)$

EXERCISES 4.2   RADICALS

Recall that, unless the directions state otherwise, all variables
represent only positive values.

A

Write this expression in radical form and simplify:

1. $32^{1/5} = \sqrt[5]{32}$ The denominator, 5, in the
                                                    exponent indicates the
                                                    fifth root.
    $= 2$ Since $2^5 = 32$.

Write this expression in radical form. Assume variables
represent non-negative real numbers:

5. $(47a)^{1/2} = \sqrt{47a}$ The denominator, 2, in the
                                                    exponent indicates the
                                                    square root.

Write this expression with a rational exponent. Assume variables
represent non-negative real numbers:

9. $\sqrt{xy} = (xy)^{1/2}$ Square root is indicated
                                                    by the exponent $\frac{1}{2}$.

Write these expressions in radical form and simplify. Assume
variables represent nonnegative real numbers:

13. $8s^{1/2} = 8\sqrt{s}$ The exponent applies only
                                                    to the variable s.

If the variable represents a real number, simplify. Use absolute signs where necessary.

17. $\sqrt{w^2} = |w|$

The absolute value is necessary, since w may be negative and the radical indicates only the positive square root.

B

Write this expression in radical form. Assume variables represent nonnegative real numbers:

21. $6x^{1/2} = 6\sqrt{x}$

The exponent applies only to the variable x.

Write using rational exponents. Assume variables represent non-negative real numbers:

25. $\sqrt{5x} = (5xy)^{1/2}$

Square root is indicated by the exponent $\frac{1}{2}$.

Find the numerical value of each expression. Use a calculator with the radical key, $\boxed{\sqrt{}}$, or a square root table (see the Appendix).

29. $\sqrt{361}$

	ENTER		DISPLAY
Using a calculator:	$\boxed{361}$	$\boxed{\sqrt{}}$	19

So $\sqrt{361} = 19$

33. $-\sqrt[4]{81}$

	ENTER				DISPLAY
Using a calculator:	$\boxed{81}$	$\boxed{\sqrt{}}$	$\boxed{\sqrt{}}$	$\boxed{+/-}$	-3

So $-\sqrt[4]{81} = -3$

If the variable represents a real number, simplify. Use absolute value signs where necessary:

37. $\sqrt{4t^2} = 2|t|$

The square root of 4 is 2, and the positive square root of x^2 is $|x|$. The radical indicates the positive square root, but since x may be negative, the absolute value signs are necessary.

C

Simplify. Assume variables represent non-negative real numbers:

41. $\sqrt{121x^4} = 11x^2$ $(11x^2)^2 = 121x^4$

Write in radical form. Assume variables represent non-negative real numbers:

45. $(w^2 + 3)^{1/3} = \sqrt[3]{w^2 + 3}$

The exponent, $\frac{1}{3}$, indicates the cube root of the quantity $(w^2 + 3)$.

49. $(p - q)^{-1/2} = \dfrac{1}{(p - q)^{1/2}}$

The negative exponent indicates the reciprocal of the expression $(p - 1)^{1/2}$

$= \dfrac{1}{\sqrt{p - q}}$

The denominator, 2, in the exponent indicates the square root.

Write using rational exponents. Assume variables represent non-negative real numbers:

53. $\sqrt{5x^3 y^5} = (5x^3 y^5)^{1/2}$

Square root is indicated by the exponent $\frac{1}{2}$.

$= 5^{1/2} x^{3/2} y^{5/2}$

Fourth law of exponents.

If the variable represents a real number, simplify. Use absolute
value signs where necessary:

57. $\sqrt{(9 - x)^2} = |9 - x|$

If x > 9, then the
quantity 9 - x represents
a negative value. But the
radical indicates the
positive square root, so
the absolute value signs
are necessary.

D

61.   Find the length of one side of a square whose area is 196m².

$s = A^{1/2}$ Formula.

$s = 196^{1/2}$ Substitute A = 196.

$s = \sqrt{196}$

$s = 14$ Since $14^2 = 196$.

So each side of the square is 14 m.

65.   A circle has an area of 2025π square inches. Find the
      radius. (Hint: Area of a circle = πr^2.)

$A = \pi r^2$ Formula.

$2025\pi = \pi r^2$ Substitute A = 2025π.

$2025 = r^2$ Divide both sides by π.

$\sqrt{2025} = r$ Find the square root of
                                     both sides.

$45 = r$

The radius of the circle is 45 in.

CHALLENGE EXERCISES

69.   Using the formula below, how fast can a car be driven
      through a curve with radius of 160 feet without skidding?

Formula:

$V = \sqrt{2.5r}$

Substitute:

$$V = \sqrt{(2.5)(160)}$$ Substitute 160 for r.
$$V = \sqrt{400}$$
$$V = 20$$

The car can be driven 20 miles an hour without skidding.

MAINTAIN YOUR SKILLS

73. Reduce: $\dfrac{98a^3 b^3 c^4}{40ab^5 c^2} = \dfrac{2 \cdot 7 \cdot 7 \cdot a^2 \cdot \cancel{a} \cdot \cancel{b^3} \cdot \cancel{c^2} \cdot c^2}{2 \cdot 2 \cdot 2 \cdot 5 \cdot \cancel{a} \cdot \cancel{b^3} \cdot b^2 \cdot \cancel{c^2}}$

Factor, and divide out the common factors.

$$= \frac{49a^2 c^2}{20b^2}$$

Multiply or divide. Reduce if possible:

77. $\dfrac{6st + 4s^2}{3} \div \dfrac{6t^2 + 4st}{9}$

$= \dfrac{2\cancel{s}(3t + 2s)}{\cancel{3}} \cdot \dfrac{\overset{3}{\cancel{9}}}{\cancel{2}t(3t + 2s)}$ Convert division to multiplication, and factor.

$= \dfrac{3s}{t}$

EXERCISES 4.3   SIMPLIFYING AND APPROXIMATING RADICALS

A

Simplify these radicals. Assume that all variables represent positive numbers:

1. $\sqrt{200} = \sqrt{2}\ \sqrt{100} = 10\sqrt{2}$ The number 200 has a perfect square factor, 100.

5. $\sqrt{27} = \sqrt{9}\ \sqrt{3} = 3\sqrt{3}$ The number 27 has a perfect square factor, 9.

9. $\sqrt{40y^2} = \sqrt{4y^2}\,\sqrt{10} = 2y\sqrt{10}$

Assume y is positive, so no restriction (absolute value) is needed.

13. $\sqrt{24st^2} = \sqrt{4t^2}\,\sqrt{6s} = 2t\sqrt{6s}$

Assume t is positive.

17. $\sqrt{49x^4} = \sqrt{49}\,\sqrt{x^4} = 7x^2$

49 is a perfect square. Also, x^2 is always positive.

B

Use a calculator or table to find the approximate values of these radicals (to the nearest thousandth). Answers below were found using a calculator:

21. $\sqrt{200} \approx 14.142$

25. $\sqrt{4428} \approx 66.543$

Simplify. Assume that all variables represent positive numbers:

29. $\sqrt{150m^{12}} = \sqrt{2 \cdot 3 \cdot 5 \cdot 5 \cdot m^2 \cdot m^2 \cdot m^2 \cdot m^2 \cdot m^2 \cdot m^2}$
$\qquad = 5m^6\sqrt{6}$

33. $\sqrt{80y^7} = \sqrt{2 \cdot 2 \cdot 2 \cdot 2 \cdot 5 \cdot y^2 \cdot y^2 \cdot y^2 \cdot y}$
$\qquad = 4y^3\sqrt{5y}$

Simplify. The variables represent real numbers. Use absolute value where necessary:

37. $\sqrt{28a^6} = \sqrt{2 \cdot 2 \cdot 7 \cdot a^2 \cdot a^2 \cdot a^2}$
$\qquad = 2\,|a^3|\,\sqrt{7}$

Use absolute value since $a^3 < 0$ when $a < 0$.

C

Simplify. Assume all variables are positive numbers:

41. $\sqrt{175a^2 b^3 c^4} = \sqrt{25a^2 b^2 c^4}\,\sqrt{7b} = 5abc^2\sqrt{7b}$

138

45. $\sqrt[3]{8x^4y^6} = \sqrt[3]{2\cdot2\cdot2\cdot x^3\cdot x\cdot y^3\cdot y^3}$

$= 2xy^2\sqrt[3]{x}$

49. $\sqrt{147\ell^{10}m^{11}} = \sqrt{49\ell^{10}m^{10}}\sqrt{3m}$

$= 7\ell^5m^5\sqrt{3m}$

Simplify. The variables represent real numbers. Use absolute value where necessary:

53. $\sqrt{48c^8d^{10}} = \sqrt{16c^8d^{10}}\sqrt{3}$

$= 4c^4|d^5|\sqrt{3}$

c^4 is always positive even if $c < 0$, so no absolute value bars are necessary. However, when $d < 0$, d^5 is negative, so absolute value bars are necessary.

57. $\sqrt[4]{16s^4t^8} = \sqrt[4]{2^4\cdot s^4\cdot t^4\cdot t^4}$

$= 2|s|t^2$

Use absolute value for s since s may be a negative number. No absolute value is needed for t^2 since even if $t < 0$ (negative), $t^2 > 0$.

Reduce the index of the following radicals if the variables represent only positive numbers:

61. $\sqrt[4]{25m^2n^2} = 25^{1/4}m^{2/4}n^{2/4}$

Change the radical form to exponential form.

$(5^2)^{1/4}m^{1/2}n^{1/2}$

Rewrite 25 as 5^2.

$= 5^{2/4}m^{2/4}n^{2/4}$

Third law of exponents.

$= 5^{1/2}m^{1/2}b^{1/2}$

Reduce the exponents.

$= \sqrt{5mn}$

Change back to radical form.

65. Find the length of one side of a square section of lawn that is watered if Ajax Hardware advertises that the sprinkler will cover an area of 192 square yards. Give the answer
   a. in simplest radical form.
   b. to the nearest tenth of a yard.

Formula:

$s = \sqrt{A}$

Substitute:

$s = \sqrt{192}$ $A = 192$

Solve:

$s = \sqrt{64} \ \sqrt{3} = 8\sqrt{3}$ Simplest radical form.

Approximate:

$s = \sqrt{192} \approx 13.9$ Use a calculator. Round to the nearest tenth.

Answer:
The length of a side is

a. $8\sqrt{3}$ yd in simplest radical form
b. 13.9 yd to the nearest tenth.

69. If the area of a circle is given, the radius can be approximated by the formula $r = \sqrt{0.318A}$. Find the radius of a circle with area 15 square meters to the nearest hundredth of a meter.

Formula:

$r \approx \sqrt{0.318A}$

Substitute:

$r \approx \sqrt{(0.318)(15)}$ $A = 15$

Simplify:

$r \approx \sqrt{4.77} \approx 2.18$ Round to the nearest hundredth.

Answer:
The radius of the circle is approximately 2.18 meters.

73.   In a radical expression in which a variable appears in the radicand, why is it sometimes necessary to restrict the variable?

   It is necessary to restrict the variable because the radicand must always be nonnegative.

CHALLENGE EXERCISES

Simplify; assume all variable exponents represent positive integers.

77. $\sqrt[4]{y^{12n}} = \sqrt[4]{y^{4n} \cdot y^{4n} \cdot y^{4n}}$

   $= \sqrt[4]{y^{4n}} \cdot \sqrt[4]{y^{4n}} \cdot \sqrt[4]{y^{4n}}$

   $= y^n \cdot y^n \cdot y^n$

   $= y^{3n}$

81.   If an auto is traveling 44 ft/sec and accelerates at the rate of 10 ft/sec^2, what is its velocity after 770 ft?

   Formula:
   $v = \sqrt{v_0 + ad}$ (See Exercise 80.)

   Substitute:
   $v = \sqrt{44 + 10(770)}$ Replace v_0 with 44,
   $V = \sqrt{7744}$ a with 10, and d with
   $v = 88$ 770.

   Answer:
   The car's velocity is 88 ft/sec.

MAINTAIN YOUR SKILLS

Add or subtract. Reduce if possible:

85. $\dfrac{10}{2x - 5} + \dfrac{6}{5 - 2x} = \dfrac{10}{2x - 5} + \dfrac{6}{-(2x - 5)}$ Recall that $5 - 2x = -(2x - 5)$.

   $= \dfrac{10}{2x - 5} + \dfrac{-6}{2x - 5}$

   $= \dfrac{4}{2x - 5}$

89. $\dfrac{5}{3x-8} - \dfrac{-4}{8-3x} = \dfrac{5}{3x-8} - \dfrac{-4}{-(3x-8)}$

$$= \dfrac{5}{3x-8} - \dfrac{4}{3x-8}$$

$$= \dfrac{1}{3x-8}$$

EXERCISES 4.4   COMBINING RADICALS

A

Combine.   Assume that all variables represent positive numbers:

1. $\sqrt{3} + \sqrt{3} = (1 + 1)\sqrt{3}$ Factor using the distributive property.

$$= 2\sqrt{3}$$

5. $4\sqrt{2} - 3\sqrt{2} = (4 - 3)\sqrt{2}$ Factor using the distributive property.

$$= 1\sqrt{2}$$
$$= \sqrt{2} \qquad \text{Simplify.}$$

9. $-5\sqrt{35} + 10\sqrt{35} - 8\sqrt{35} = (-5 + 10 - 8)\sqrt{35}$

$$= -3\sqrt{35}$$

13. $\sqrt{18} - 4\sqrt{2} = \sqrt{9 \cdot 2} - 4\sqrt{2}$

$$= 3\sqrt{2} - 4\sqrt{2} \qquad \text{Simplify the first}$$
$$= (3 - 4)\sqrt{2} \qquad \text{radical.}$$
$$= -1\sqrt{2} \qquad \text{Use the distributive}$$
$$= -\sqrt{2} \qquad \text{property to factor.}$$

17. $\sqrt{13} + \sqrt{52} - 3\sqrt{13}$

$= \sqrt{13} + \sqrt{4 \cdot 13} - 3\sqrt{13}$ Simplify.

$= \sqrt{13} + 2\sqrt{13} - 3\sqrt{13}$

$= (1 + 2 - 3)\sqrt{13}$
$= 0$

B

21. $\sqrt{24} - 6\sqrt{6} + \sqrt{54} = 2\sqrt{6} - 6\sqrt{6} + 3\sqrt{6}$ Simplify each radical.
    $\qquad\qquad\qquad\qquad = -\sqrt{6}$ Combine like radicals.

25. $6\sqrt{3y} + 3\sqrt{3y} = (6 + 3)\sqrt{3y}$ Factor.
    $\qquad\qquad\qquad = 9\sqrt{3y}$

29. $\sqrt{25x} + \sqrt{25x} = \sqrt{25}\,\sqrt{x} + \sqrt{25}\,\sqrt{x}$ Simplify each radical.
    $\qquad\qquad\qquad = 5\sqrt{x} + 5\sqrt{x}$
    $\qquad\qquad\qquad = 10\sqrt{x}$ Factor and simplify.

33. $\quad\sqrt{125x^3} - x\sqrt{45x} - \sqrt{20x^3}$
    $= 5x\sqrt{5x} - 3x\sqrt{5x} - 2x\sqrt{5x}$ Simplify each radical
    $= (5x - 3x - 2x)\sqrt{5x}$ Factor.
    $= 0\sqrt{5x}$
    $= 0$

37. $\quad\sqrt[3]{4} + \sqrt[3]{32} - 2\sqrt[3]{4}$
    $= \sqrt[3]{4} + \sqrt[3]{8\cdot4} - 2\sqrt[3]{4}$ Simplify.
    $= \sqrt[3]{4} + 2\sqrt[3]{4} - 2\sqrt[3]{4}$
    $= (1 + 2 - 2)\sqrt[3]{4} = \sqrt[3]{4}$

C

41. $\sqrt{98x^3} + 7x\sqrt{18x} = 7x\sqrt{2x} + 21x\sqrt{2x}$ Simplify each radical.
    $\qquad\qquad\qquad\quad = 28x\sqrt{2x}$ Combine like radicals.

45. $\quad 5y\sqrt{63y^3} + 7y\sqrt{28y^3}$
    $= 5y\sqrt{9y^2}\,\sqrt{7y} + 7y\sqrt{4y^2}\,\sqrt{7y}$ Simplify each radical.
    $= 5y\cdot3y\sqrt{7y} + 7y\cdot2y\sqrt{7y}$
    $= 15y^2\sqrt{7y} + 14y^2\sqrt{7y} = 29y^2\sqrt{7y}$ Factor and simplify.

49. $2\sqrt{40m^2 n} - 3\sqrt{90m^2 n}$ Simplify each radical.

$= 2\sqrt{4m}\ \sqrt{10n} - 3\sqrt{9m^2}\ \sqrt{10n}$

$= 2 \cdot 2m\sqrt{10n} - 3 \cdot 3m\sqrt{10n}$

$= 4m\sqrt{10n} - 9m\sqrt{10n} = -5m\sqrt{10n}$ Factor and simplify.

53. $13\sqrt{9x^2} - \sqrt{4x^2} - 3\sqrt{121x^2}$ Simplify each radical.

$= 13 \cdot 3x - 2x - 3 \cdot 11x = 39x - 2x - 33x = 4x$

57. $\sqrt[4]{16x^{10}} - x\sqrt[4]{x^6}$

$= \sqrt[4]{2^4 x^8 x^2} - x\sqrt[4]{x^4 x^2}$ Simplify each radical.

$= 2x^2 \sqrt[4]{x^2} - x^2 \sqrt[4]{x^2}$

$= (2x^2 - x^2)\sqrt[4]{x^2}$ Factor and simplify.

$= x^2 \sqrt[4]{x^2}$

D

49. Find the perimeter of a triangle with sides $\sqrt{80}$ in.,
$\sqrt{125}$ in., and $\sqrt{45}$ in.
a. in simplest radical form
b. to the nearest tenth of an inch.

Formula:
$P = a + b + c$

Substitution:

$P = \sqrt{80} + \sqrt{125} + \sqrt{45}$ Replace a, b, and c with $\sqrt{80}$, $\sqrt{125}$, and $\sqrt{45}$.

Solve:

$P = \sqrt{16 \cdot 5} + \sqrt{25 \cdot 5} + \sqrt{9 \cdot 5}$ Simplify each radical.

$P = 4\sqrt{5} + 5\sqrt{5} + 3\sqrt{5}$

$P = 12\sqrt{5}$ Simplest radical form.
$P \approx 26.8$ Use a calculator.

Answer:
The perimeter of the triangle is:

a. $12\sqrt{5}$ in. in simplest radical form.
b. 26.8 in. rounded to the nearest tenth.

STATE YOUR UNDERSTANDING

65. Are $\sqrt{2}$ and $\sqrt[3]{2}$ like or unlike radical expressions? State the reason(s) for your answer.

The two expressions have the same radicand, 2, but do <u>not</u> have the same index. Therefore they are unlike radical expressions.

CHALLENGE EXERCISES

Combine; assume all variable exponents represent positive integers:

69. $\sqrt[n]{x^{2n}} + \sqrt[n]{2^n x^{2n}} + \sqrt[n]{3^{2n} x^{2n}}$

$= \sqrt[n]{(x^2)^n} + \sqrt[n]{(2)^n (x^2)^n} + \sqrt[n]{(3^2)^n (x^2)^n}$
$= x^2 + 2x^2 + 3^2 x^2$
$= x^2 + 2x^2 + 9x^2$
$= 12x^2$

MAINTAIN YOUR SKILLS

Perform the indicated operations. Reduce, if possible:

73. $\dfrac{1}{4} - \dfrac{1}{x} + \dfrac{x + 2}{x - 2}$ The LCD is $4x(x - 2)$.

$= \dfrac{1}{4} \cdot \dfrac{x(x - 2)}{x(x - 2)} - \dfrac{1}{x} \cdot \dfrac{4(x - 2)}{4(x - 2)} + \dfrac{x + 2}{x - 2} \cdot \dfrac{4x}{4x}$

$= \dfrac{x^2 - 2x}{4x(x - 2)} - \dfrac{4x - 8}{4x(x - 2)} + \dfrac{4x^2 + 8x}{4x(x - 2)}$

$= \dfrac{x^2 - 2x - (4x - 8) + 4x^2 + 8x}{4x(x - 2)}$

$= \dfrac{x^2 - 2x - 4x + 8 + 4x^2 + 8x}{4x(x - 2)}$

$= \dfrac{5x^2 + 2x + 8}{4x(x - 2)}$

77. $\dfrac{x}{x - 5} - \dfrac{3}{x + 4}$

$= \dfrac{x}{x - 5} \cdot \dfrac{x + 4}{x + 4} - \dfrac{3}{x + 4} \cdot \dfrac{x - 5}{x - 5}$ The LCD is $(x - 5)(x + 4)$.

$= \dfrac{x(x + 4) - 3(x - 5)}{(x - 5)(x + 4)}$

$= \dfrac{x^2 + 4x - 3x + 15}{(x - 5)(x + 4)}$

$= \dfrac{x^2 + x + 15}{(x - 5)(x + 4)}$

EXERCISES 4.5 MULTIPLYING RADICALS

A

Multiply and simplify. All variables represent positive numbers:

1. $\sqrt{2}\,\sqrt{8} = \sqrt{16} = 4$ Multiply, then simplify.

5. $2\sqrt{6}\,\sqrt{10} = 2\sqrt{60}$ Multiply.
    $\qquad\qquad = 2\sqrt{4 \cdot 15}$ Simplify.
    $\qquad\qquad = 4\sqrt{15}$

9. $\sqrt[3]{2x}\,\sqrt[3]{5x} = \sqrt[3]{10x^2}$ Multiply. Simplest radical form, since there are no cubes in the radicand.

13. $(5\sqrt{x})(\sqrt{5x}) = 5(\sqrt{x} \cdot \sqrt{5x})$
     $\qquad\qquad\quad = 5\sqrt{5x^2}$ Multiply.
     $\qquad\qquad\quad = 5x\sqrt{5}$ Simplify.

17. $3\sqrt{5}(5\sqrt{3} - 3\sqrt{5}) = 15\sqrt{15} - 9\sqrt{25}$ Multiply.
     $\qquad\qquad\qquad\quad = 15\sqrt{15} - 9 \cdot 5$ Simplify
     $\qquad\qquad\qquad\quad = 15\sqrt{15} - 45$

B

21. $(3\sqrt{y})(2\sqrt{10}) = (3 \cdot 2)(\sqrt{y} \cdot \sqrt{10})$

    $= 6\sqrt{10y}$ Multiply.

25. $(6s\sqrt{5t})(s\sqrt{10}) = (6s \cdot s)(\sqrt{5t} \cdot \sqrt{10})$

    $= 6s^2\sqrt{50t}$ Multiply.

    $= 6s^2\sqrt{25}\sqrt{2t}$ Simplify.

    $= 6s^2 \cdot 5\sqrt{2t} = 30s^2\sqrt{2t}$

29. $\left(5\sqrt{10t^2}\right)\left(\sqrt{6t^3}\right) = 5\sqrt{60^5}$ Multiply.

    $= 5\sqrt{4t^4}\sqrt{15t}$ Simplify.

    $= 5 \cdot 2t^2\sqrt{15t} = 10t^2\sqrt{15t}$

33. $\sqrt{3}(\sqrt{27} - \sqrt{12}) = \sqrt{81} - \sqrt{36}$ Multiply and simplify.

    $= 9 - 6$

    $= 3$

37. $3\sqrt{7}(\sqrt{7} - \sqrt{14}) = 3\sqrt{49} - 3\sqrt{98}$ Multiply

    $= 3 \cdot 7 - 3\sqrt{49}\sqrt{2}$

    $= 21 - 3 \cdot 7\sqrt{2}$

    $= 21 - 21\sqrt{2}$

C

41. $(\sqrt{10} + \sqrt{2})(\sqrt{10} + \sqrt{2}) = \sqrt{100} + 2\sqrt{20} + \sqrt{4}$

    $= 10 + 4\sqrt{5} + 2$

    $= 12 + 4\sqrt{5}$

45. $(2\sqrt{x} + 1)(\sqrt{x} - 4) = 2\sqrt{x^2} - 7\sqrt{x} - 4$

    $= 2x - 7\sqrt{x} - 4$

49. $(2\sqrt{b} - \sqrt{c})(2\sqrt{b} + \sqrt{c})$ The factors are conjugate pairs.

    $= 4\sqrt{b^2} - \sqrt{c^2}$

    $= 4b - c$

53. $2\sqrt{3} \cdot 5\sqrt[4]{2} = 2 \cdot 3^{1/2} \cdot 5 \cdot 2^{1/4}$ **Change to exponential form.**

$\qquad\qquad = 2 \cdot 5 \cdot 3^{2/4} \cdot 2^{1/4}$ **Express exponents with a common denominator.**

$\qquad\qquad = 10\sqrt[4]{3^2 \cdot 2^1}$ **Change back to radical form.**

$\qquad\qquad = 10\sqrt[4]{18}$ **Multiply.**

57. $\sqrt{2x} \cdot \sqrt[3]{3x} \cdot \sqrt[4]{4x}$ **Change to exponential form.**

$\qquad = (2x)^{1/2} \cdot (3x)^{1/3} \cdot (4x)^{1/4}$

$\qquad = (2x)^{6/12} \cdot (3x)^{4/12} \cdot (4x)^{3/12}$

    Express each exponent with a common denominator.

$\qquad = \sqrt[12]{(2x)^6} \cdot \sqrt[12]{(3x)^4} \cdot \sqrt[12]{(4x)^3}$ **Change back to radical form.**

$\qquad = \sqrt[12]{64x^6} \cdot \sqrt[12]{81x^4} \cdot \sqrt[12]{64x^3}$ **Simplify each radicand.**

$\qquad = \sqrt[12]{2^{12}3^4x^{13}} = \sqrt[12]{2^{12}x^{12}}\sqrt[12]{3^4 \cdot x}$ **Multiply and factor.**

$\qquad = 2x\sqrt[12]{81x}$ **Simplify.**

D

53. Find the area of a triangle with base $\sqrt{20}$ ft and height $\sqrt{15}$ ft. Give the answer
    a. in simplified radical form.
    b. to the nearest tenth of a square foot.

Formula:
$A = \frac{1}{2}bh$

Substitute:
$A = \frac{1}{2}(\sqrt{20})(\sqrt{15})$ **Replace b with $\sqrt{20}$ and h**

    with $\sqrt{15}$.

Simplify:

$A = \frac{1}{2}(\sqrt{4} \; \sqrt{5})(\sqrt{15})$

$A = \frac{1}{2}(2)\sqrt{5} \; \sqrt{15}$

$A = \sqrt{75}$

$A = \sqrt{25} \; \sqrt{3}$

$A = 5\sqrt{3}$

$A \approx 8.7$ Use a calculator.

Answer:

The area of the triangle is:

a. $5\sqrt{3}$ ft^2 in simplest radical form.
b. 8.7 ft^2 rounded to the nearest tenth.

65. Find the volume of a pyramid with a square base if the base is $\sqrt{30}$ m on a side and the height is $\sqrt{50}$ m. Express the answer in simplifed radical form. The formula for the volume of a pyramid is $V = \frac{1}{3}b^2 h$.

Formula:

$V = \frac{1}{3}b^2 h$

Substitute:

$V = \frac{1}{3}(\sqrt{30})^2 (\sqrt{50})$ Replace b with $\sqrt{30}$ and h with $\sqrt{50}$.

Simplify:

$V = \frac{1}{3}\sqrt{30^2} \cdot \sqrt{25} \cdot \sqrt{2}$

$V = \frac{1}{3}(30)(5)\sqrt{2}$

$V = 50\sqrt{2}$

Answer:

Expressed in simplified radical form, the volume of the pyramid is $50\sqrt{2}$ m^3.

STATE YOUR UNDERSTANDING

69. What conditions must be satisfied before two radical expressions can be multiplied?

The indices of the two radical expressions must be the same.

Multiply; assume all variables represent positive integers.

73. $\sqrt[n]{2x^{n+2}} \cdot \sqrt[n]{2^{n-1}x^{n-2}}$

$= \sqrt[n]{2x^{n+2} \cdot 2^{n-1} \cdot x^{n-2}}$

$= \sqrt[n]{(2 \cdot 2^{n-1})(x^{n+2} \cdot x^{n-2})}$

$= \sqrt[n]{2^{1+(n-1)} \, x^{(n+2)+(n-2)}}$

$= \sqrt[n]{2^n \, x^{2n}}$

$= \sqrt[n]{(2)^n \, (x^2)^n}$

$= 2x^2$

77. $\left(\sqrt[3]{a} + \sqrt[3]{b}\right)\left(\sqrt[3]{a^2} - \sqrt[3]{ab} + \sqrt[3]{b^2}\right)$

$= \left(\sqrt[3]{a} + \sqrt[3]{b}\right)\left(\sqrt[3]{a^2}\right) + \left(\sqrt[3]{a} + \sqrt[3]{b}\right)\left(-\sqrt[3]{ab}\right)$

$\quad + \left(\sqrt[3]{a} + \sqrt[3]{b}\right)\left(\sqrt[3]{b^2}\right)$

$= \sqrt[3]{a^3} + \sqrt[3]{a^2 b} - \sqrt[3]{a^2 b} - \sqrt[3]{ab^2} + \sqrt[3]{ab^2} + \sqrt[3]{b^3}$

$= \sqrt[3]{a^3} + \sqrt[3]{b^3}$

$= a + b$

MAINTAIN YOUR SKILLS

Simplify:

81. $\dfrac{3 - \dfrac{1}{y}}{\dfrac{1}{y} + 4} = \dfrac{y\left[3 - \dfrac{1}{y}\right]}{y\left[\dfrac{1}{y} + 4\right]}$

$\qquad\qquad = \dfrac{3y - 1}{1 + 4y}$

$\qquad\qquad$ or $\dfrac{3y - 1}{4y + 1}$

85. $5(y + 4)^{-1} - 3(y + 4)^{-1} = 15$

$$\frac{5}{y + 4} - \frac{3}{y + 4} = 15$$

$$(y + 4)\left[\frac{5}{y + 4} - \frac{3}{y + 4}\right] = (y + 4)(15)$$

$$5 - 3 = 15y + 60$$
$$2 = 15y + 60$$
$$-58 = 15y$$
$$-\frac{58}{15} = y$$

The solution set is $\left\{-\frac{58}{15}\right\}$.

EXERCISES 4.6   DIVIDING RADICALS

A

Rationalize the denominator:

1. $\dfrac{2}{\sqrt{5}} = \dfrac{2}{\sqrt{5}} \cdot \dfrac{\sqrt{5}}{\sqrt{5}} = \dfrac{2\sqrt{5}}{5}$

Multiply both numerator and denominator by $\sqrt{5}$.

5. $\sqrt{\dfrac{1}{8}} = \dfrac{\sqrt{1}}{\sqrt{8}} \cdot \dfrac{\sqrt{2}}{\sqrt{2}} = \dfrac{\sqrt{2}}{4}$

Multiply numerator and denominator by $\sqrt{2}$, since $\sqrt{16} = 4$.

9. $\dfrac{-3}{\sqrt{6}} = \dfrac{-3}{\sqrt{6}} \cdot \dfrac{\sqrt{6}}{\sqrt{6}} = \dfrac{-3\sqrt{6}}{6} = -\dfrac{\sqrt{6}}{2}$

Multiply numerator and denominator by $\sqrt{6}$. Reduce.

13. $\sqrt{\dfrac{3}{8}} = \dfrac{\sqrt{3}}{\sqrt{8}} \cdot \dfrac{\sqrt{2}}{\sqrt{2}} = \dfrac{\sqrt{6}}{4}$

Multiply numerator and denominator by $\sqrt{2}$.

17. $\dfrac{7}{\sqrt{14}} = \dfrac{7}{\sqrt{14}} \cdot \dfrac{\sqrt{14}}{\sqrt{14}} = \dfrac{7\sqrt{14}}{14}$

$$= \dfrac{\sqrt{14}}{2}$$

Multiply numerator and denominator by $\sqrt{14}$.

Reduce.

B

21. $\sqrt{\dfrac{a}{2}} = \dfrac{\sqrt{a}}{\sqrt{2}} \cdot \dfrac{\sqrt{2}}{\sqrt{2}} = \dfrac{\sqrt{2a}}{2}$ Multiply numerator and denominator by $\sqrt{2}$.

25. $\dfrac{2\sqrt{5}}{5\sqrt{2}} = \dfrac{2\sqrt{5}}{5\sqrt{2}} \cdot \dfrac{\sqrt{2}}{\sqrt{2}} = \dfrac{2\sqrt{10}}{5 \cdot 2}$ Multiply numerator and denominator by $\sqrt{2}$.

$= \dfrac{2\sqrt{10}}{10} = \dfrac{\sqrt{10}}{5}$ Reduce.

29. $\dfrac{-2\sqrt{8}}{\sqrt{6}} = \dfrac{-2\sqrt{8}}{\sqrt{6}} \cdot \dfrac{\sqrt{6}}{\sqrt{6}}$ Multiply numerator and denominator by $\sqrt{6}$.

$= \dfrac{-2\sqrt{48}}{6}$

$= \dfrac{-2 \cdot 4\sqrt{3}}{6}$ Simplify the radical.

$= \dfrac{-8\sqrt{3}}{6} = -\dfrac{4\sqrt{3}}{3}$ Reduce.

or

$\dfrac{-2\sqrt{8}}{\sqrt{6}} = \dfrac{-4\sqrt{2}}{\sqrt{6}} \cdot \dfrac{\sqrt{6}}{\sqrt{6}}$ Simplify the radical in the numerator, ($\sqrt{8} = 2\sqrt{2}$) and then multiply numerator and denominator by $\sqrt{6}$.

$= \dfrac{-4\sqrt{12}}{6}$

$= \dfrac{-8\sqrt{3}}{6} = -\dfrac{4\sqrt{3}}{3}$

33. $\sqrt[3]{\dfrac{3}{4}} = \dfrac{\sqrt[3]{3}}{\sqrt[3]{4}} \cdot \dfrac{\sqrt[3]{2}}{\sqrt[3]{2}}$ Multiply by the smallest radical factor that will result in a perfect cube, $\sqrt[3]{2}$

$= \dfrac{\sqrt[3]{6}}{\sqrt[3]{8}} = \dfrac{\sqrt[3]{6}}{2}$

152

37. $\sqrt{\dfrac{3}{4}} + \sqrt{\dfrac{4}{3}} - \sqrt{\dfrac{1}{12}}$

    $= \dfrac{\sqrt{3}}{\sqrt{4}} + \dfrac{\sqrt{4}}{\sqrt{3}} - \dfrac{\sqrt{1}}{\sqrt{12}}$ Simplify, and rationalize denominators.

    $= \dfrac{\sqrt{3}}{2} + \dfrac{2}{\sqrt{3}} \cdot \dfrac{\sqrt{3}}{\sqrt{3}} - \dfrac{1}{\sqrt{4} \cdot \sqrt{3}}$

    $= \dfrac{\sqrt{3}}{2} + \dfrac{2\sqrt{3}}{3} - \dfrac{1}{2\sqrt{3}} \cdot \dfrac{\sqrt{3}}{\sqrt{3}}$

    $= \dfrac{\sqrt{3}}{2} + \dfrac{2\sqrt{3}}{3} - \dfrac{\sqrt{3}}{6}$

    $= \dfrac{3\sqrt{3}}{6} + \dfrac{4\sqrt{3}}{6} - \dfrac{\sqrt{3}}{6}$ Write each fraction with a common denominator.

    $= \dfrac{6\sqrt{3}}{6}$

    $= \sqrt{3}$

c

Rationalize the numerator:

41. $\sqrt{\dfrac{3}{2}} = \dfrac{\sqrt{3}}{\sqrt{2}} \cdot \dfrac{\sqrt{3}}{\sqrt{3}} = \dfrac{3}{\sqrt{6}}$

Rationalize the denominator:

45. $\dfrac{2 + \sqrt{3}}{\sqrt{3}} = \dfrac{(2 + \sqrt{3})}{\sqrt{3}} \cdot \dfrac{\sqrt{3}}{\sqrt{3}}$

          $= \dfrac{2\sqrt{3} + 3}{3}$

49. $\dfrac{\sqrt{5} + \sqrt{2}}{\sqrt{3} - \sqrt{5}} = \dfrac{(\sqrt{5} + \sqrt{2})}{(\sqrt{3} - \sqrt{5})} \cdot \dfrac{(\sqrt{3} + \sqrt{5})}{(\sqrt{3} + \sqrt{5})}$

          $= \dfrac{\sqrt{15} + 5 + \sqrt{6} + \sqrt{10}}{-2}$

    or $- \dfrac{\sqrt{15} + 5 + \sqrt{6} + \sqrt{10}}{2}$

53. $\dfrac{\sqrt{a}}{\sqrt{2a} + \sqrt{b}} = \dfrac{\sqrt{a}}{\sqrt{2a} + \sqrt{b}} \cdot \dfrac{\sqrt{2a} - \sqrt{b}}{\sqrt{2a} - \sqrt{b}}$

$\qquad = \dfrac{a\sqrt{2} - \sqrt{ab}}{2a - b}$

57. $\dfrac{\sqrt{xy} - \sqrt{x}}{\sqrt{x} + \sqrt{xy}} = \dfrac{\sqrt{xy} - \sqrt{x}}{\sqrt{x} + \sqrt{xy}} \cdot \dfrac{\sqrt{x} - \sqrt{xy}}{\sqrt{x} - \sqrt{xy}}$

$\qquad = \dfrac{\sqrt{x^2 y} - \sqrt{x^2 y^2} - \sqrt{x^2} + \sqrt{x^2 y}}{x - xy}$

$\qquad = \dfrac{x\sqrt{y} - xy - x + x\sqrt{y}}{x - xy}$

$\qquad = \dfrac{-xy + 2x\sqrt{y} - x}{x - xy}$

$\qquad = \dfrac{-x(y - 2\sqrt{y} + 1)}{-x(y - 1)}$

$\qquad = \dfrac{y - 2\sqrt{y} + 1}{y - 1}$

Rationalize the numerator:

61. $\dfrac{2 + \sqrt{3}}{\sqrt{3}} = \dfrac{2 + \sqrt{3}}{\sqrt{3}} \cdot \dfrac{2 - \sqrt{3}}{2 - \sqrt{3}}$

$\qquad = \dfrac{4 - 3}{2\sqrt{3} - 3}$

$\qquad = \dfrac{1}{2\sqrt{3} - 3}$

65. A pendulum is 1 foot long. Find the time it takes to complete one cycle. Give the answer
    a. in simplified radical form.
    b. to the nearest tenth of a second.

Formula:

$$T = 2\pi \frac{\sqrt{L}}{\sqrt{32}}$$

Substitute:

$$T = 2\pi \frac{\sqrt{1}}{\sqrt{32}}$$ Substitute L = 1.

Simplify:

$$T = 2\pi \frac{1}{\sqrt{32}} \cdot \frac{\sqrt{2}}{\sqrt{2}}$$ Rationalize the denominator.

$$T = \frac{2\pi\sqrt{2}}{8}$$

$$T = \frac{\pi\sqrt{2}}{4}$$ Simplest radical form.

$$T \approx \frac{(3.14)(1.41)}{4}$$

$$T \approx 1.1$$ To the nearest tenth.

Answer:

The pendulum takes $\frac{\pi\sqrt{2}}{4}$ seconds, or approximately 1.1 seconds to complete one cycle.

STATE YOUR UNDERSTANDING

69. Explain what is meant by "rationalize the denominator."

    Division is referred to as rationalizing the denominator. The new fraction that results from this process will always have a denominator with no radicals.

Use the results of Exercises 76 and 77 of Section 4.5 to rationalize the denominator:

73. $\dfrac{1}{\sqrt[3]{8} - \sqrt[3]{4}} = \dfrac{1}{\sqrt[3]{8} - \sqrt[3]{4}} \cdot \dfrac{\sqrt[3]{64} + \sqrt[3]{32} + \sqrt[3]{16}}{\sqrt[3]{64} + \sqrt[3]{32} + \sqrt[3]{16}}$

$\qquad = \dfrac{4 + 2\sqrt[3]{4} + 2\sqrt[3]{2}}{8 - 4}$

$\qquad = \dfrac{2\left(2 + \sqrt[3]{4} + \sqrt[3]{2}\right)}{2(2)}$

$\qquad = \dfrac{2 + \sqrt[3]{4} + \sqrt[3]{2}}{2}$

MAINTAIN YOUR SKILLS

Simplify:

77. $\dfrac{\dfrac{3}{1} - \dfrac{2}{b}}{b + a} = \dfrac{ab\left(\dfrac{3}{a} - \dfrac{2}{b}\right)}{ab(b + a)}$

$\qquad = \dfrac{3b - 2a}{ab(b + a)}$

Solve:

81. $\dfrac{6}{y - 5} + \dfrac{6}{2y + 1} = \dfrac{-1}{2y^2 - 9y - 15}$

$6(2y + 1) + 5(y - 5) = -1$ Multiply by $2y^2 - 9y - 15$
$12y + 6 + 5y - 25 = -1$ since it factors into
$17y - 19 = -1$ $(y - 5)(2y + 1)$ and is the
$17y = 18$ LCD.
$y = \dfrac{18}{17}$

The solution set is $\left\{\dfrac{18}{17}\right\}$.

A

Solve:

1. $\sqrt{a} = 11$

$(\sqrt{a})^2 = (11)^2$ Power property.

$a = 121$

Check:

$\sqrt{121} = 11$

$11 = 11$

The solution set is {121}.

5. $\sqrt{x} + 4 = 5$

$\sqrt{x} = 1$ Isolate the radical on the left by subtracting 4 from both sides.

$(\sqrt{x})^2 = (1)^2$ Square both sides.

$x = 1$

Check:

$\sqrt{1} + 4 = 5$

$1 + 4 = 5$

$5 = 5$

The solution set is {1}.

9. $\sqrt{2x + 1} = 3$

$(\sqrt{2x + 1})^2 = (3)^2$ Square both sides.

$2x + 1 = 9$

$2x = 8$

$x = 4$

Check:

$\sqrt{2 \cdot 4 + 1} = 3$

$\sqrt{9} = 3$

$3 = 3$

The solution set is {4}.

13. $5 - \sqrt{2x} = -11$

$\quad\quad -\sqrt{2x} = -16$ Isolate the radical on the left.

$\quad (-\sqrt{2x})^2 = (-16)^2$ Square both sides.

$\quad\quad\quad\quad 2x = 256$

$\quad\quad\quad\quad\ x = 128$

Check:

$5 - \sqrt{2 \cdot 128} = -11$

$\quad 5 - \sqrt{256} = -11$

$\quad\quad 5 - 16 = -11$

$\quad\quad\quad -11 = -11$

The solution set is {128}.

17. $\quad \sqrt[3]{x} = 5$

$\quad \left(\sqrt[3]{x}\right)^3 = (5)^3$ Since the index is 3, cube both sides.

$\quad\quad\quad\ x = 125$

Check:

$\sqrt[3]{125} = 5$

$\quad\quad\ 5 = 5$

The solution set is {125}.

B

21. $\sqrt{3 - 4x} = 2x$

$\quad 3 - 4x = 4x^2$ Square both sides.

$\quad\quad\quad 0 = 4x^2 + 4x - 3$ A quadratic equation.

$\quad\quad\quad 0 = (2x - 1)(2x + 3)$ Factor.

$2x - 1 = 0$ or $2x + 3 = 0$ Zero-product property.

$\quad\quad x = \frac{1}{2}$ or $\quad\quad x = -\frac{3}{2}$

Check:

$$\sqrt{3 - 4\left(\frac{1}{2}\right)} = 2\left(\frac{1}{2}\right)$$

$$\sqrt{3 - 2} = 1$$

$$\sqrt{1} = 1$$

$$1 = 1 \qquad \text{Checks.}$$

$$\sqrt{3 - 4\left(-\frac{3}{2}\right)} = 2\left(-\frac{3}{2}\right)$$

$$\sqrt{3 + 6} = -3$$

$$\sqrt{9} = -3$$

$$3 = -3 \qquad \text{Does not check.}$$

The solution set is $\left\{\frac{1}{2}\right\}$.

25. $\sqrt{x + 2} = x + 2$

$\quad x + 2 = x^2 + 4x + 4$ Square both sides.

$\quad\quad\quad 0 = x^2 + 3x + 2$ Quadratic equation.

$\quad\quad\quad 0 = (x + 2)(x + 1)$ Factor.

$\quad x + 2 = 0$ or $x + 1 = 0$ Zero-product property.

$\quad\quad\quad x = -2$ or $\quad\quad x = -1$

Check:

$$\sqrt{-2 + 2} = -2 + 2$$

$$\sqrt{0} = 0$$

$$0 = 0 \qquad \text{Checks.}$$

$$\sqrt{-1 + 2} = -1 + 2$$

$$\sqrt{1} = 1$$

$$1 = 1 \qquad \text{Checks.}$$

The solution set is $\{-2, -1\}$.

29. $\sqrt{12x + 13} = 3x - 2$

$\quad 12x + 13 = 9x^2 - 12x + 4$ Square both sides.

$\quad\quad\quad\quad 0 = 9x^2 - 24x - 9$

$\quad\quad\quad\quad 0 = 3(3x^2 - 8x - 3)$ Factor completely.

$\quad\quad\quad\quad 0 = 3(x - 3)(3x + 1)$

$\quad x - 3 = 0$ or $3x + 1 = 0$ Zero-product property.

$\quad\quad\quad x = 3$ or $\quad\quad x = -\frac{1}{3}$

Check.

$$\sqrt{12 \cdot 3 + 13} = 3(3) - 2$$

$$\sqrt{49} = 9 - 2$$

$$7 = 7 \qquad \text{Checks.}$$

$$\sqrt{12\left(-\frac{1}{3}\right) + 13} = 3\left(-\frac{1}{3}\right) - 2$$

$$\sqrt{9} = -1 - 2$$

$$3 = -3 \qquad \text{Does not check.}$$

The solution set is {3}.

33. $\sqrt{2x + 5} + 5 = x$

$$\sqrt{2x + 5} = x - 5 \qquad \text{Isolate the radical.}$$

$$2x + 5 = x^2 - 10x + 25 \qquad \text{Square both sides.}$$

$$0 = x^2 - 12x + 20$$

$$0 = (x - 10)(x - 2) \qquad \text{Factor.}$$

$$x - 10 = 0 \text{ or } x - 2 = 0 \qquad \text{Zero-product property.}$$

$$x = 10 \text{ or } \qquad x = 2$$

Check:

$$\sqrt{2(10) + 5} + 5 = 10$$

$$\sqrt{25} + 5 = 10$$

$$5 + 5 = 10$$

$$10 = 10 \qquad \text{Checks.}$$

$$\sqrt{2(2) + 5} + 5 = 2$$

$$\sqrt{9} + 5 = 2$$

$$3 + 5 = 2$$

$$8 \neq 2$$

The solution set is {10}.

37. $\sqrt[4]{2x - 7} = \sqrt[4]{x + 2}$

$$\left(\sqrt[4]{2x - 7}\right)^4 = \left(\sqrt[4]{x + 2}\right)^4 \qquad \text{Since the index is 4,}$$

$$2x - 7 = x + 2 \qquad \text{raise both sides to}$$

$$x = 9 \qquad \text{the fourth power.}$$

Check:

$$\sqrt[4]{2 \cdot 9 - 7} = \sqrt[4]{9 + 2}$$

$$\sqrt[4]{11} = \sqrt[4]{11} \qquad \text{Checks.}$$

The solution set is {9}.

C

41. $\sqrt{x} - \sqrt{x + 9} = -1$

$\sqrt{x} = \sqrt{x + 9} - 1$ — Isolate $\sqrt{x}$ on the left side. Square both sides to eliminate the radical on the left.

$x = x + 9 - 2\sqrt{x + 9} + 1$

$-10 = -2\sqrt{x + 9}$ — Simplify.

$5 = \sqrt{x + 9}$ — Divide both sides by -2.
$25 = x + 9$ — Square both sides to eliminate the second radical.

$16 = x$

Check:

$\sqrt{16} - \sqrt{16 + 9} = -1$
$4 - \sqrt{25} = -1$
$4 - 5 = -1$
$-1 = -1$ — Checks.

The solution set is $\{16\}$.

45. $\sqrt{x + 1} + \sqrt{x} = -7$

$\sqrt{x + 1} = -\sqrt{x} - 7$ — Isolate $\sqrt{x + 1}$ on the left.

$(\sqrt{x + 1})^2 = (-\sqrt{x} - 7)^2$ — Square both sides to eliminate the radical on the left.

$x + 1 = x + 14\sqrt{x} + 49$

$-48 = 14\sqrt{x}$ — Simplify.

$-\dfrac{24}{7} = \sqrt{x}$ — Divide both sides by 14.

$\dfrac{576}{49} = x$ — Square both sides to eliminate the second radical.

Check:

$\sqrt{\dfrac{576}{49} + 1} + \sqrt{\dfrac{576}{49}} = -7$

$\sqrt{\dfrac{625}{49}} + \sqrt{\dfrac{576}{49}} = -7$

$\dfrac{25}{7} + \dfrac{24}{7} = -7$

$\dfrac{49}{7} = -7$

$7 \neq -7$ — Does not check.

The solution set is the empty set, $\emptyset$.

49.
$$\sqrt{2y} + 3 = \sqrt{y-7}$$

$2y + 6\sqrt{2y} + 9 = y - 7$ Square both sides.

$6\sqrt{2y} = -y - 16$ Simplify.
$36(2y) = y^2 + 32y + 256$ Square both sides.
$72y = y^2 + 32y + 256$
$0 = y^2 - 40y + 256$
$0 = (y - 32)(y - 8)$
$y - 32 = 0$ or $y - 8 = 0$
$y = 32$ or $\quad$ $y = 8$

Check:

$$\sqrt{2 \cdot 32} + 3 = \sqrt{32 - 7}$$
$$\sqrt{64} + 3 = \sqrt{25}$$
$$8 + 3 = 5$$
$$11 \neq 5$$ Does not check.

$$\sqrt{2 \cdot 8} + 3 = \sqrt{8 - 7}$$
$$\sqrt{16} + 3 = \sqrt{1}$$
$$4 + 3 = 1$$
$$7 \neq 1$$ Does not check

The solution is the empty set, $\emptyset$.

53.
$$\sqrt{5w + 9} = 3 + \sqrt{w}$$

$5w + 9 = 9 + 6\sqrt{w} + w$ Square both sides.

$4w = 6\sqrt{w}$ Simplify.
$16w^2 = 36w$ Square both sides.
$16w^2 - 36w = 0$
$4w(4w - 9) = 0$ Factor.
$4w = 0$ or $4w - 9 = 0$ Zero-product property.
$w = 0$ or $\quad$ $w = \dfrac{9}{4}$

Check:

$$\sqrt{5\left(\dfrac{9}{4}\right) + 9} = 3 + \sqrt{\dfrac{9}{4}}$$

$$\sqrt{\dfrac{81}{4}} = 3 + \dfrac{3}{2}$$

$$\dfrac{9}{2} = \dfrac{6}{2} + \dfrac{3}{2}$$

$$\dfrac{9}{2} = \dfrac{9}{2}$$ Checks.

$$\sqrt{5 \cdot 0 + 9} = 3 + \sqrt{0}$$
$$\sqrt{9} = 3$$
$$3 = 3$$ Checks.

The solution set is $\left\{0, \dfrac{9}{4}\right\}$.

57.
$$\sqrt{2x - 1} - 2 = \sqrt{x - 4}$$

$2x - 1 - 4\sqrt{2x - 1} + 4 = x - 4$ Square both sides.

$\qquad\qquad -4\sqrt{2x - 1} = -x - 7$ Simplify.

$\qquad\qquad 16(2x - 1) = x^2 + 14x + 49$ Square both sides.

$\qquad\qquad 32x - 16 = x^2 + 14x + 49$

$\qquad\qquad 0 = x^2 - 18x + 65$

$\qquad\qquad 0 = (x - 5)(x - 13)$

$x - 5 = 0$ or $x - 13 = 0$

$\quad x = 5$ or $\qquad x = 13$

Check:

$$\sqrt{2 \cdot 5 - 1} - 2 = \sqrt{5 - 4}$$

$\qquad\quad \sqrt{9} - 2 = \sqrt{1}$

$\qquad\quad 3 - 2 = 1$

$\qquad\qquad 1 = 1$ Checks.

$$\sqrt{2 \cdot 13 - 1} - 2 = \sqrt{13 - 4}$$

$\qquad\quad \sqrt{25} - 2 = \sqrt{9}$

$\qquad\quad 5 - 2 = 3$

$\qquad\qquad 3 = 3$ Checks.

The solution set is {5, 13}.

61. The pressure of the steam from a fire hydrant can be calculated from the formula $G = 26.8d^2 \sqrt{p}$, where G represents the discharge in gallons per minute (gpm), d represents the diameter of the outlet in inches, and p represents the pressure in pounds per square inch (psi). Find the pressure (psi) of water from an outlet that is 4.5 inches in diameter and that discharges at 900 gallons per minute (round to the nearest tenth).

Formula:

$G = 26.8d^2 \sqrt{p}$

Substitute:

$\quad 900 = 26.8(4.5)^2 \sqrt{p}$ Replace d with 4.5, and G

$\quad 900 = (26.8)(20.25)\sqrt{p}$ with 900.

$\quad 900 = 542.70\sqrt{p}$

$\quad 1.66 = \sqrt{p}$ Use a calculator. Divide both sides by 542.70. Round to the nearest hundredth.

$\quad 2.75 = p$ Square both sides.

Answer:

Rounded to the nearest tenth, the pressure in pounds per square inch (psi) is 2.75.

65. The interest rate (r) (compounded annually) needed to have P dollars grow to A dollars at the end of two years is given by $r = \sqrt{\dfrac{A}{P}} - 1$. Find the value of A if P = 2000 and r = 0.1.

Formula:

$$r = \sqrt{\dfrac{A}{P}} - 1$$

Substitute:

$0.1 = \sqrt{\dfrac{A}{2000}} - 1$	Replace r with 0.1, and P with 2000.
$1.1 = \sqrt{\dfrac{A}{2000}}$	Isolate the radical.
$1.21 = \dfrac{A}{2000}$	Square both sides.
$2420 = A$	Multiply both sides by 2000.

CHALLENGE EXERCISES

69. Using the formula below, what is the length (to the nearest foot) of a pendulum with a period of 3 seconds? Let $\pi \approx 3.14$.

Formula:

$$T = 2\pi\sqrt{\dfrac{L}{g}} \text{ , where } g = 32 \text{ ft/sec}^2$$

Substitute:

$3 = 2\pi\sqrt{\dfrac{L}{32}}$	Replace T with 3, and g with 32.
$\dfrac{3}{2\pi} = \sqrt{\dfrac{L}{32}}$	Divide both sides by 2π.
$\dfrac{9}{4\pi^2} = \dfrac{L}{32}$	Square both sides.
$32\left(\dfrac{9}{4\pi^2}\right) = L$	Multiply both sides by 32.
$\dfrac{72}{\pi^2} = L$	
$7.30 \approx L$	Evaluate the left side using a calculator. Let $\pi \approx 3.14$.
$7 = L$	Round to the nearest foot.

Answer:
The length of the pendulum is approximately equal to 7 ft.

Solve:

73. $-15 < 4x + 1 < 29$
     $-16 < 4x < 28$ Subtract 1.
     $-4 < x < 7$ Divide by 4.

The solution set is $(-4, 7)$.

77. $-7 \leq 4x + 9 \leq 13$
     $-16 \leq 4x \leq 4$ Subtract 9.
     $-4 \leq x \leq 1$ Divide by 4.

The solution set is $[-4, 1]$.

EXERCISES 4.8   COMPLEX NUMBERS

A

Write in standard form:

1. $\sqrt{-16} = \sqrt{16 \cdot -1} = \sqrt{16}\,\sqrt{-1}$
    $= 4\sqrt{-1}$ Since $\sqrt{16} = 4$
    $= 4i$ Since $\sqrt{-1} = i$.

5. $\sqrt{-18} = \sqrt{9 \cdot 2 \cdot -1}$
    $= \sqrt{9} \cdot \sqrt{-1} \cdot \sqrt{2}$
    $= 3i\sqrt{2}$

9. $\sqrt{-75} - \sqrt{-3} = \sqrt{75 \cdot -1} - \sqrt{3 \cdot -1}$
    $= 5i\sqrt{3} - i\sqrt{3}$ Since $\sqrt{75} = 5\sqrt{3}$, and
           $\sqrt{-1} = i$.

    $= 4i\sqrt{3}$

Simplify:

13. $i^{54} = i^{52} \cdot i^2$ Factor: 52 is the largest
     $= i^{4 \cdot 13} \cdot i^2$ multiple of four that is
            less than 54.
     $= (i^4)^{13} \cdot i^2$
     $= 1^{13} \cdot i^2$ $i^4 = 1$.
     $= -1$ Since $i^2 = -1$.

Add or subtract:

17.
$$(4 + 2i) + (3 + 4i)$$
$$= (4 + 3) + (2i + 4i)$$
$$= 7 + 6i$$

Regroup the real parts and imaginary parts of the two complex numbers.

21.
$$(4 + \sqrt{-1}) + (2 + \sqrt{-9})$$
$$= (4 + i) + (2 + 3i)$$

$$= (4 + 2) + (i + 3i)$$
$$= 6 + 4i$$

Write each complex number in standard form.
Regroup.

B

Perform the indicated operations:

25.
$$7 - (-8 - i) = 7 + (8 + i)$$
$$= 15 + i$$

Subract by adding the opposite.

29.
$$(8 + 3i) - (-3 + i) = (8 + 3i) + (3 - i)$$
$$= (8 + 3) + (3i - i)$$
$$= 11 + 2i$$

33.
$$3i(2 + 5i) = 6i + 15i^2$$

$$= -15 + 6i$$

When written in standard form, complex numbers can be multiplied in the same way as polynomials.
Since $i^2 = -1$.

37.
$$\frac{i^{16}}{i^{21}} = \frac{1}{i^5}$$

Use laws of exponents to reduce the fraction.

$$= \frac{1}{i^5} \cdot \frac{i^3}{i^3} = \frac{i^3}{i^8}$$

Multiply by one, $\frac{i^3}{i^3}$, so that the denominator will be a power of i that is a multiple of 4.

$$= \frac{i^3}{(i^4)^2} = \frac{i^3}{1^2} = i^3 = -i$$

Since $i^3 = -i$.

41.
$$\frac{i^{21}}{i^{-10}} = i^{31}$$

Second law of exponents.

$$= i^{28} \cdot i^3$$

The largest multiple of four that is less than 31 is 28.

$$= (i^4)^7 \cdot i^3$$
$$= 1^7 \cdot i^3$$
$$= i^3$$
$$= -i$$

$i^4 = 1$

$i^3 = i^2 \cdot i = -1 \cdot i = -i$.

C

Perform the indicated operations:

45. $(7 + \sqrt{-16})(-2 - \sqrt{-25})$
$= (7 + 4i)(-2 - 5i)$
$= -14 - 35i - 8i - 20i^2$

In standard form, complex numbers can be multiplied like polynomials. Substitute -1 for i^2.

$= -14 - 43i - 20(-1)$
$= -14 - 43i + 20 = 6 - 43i$

49. $\dfrac{1}{2i} = \dfrac{1}{2i} \cdot \dfrac{i^3}{i^3} = \dfrac{i^3}{2i^4}$

Multiply by one so that the denominator will be a power of i that is a multiple of 4.

$= \dfrac{-i}{2(1)} = -\dfrac{i}{2}$ or $-\dfrac{1}{2}i$

Since $i^3 = -i$ and $i^4 = 1$.

53. $(\sqrt{8} + i\sqrt{12})(\sqrt{2} - i\sqrt{3})$
$= \sqrt{16} - i\sqrt{24} + i\sqrt{24} - i^2\sqrt{36}$ Multiply.
$= 4 - 6i^2$ Simplify.
$= 4 - 6(-1) = 4 + 6 = 10$ Replace i^2 with -1.

57. $\dfrac{-5 - 4i}{-2 + 3i} = \dfrac{(-5 - 4i)}{(-2 + 3i)} \cdot \dfrac{(-2 - 3i)}{(-2 - 3i)}$

Multiply numerator and denominator by the conjugate of the denominator.

$= \dfrac{10 + 15i + 8i + 12i^2}{4 - 9i^2}$

$= \dfrac{10 + 23i + 12(-1)}{4 - 9(-1)}$

Replace i^2 with -1, and combine like terms.

$= \dfrac{-2 + 23i}{13}$ or $-\dfrac{2}{13} + \dfrac{23}{13}i$ Standard form.

61. $\dfrac{3}{(2 - i)^2} = \dfrac{3}{4 - 4i + i^2} = \dfrac{3}{4 - 4i - 1}$

Multiply and simplify, in the denominator.

$= \dfrac{3}{3 - 4i} \cdot \dfrac{3 + 4i}{3 + 4i}$

Multiply by the conjugate of the denominator.

$= \dfrac{9 + 12i}{9 - 16i^2}$

$= \dfrac{9 + 12i}{9 + 16} = \dfrac{9 + 12i}{25}$ Fraction form.

$= \dfrac{9}{25} + \dfrac{12}{25}i$ Standard form.

65. $\dfrac{-6i}{3 + \sqrt{-3}} = \dfrac{-6i}{3 + i\sqrt{3}}$ Simplify in the denominator.

$= \dfrac{-6i}{3 + i\sqrt{3}} \cdot \dfrac{3 - i\sqrt{3}}{3 - i\sqrt{3}}$ Multiply by the conjugate.

$= \dfrac{-18i + 6\sqrt{3}i^2}{9 - 3i^2}$

$= \dfrac{-18i - 6\sqrt{3}}{9 + 3}$ $i^2 = -1$.

$= \dfrac{-6\sqrt{3} - 18i}{12}$

$= \dfrac{-\sqrt{3} - 3i}{2}$ Reduced fraction form.

$= -\dfrac{\sqrt{3}}{2} - \dfrac{3}{2}i$ Standard form.

D

A formula to find the total series of impedance is $Z_T = Z_1 + Z_2$ where Z_T represents the total impedance, $Z_1 = R_1 + jX_{c1}$ and $Z_2 = R_2 + jX_{c2}$. Find the total series impedance of resistance and capacitance if:

69. $R_1 = 15700\Omega$, $X_{c1} = 18900\Omega$, $R_2 = 23800\Omega$, and $X_{c2} = 30500\Omega$.

Formulas:
$Z_T = Z_1 + Z_2$
$Z_1 = R_1 + jX_{c1}$
$Z_2 = R_2 + jX_{c2}$

Substitute:
$Z_T = (15700 + j18900) + (23800 + j30500)$

Solve:
$Z_T = 39500 + j49400$

The total impedance is 39500 + j49400 ohms.

73. Evaluate $4x^2 + 4$ when $x = i$.

$$4(i)^2 + 4$$
$$= 4(-1) + 4 \qquad\qquad\qquad$$
$$= 0$$

Replace x with i.
$i^2 = -1$.

77. Factor $x^2 + 25$.
$$\begin{aligned} x^2 + 25 &= (x)^2 - (-25) \\ &= (x)^2 - (-5i)^2 \\ &= [x - (-5i)][x + (-5i)] \\ &= (x + 5i)(x - 5i) \end{aligned}$$

Difference of squares.

MAINTAIN YOUR SKILLS

Simplify:

81.
$$1 - \cfrac{1}{1 - \cfrac{1}{1 - x}} = 1 - \cfrac{1}{\cfrac{1 - x - 1}{1 - x}}$$

$$= 1 - \cfrac{1}{\cfrac{-x}{1 - x}}$$

$$= 1 - \cfrac{1 - x}{-x}$$

$$= \frac{-x - (1 - x)}{-x}$$

$$= \frac{-x - 1 + x}{-x}$$

$$= \frac{-1}{-x}$$

$$= \frac{1}{x}$$

85. Divide: $(4x^3 + 2x^2 + 2x + 2) \div 2x + 3$

```
                    2x² - 2x + 4
        2x + 3) 4x³ + 2x² + 2x + 2
                4x³ + 6x²
                     -4x² + 2x
                     -4x² - 6x
                           8x +  2
                           8x + 12
                               -10
```

The quotient is $2x^2 - 2x + 4 - \dfrac{10}{2x + 3}$

CHAPTER 5

QUADRATIC EQUATIONS

EXERCISES 5.1   QUADRATIC EQUATIONS SOLVED BY COMPLETING THE SQUARE

A

Solve:

1. $x^2 = 1$

    $x = \pm\sqrt{1} = \pm 1$ If $x^2 = k$, then $x = \pm\sqrt{k}$.

    The solution set is $\{\pm 1\}$.

5. $a^2 = 0.16$

    $a = \pm\sqrt{0.16} = \pm 0.4$

    The solution set is $\{\pm 0.4\}$.

9. $(x + 1)^2 = 1$

    $x + 1 = \pm\sqrt{1}$ If $x^2 = k$, then $x = \pm\sqrt{k}$.
    $x + 1 = \pm 1$
    $x + 1 = 1$ or $x + 1 = -1$ Rewrite as two equations.
    $x = 0$ or $x = -2$

    The solution set is $\{-2, 0\}$.

Solve by completing the square; check by factoring when possible.

13. $x^2 - 10x - 24 = 0$
     $x^2 - 10x = 24$ Add 24 to both sides.
     $x^2 - 10x + 25 = 24 + 25$ Add the square of half the coefficient of x: $\left[\frac{1}{2}(-10)\right]^2$.

     $(x - 5)^2 = 49$ Factor on the left. Simplify on the right.

     $x - 5 = \pm 7$
     $x - 5 = 7$ or $x - 5 = -7$
     $x = 12$ or $x = -2$

     $x^2 - 10x - 24 = 0$ Check by factoring.
     $(x - 12)(x + 2) = 0$
     $x - 12 = 0$ or $x + 2 = 0$
     $x = 12$ or $x = -2$

     The solution set is $\{-2, 12\}$.

171

17. $z^2 + 2z - 3 = 0$

    $z^2 + 2z = 3$ Add 3 to both sides.

    $z^2 + 2z + 1 = 3 + 1$ Add $\left[\frac{1}{2}(2)\right]^2$.

    $(z + 1)^2 = 4$ Factor on the left.

    $z + 1 = \pm 2$

    $z + 1 = -2$ or $z + 1 = 2$

    $z = -3$ or $z = 1$

    $z^2 + 2z - 3 = 0$ Factor the original

    $(z + 3)(z - 1) = 0$ equation to check.

    $z = -3$ or $z = 1$ Checks.

    The solution set is $\{-3, 1\}$.

B

Solve:

21. $(y + 5)^2 = -1$

    $y + 5 = \pm\sqrt{-1}$ If $x^2 = k$, then $x = \pm\sqrt{k}$.

    $y + 5 = \pm i$

    $y + 5 = i$ or $y + 5 = -i$

    $y = -5 + i$ or $y = -5 - i$

    The solution set is $\{-5 \pm i\}$.

25. $(x + 3)^2 = 2$

    $x + 3 = \pm\sqrt{2}$

    $x + 3 = \sqrt{2}$ or $x + 3 = -\sqrt{2}$

    $x = -3 + \sqrt{2}$ or $x = -3 - \sqrt{2}$

    The solution set is $\{-3 \pm \sqrt{2}\}$.

29. $(c - 6)^2 = -18$

    $c - 6 = \pm\sqrt{-18}$ $x = \pm\sqrt{k}$

    $c - 6 = \pm 3i\sqrt{2}$ Simplify.

    $c - 6 = 3i\sqrt{2}$ or $c - 6 = -3i\sqrt{2}$

    $c = 6 + 3i\sqrt{2}$ or $c = 6 - 3i\sqrt{2}$

    The solution set is $\{6 \pm 3i\sqrt{2}\}$.

Solve by completing the square.

33. $x^2 - x - 1 = 0$

    $x^2 - x = 1$ Add 1 to both sides.

$x^2 - x + \frac{1}{4} = 1 + \frac{1}{4}$ Add $\left(\frac{1}{2} \cdot -1\right)^2$ to both sides.

$$\left(x - \frac{1}{2}\right)^2 = \frac{5}{4}$$

Factor the left side and simplify the right.

$$x - \frac{1}{2} = \pm\sqrt{\frac{5}{4}}$$

$$x - \frac{1}{2} = \pm\frac{\sqrt{5}}{2}$$

$$x - \frac{1}{2} = \frac{\sqrt{5}}{2} \text{ or } x - \frac{1}{2} = -\frac{\sqrt{5}}{2}$$

$$x = \frac{1}{2} + \frac{\sqrt{5}}{2} \text{ or } x = \frac{1}{2} - \frac{\sqrt{5}}{2}$$

$$x = \frac{1 + \sqrt{5}}{2} \text{ or } x = \frac{1 - \sqrt{5}}{2}$$

The solution set is $\left\{\frac{1 \pm \sqrt{5}}{2}\right\}$.

37. $-2y^2 + 14y + 4 = 0$

    $y^2 - 7y - 2 = 0$ Divide both sides by −2.

       $y^2 - 7y = 2$ Add 2 to both sides.

$y^2 - 7y + \frac{49}{4} = 2 + \frac{49}{4}$ Add $\left(\frac{1}{2} \cdot -7\right)^2 = \frac{49}{4}$ to both sides.

$$\left(y - \frac{7}{2}\right)^2 = \frac{57}{4}$$

Factor on the left and simplify on the right.

$$y - \frac{7}{2} = \pm\sqrt{\frac{57}{4}}$$

$$y - \frac{7}{2} = \pm\frac{\sqrt{57}}{2}$$

$$y - \frac{7}{2} = \frac{\sqrt{57}}{2} \text{ or } y - \frac{7}{2} = -\frac{\sqrt{57}}{2}$$

$$y = \frac{7}{2} + \frac{\sqrt{57}}{2} \text{ or } y = \frac{7}{2} - \frac{\sqrt{57}}{2}$$

The solution set is $\left\{\frac{7 \pm \sqrt{57}}{2}\right\}$.

C

41. $4x^2 - 12x + 9 = 25$
        $4x^2 - 12x = 16$
           $x^2 - 3x = 4$

Subtract 9 from both sides.
Divide both sides by 4.

$$x^2 - 3x + \frac{9}{4} = 4 + \frac{9}{4}$$

Add $\left(\frac{1}{2} \cdot -3\right)^2 = \frac{9}{4}$ to both sides.

$$\left(x - \frac{3}{2}\right)^2 = \frac{25}{4}$$

Factor on the left and simplify on the right.

$$x - \frac{3}{2} = \pm \frac{25}{4}$$

$$x - \frac{3}{2} = \frac{5}{2} \text{ or } x - \frac{3}{2} = -\frac{5}{2}$$

$$x = \frac{3}{2} + \frac{5}{2} \text{ or } x = \frac{3}{2} - \frac{5}{2}$$

$$x = 4 \text{ or } \qquad x = -1$$

The solution set is $\{-1, 4\}$.

45. $9x^2 - 12x + 4 = -81$
        $9x^2 - 12x = -85$

Subtract 4 from both sides.

$$x^2 - \frac{4}{3}x = -\frac{85}{9}$$

Divide both sides by 9.

$$x^2 - \frac{4}{3}x + \frac{4}{9} = \frac{4}{9} - \frac{85}{9}$$

Add $\left(\frac{1}{2} \cdot -\frac{4}{3}\right)^2 = \frac{4}{9}$ to both sides.

$$\left(x - \frac{2}{3}\right)^2 = -\frac{81}{9}$$

$$x - \frac{2}{3} = \pm \sqrt{-9}$$

$$x - \frac{2}{3} = \pm 3i$$

$$x - \frac{2}{3} = 3i \text{ or } x - \frac{2}{3} = -3i$$

$$x = \frac{2}{3} + 3i \text{ or } x = \frac{2}{3} - 3i$$

The solution set is $\left\{\frac{2}{3} \pm 3i\right\}$.

53. $3x^2 + 2x = 6$

$x^2 + \frac{2}{3}x = 2$ Divide both sides by 3.

$x^2 + \frac{2}{3}x + \frac{1}{9} = 2 + \frac{1}{9}$ Add $\left[\frac{1}{2}\left(\frac{2}{3}\right)\right]^2$ to both sides.

$\left(x + \frac{1}{3}\right)^2 = \frac{19}{9}$

$x + \frac{1}{3} = \pm \frac{\sqrt{19}}{3}$ Solve.

$x + \frac{1}{3} = \frac{\sqrt{19}}{3}$ or $x + \frac{1}{3} = -\frac{\sqrt{19}}{3}$

$x = -\frac{1}{3} + \frac{\sqrt{19}}{3}$ or $x = -\frac{1}{3} - \frac{\sqrt{19}}{3}$

$x = \frac{-1 + \sqrt{19}}{3}$ or $x = \frac{-1 - \sqrt{19}}{3}$

The solution set is $\left\{\frac{-1 \pm \sqrt{19}}{3}\right\}$.

57. $6x^2 = -3x - 1$

$6x^2 + 3x + 1 = 0$ Write the equation in standard form.

$6x^2 + 3x = -1$ Subtract 1 from both sides.

$x^2 + \frac{1}{2}x = -\frac{1}{6}$ Divide both sides by 6.

$x^2 + \frac{1}{2}x + \frac{1}{16} = -\frac{1}{6} + \frac{1}{16}$ Add $\left[\frac{1}{2}\left(\frac{1}{2}\right)\right]^2$ to both sides.

$\left(x + \frac{1}{4}\right)^2 = -\frac{5}{48}$ Factor on the left, and simplify on the right.

$x + \frac{1}{4} = \pm \sqrt{-\frac{5}{48}}$ Solve.

$= \pm \sqrt{-\frac{15}{144}}$ Rationalize the denominator.

$x + \frac{1}{4} = \pm \frac{\sqrt{15}}{12}i$

$x + \frac{1}{4} = \frac{\sqrt{15}}{12}i$ or $x + \frac{1}{4} = -\frac{\sqrt{15}}{12}i$

$x = -\frac{1}{4} + \frac{\sqrt{15}}{12}i$ or $x = -\frac{1}{4} - \frac{\sqrt{15}}{12}i$ Write the complex roots in the form $a + bi$.

The solution set is $\left\{-\frac{1}{4} \pm \frac{\sqrt{15}}{12}i\right\}$.

61. $3x^2 - 4x + 5 = 0$

      $x^2 - \dfrac{4}{3}x + \dfrac{5}{3} = 0$ Divide both sides by 3.

           $x^2 - \dfrac{4}{3}x = -\dfrac{5}{3}$ Subtract $\dfrac{5}{3}$ from both sides.

           $x^2 - \dfrac{4}{3}x + \dfrac{4}{9} = -\dfrac{5}{3} + \dfrac{4}{9}$ Add $\left[\dfrac{1}{2}\left(-\dfrac{4}{3}\right)\right]^2$ to both sides.

      $\left(x - \dfrac{2}{3}\right)^2 = -\dfrac{11}{9}$ Factor on the left; simplify on the right.

        $x - \dfrac{2}{3} = \pm\sqrt{-\dfrac{11}{9}}$ Solve.

        $x - \dfrac{2}{3} = \pm\dfrac{\sqrt{11}}{3}i$

    $x - \dfrac{2}{3} = \dfrac{\sqrt{11}}{3}i$ or $x - \dfrac{2}{3} = -\dfrac{\sqrt{11}}{3}i$

    $x = \dfrac{2}{3} + \dfrac{\sqrt{11}}{3}i$ or $x = \dfrac{2}{3} - \dfrac{\sqrt{11}}{3}i$ Standard form for complex roots.

    The solution set is $\left\{\dfrac{2}{3} \pm \dfrac{\sqrt{11}}{3}i\right\}$.

D

Solve using the Pythagorean theorem:

65.   If $a = 3$, and $b = 4$, then $c = ?$

      $c^2 = a^2 + b^2$ Pythagorean Theorem.
      $c^2 = 3^2 + 4^2$ Let $a = 3$ and $b = 4$.
      $c^2 = 9 + 16$
      $c^2 = 25$

        $c = \pm\sqrt{25} = \pm 5$ Disregard the negative root since a length cannot be negative.

    The length of side c is 5.

69.   What is the rise of a rafter that is 20 feet in length and has a run of 16 feet?

      $a^2 = c^2 - b^2$ Pythagorean Theorem.
      $a^2 = 20^2 - 16^2$ The "rise" corresponds to the length of a, so
      $a^2 = 400 - 256$ replace c with 20 and b
      $a^2 = 144$ with 16.

        $a = \pm\sqrt{144} = \pm 12$

    The rise of the rafter is 12 feet.

73. The number of two-station telephone connections (D) that can be made between N different stations is given by the formula $2(D - N) = N(N - 3)$. If there are 66 possible connections on the operator's board, how many stations are in the system?

Formula:
$2(D - N) = N(N - 3)$

Substitute:
$2(66 - N) = N(N - 3)$ $D = 66$.

Solve:
$$132 - 2N = N^2 - 3N$$
$$N^2 - N - 132 = 0$$
$$N^2 - N = 132$$
$$N^2 - N + \frac{1}{4} = 132 + \frac{1}{4}$$
$$\left(N - \frac{1}{2}\right)^2 = \frac{529}{4}$$
$$N - \frac{1}{2} = \pm \frac{23}{2}$$
$$N - \frac{1}{2} = \frac{23}{2} \text{ or } N - \frac{1}{2} = -\frac{23}{2}$$
$$N = \frac{24}{2} \text{ or } \qquad N = -\frac{22}{2}$$
$$N = 12 \text{ or } \qquad N = -11$$

Reject -11 because the solution represents a number of stations.

Answer:
There are 12 stations in the system.

77. The sum of two numbers is 21 and their product is 108. What are the numbers?

Select a variable:
Let x represent one of the numbers. Then 21 - x represents the other.

Translate to algebra:
$x(21 - x) = 108$ The product of the two numbers is 108.

Solve:

$$21x - x^2 = 108$$
$$x^2 - 21x + 108 = 0 \qquad \text{Standard form.}$$
$$x^2 - 21x = -108$$

$$x^2 - 21x + \frac{441}{4} = -108 + \frac{441}{4} \qquad \text{Add } \left(\frac{1}{2} \cdot -21\right)^2 = \frac{441}{4} \text{ to}$$

both sides.

$$\left(x - \frac{21}{2}\right)^2 = \frac{9}{4}$$

$$x - \frac{21}{2} = \pm\frac{3}{2}$$

$$x - \frac{21}{2} = \frac{3}{2} \text{ or } x - \frac{21}{2} = -\frac{3}{2}$$

$$x = 12 \text{ or } \qquad x = 9$$

$$21 - x = 9 \text{ or } \quad 21 - x = 12$$

Answer:
The two numbers are 12 and 9.

In Exercises 81-88, assume $a \geq 0$, $b \geq 0$, and $c \geq 0$.

81. Solve $a^2 + b^2 = c^2$ for c
$$c^2 = a^2 + b^2$$

$$c = \pm\sqrt{a^2 + b^2} \qquad \text{If } x^2 = k, \text{ then } x = \pm\sqrt{k}.$$

85. Solve $x^2 = 16c$ for x.
$$x^2 = 16c$$

$$x = \pm\sqrt{16c}$$

$$x = \pm 4\sqrt{c}$$

89. Solve the formula $V = \pi r^2 h$ for r.
$$V = \pi r^2 h$$

$$\frac{V}{\pi h} = r^2 \qquad \text{Divide both sides by } \pi h.$$

$$\pm\sqrt{\frac{V}{\pi h}} = r \qquad \text{If } x^2 = k, \text{ then } x = \pm\sqrt{k}.$$

STATE YOUR UNDERSTANDING

93. Explain how to use the completing the square method for solving a quadratic equation.

When the equation is of the form $x^2 + bx$, a constant term $\left(\frac{1}{2} \cdot b\right)^2$ is added so that the resulting trinomial is a perfect square trinomial. Then the trinomial can be factored as the square of a binomial, which leads to the solution using $x = \pm\sqrt{k}$.

Solve for x by completing the square.

97. $x^2 + 4ax + a = 0$
          $x^2 + 4ax = -a$ Subtract a from both sides.

      $x^2 + 4ax + 4a^2 = 4a^2 - a$ Add $\left(\frac{1}{2} \cdot 4a\right)^2 = 4a^2$ to both sides.

          $(x + 2a)^2 = 4a^2 - a$

              $x + 2a = \pm\sqrt{4a^2 - a}$

      $x + 2a = \sqrt{4a^2 - a}$ or $x + 2a = -\sqrt{4a^2 - a}$

          $x = -2a + \sqrt{4a^2 - a}$ or $x = -2a - \sqrt{4a^2 - a}$

      The solution set is $\left\{-2a \pm \sqrt{4a^2 - a}\right\}$

MAINTAIN YOUR SKILLS

Write using rational exponents:

101. $\sqrt[3]{a^2 b} = (a^2 b)^{1/3} = a^{2/3} b^{1/3}$

Simplify. Assume all variables represent positive numbers:

105. $\sqrt{164 c^4 d^7} = \sqrt{4 \cdot 41 \cdot c^4 d^6 d}$
           $= \sqrt{4 c^4 d^6} \cdot \sqrt{41 d}$
           $= 2 c^2 d^3 \sqrt{41 d}$

EXERCISES 5.2   THE QUADRATIC FORMULA

A

Solve for x using the quadratic formula. Assume all other variables represent positive numbers:

1. $x^2 + 9x + 14 = 0$ — The equation is in standard form.

$a = 1$, $b = 9$, $c = 14$ — Identify a, b, and c.

$$x = \frac{-b \pm \sqrt{b^2 - 4ac}}{2a}$$ — Quadratic formula.

$$x = \frac{-9 \pm \sqrt{(9)^2 - 4(1)(14)}}{2(1)}$$ — Substitute.

$$x = \frac{-9 \pm \sqrt{81 - 56}}{2}$$ — Simplify.

$$x = \frac{-9 \pm \sqrt{25}}{2}$$

$$x = \frac{-9 \pm 5}{2}$$

$$x = \frac{-9 + 5}{2} \text{ or } x \frac{-9 - 5}{2}$$

$x = -2 \qquad\qquad x = -7$

The solution set is $\{-7, -2\}$.

5. $2x^2 - 10x - 72 = 0$

$a = 2$, $b = -10$, $c = -72$ — Identify a, b, and c.

$$x = \frac{-b \pm \sqrt{b^2 - 4ac}}{2a}$$ — Quadratic formula.

$$x = \frac{10 \pm \sqrt{(-10)^2 - 4(2)(-72)}}{2(2)}$$ — Substitute.

$$x = \frac{10 \pm \sqrt{100 + 576}}{4}$$

$$x = \frac{10 \pm \sqrt{676}}{4}$$

$$x = \frac{10 \pm 26}{4}$$

$$x = \frac{10 + 26}{4} \text{ or } x = \frac{10 - 26}{4}$$

$x = 9 \qquad\qquad x = -4$

The solution set is $\{-4, 9\}$.

9. $x^2 - 20x = -96$ Standard form.
    $a = 1, \ b = -20, \ c = 96$

$$x = \frac{-b \pm \sqrt{b^2 - 4ac}}{2a}$$

$$x = \frac{-(-20) \pm \sqrt{(-20)^2 - 4(1)(96)}}{2(1)}$$

$$x = \frac{20 \pm \sqrt{400 - 384}}{2}$$

$$x = \frac{20 \pm \sqrt{16}}{2}$$

$$x = \frac{20 \pm 4}{2}$$

$$x = \frac{20 + 4}{2} \quad \text{or} \quad x = \frac{20 - 4}{2}$$

$$x = 12 \qquad\qquad x = 8$$

The solution set is $\{8, \ 12\}$.

13. $4x^2 = -8x - 3$
    $4x^2 + 8x + 3 = 0$ Standard form.
    $a = 4, \ b = 8, \ c = 3$

$$x = \frac{-b \pm \sqrt{b^2 - 4ac}}{2a}$$ Quadratic formula.

$$x = \frac{-(8) \pm \sqrt{(8)^2 - 4(4)(3)}}{2(4)}$$ Substitute.

$$x = \frac{-8 \pm \sqrt{64 - 48}}{8}$$ Simplify.

$$x = \frac{-8 \pm \sqrt{16}}{8}$$

$$x = \frac{-8 \pm 4}{8}$$

$$x = -\frac{1}{2} \qquad\qquad x = -\frac{3}{2}$$

The solution set is $\left\{-\frac{3}{2}, \ -\frac{1}{2}\right\}$.

17. $3x^2 + 12x + 6 = 0$
$a = 3, \ b = 12, \ c = 6$

$$x = \frac{-b \pm \sqrt{b^2 - 4ac}}{2a}$$

$$x = \frac{-12 \pm \sqrt{(12)^2 - 4(3)(6)}}{2(3)}$$

$$x = \frac{-12 \pm \sqrt{144 - 72}}{6}$$

$$x = \frac{-12 \pm \sqrt{72}}{6}$$

$$x = \frac{-12 \pm 6\sqrt{2}}{6}$$

$$x = \frac{6(-2 \pm \sqrt{2})}{6}$$

$$x = -2 \pm \sqrt{2}$$

The solution set is $\{-2 \pm \sqrt{2}\}$.

B

21. $x^2 - 4x - 7 = 0$ Standard form.
$a = 1, \ b = -4, \ c = -7$

$$x = \frac{-b \pm \sqrt{b^2 - 4ac}}{2a}$$

$$x = \frac{-(-4) \pm \sqrt{(-4)^2 - 4(1)(-7)}}{2(1)}$$

$$x = \frac{4 \pm \sqrt{16 + 28}}{2}$$

$$= \frac{4 \pm \sqrt{44}}{2}$$

$$x = \frac{4 \pm 2\sqrt{11}}{2} = \frac{2(2 + \sqrt{11})}{2} \qquad \text{Divide out the common factor, 2.}$$

$$x = 2 \pm \sqrt{11}$$

The solution set is $\{2 \pm \sqrt{11}\}$.

182

25. $3x^2 - 15x + 21 = 0$
    $a = 3, \ b = -15, \ c = 21$

$$x = \frac{-b \pm \sqrt{b^2 - 4ac}}{2a}$$

$$x = \frac{15 \pm \sqrt{225 - 4(3)(21)}}{6}$$

$$x = \frac{15 \pm \sqrt{225 - 252}}{6} = \frac{15 \pm \sqrt{-27}}{6}$$

$$x = \frac{15 \pm 3i\sqrt{3}}{6}$$

$$x = \frac{3(5 \pm i\sqrt{3})}{3(2)}$$ Factor the numerator to reduce.

$$x = \frac{5 \pm i\sqrt{3}}{2}$$

or $\frac{5}{2} \pm \frac{\sqrt{3}}{2}i$ Standard form for complex roots: $a + bi$.

The solution set is $\left\{ \frac{5}{2} \pm \frac{\sqrt{3}}{2}i \right\}$.

29. $-2x^2 + 3x + 7 = 0$
    $a = -2, \ b = 3, \ c = 7$ Identify a, b, and c.

$$x = \frac{-b \pm \sqrt{b^2 - 4ac}}{2a}$$

$$x = \frac{-3 \pm \sqrt{9 - 4(-2)(7)}}{2(-2)}$$

$$x = \frac{-3 \pm \sqrt{65}}{-4}$$

or $\frac{-1(3 \pm \sqrt{65})}{-1(4)}$ Factor out −1 in the numerator and denominator.

The solution set is $\left\{ \frac{3 \pm \sqrt{65}}{4} \right\}$.

33. $6x^2 - 19ax + 15a^2 = 0$ Standard form.

     $a = 6, \; b = -19a, \; c = 15a^2$ Identify a, b, and c.

$$x = \frac{-b \pm \sqrt{b^2 - 4ac}}{2a}$$ Quadratic formula.

$$x = \frac{-(-19a) \pm \sqrt{(-19a)^2 - 4(6)(15a^2)}}{2(6)}$$ Substitute.

$$x = \frac{19a \pm \sqrt{361a^2 - 360a^2}}{12}$$

$$x = \frac{19a \pm \sqrt{a^2}}{12}$$

$$x = \frac{19a \pm a}{12}$$

$$x = \frac{19a + a}{12} \text{ or } x = \frac{19a - a}{12}$$

$$x = \frac{5a}{3} \qquad\qquad x = \frac{3a}{2}$$

The solution set is $\left\{ \dfrac{3a}{2}, \; \dfrac{5a}{2} \right\}$.

37. $3x^2 + 7x + 2 = 0$

     $a = 3, \; b = 7, \; c = 2$

$$x = \frac{-b \pm \sqrt{b^2 - 4ac}}{2a}$$

$$x = \frac{-7 \pm \sqrt{(-7)^2 - 4(3)(2)}}{2(3)}$$

$$x = \frac{-7 \pm \sqrt{49 - 24}}{6}$$

$$x = \frac{-7 \pm \sqrt{25}}{6}$$

$$x = \frac{-7 + 5}{6} \text{ or } x = \frac{-7 - 5}{6}$$

$$x = -\frac{1}{3} \text{ or } \qquad x = -2$$

The solution set is $\left\{ -\dfrac{1}{3}, \; -2 \right\}$.

C

41. $$6x^2 = -19x - 10$$
$$6x^2 + 19x + 10 = 0$$ Write the equation in standard form.

$$a = 6, \ b = 19, \ c = 10$$ Identify a, b, and c.

$$x = \frac{-(19) \pm \sqrt{(19)^2 - 4(6)(10)}}{2(6)}$$ Substitute into the formula from memory.

$$x = \frac{-19 \pm \sqrt{361 - 240}}{12}$$ Simplify.

$$x = \frac{-19 \pm \sqrt{121}}{12}$$

$$x - \frac{-19 \pm 11}{12}$$

$$x = \frac{-19 + 11}{12} \ \text{or} \ x = \frac{-19 - 11}{12}$$

$$x = -\frac{2}{3} \qquad x = -\frac{5}{2}$$

The solution set is $\left\{ -\frac{5}{2}, \ -\frac{2}{3} \right\}$.

45. $$-6x^2 = 13x - 5$$
$$6x^2 + 13x - 5 = 0$$ Standard form.
$$a = 6, \ b = 13, \ c = -5$$

$$x = \frac{-(13) \pm \sqrt{(13)^2 - 4(6)(-5)}}{2(6)}$$ Substitute into the formula from memory.

$$x = \frac{-13 \pm \sqrt{169 + 120}}{12}$$

$$x = \frac{-13 \pm \sqrt{289}}{12}$$

$$x = \frac{-13 \pm 17}{12}$$

$$x = \frac{-13 + 17}{12} \ \text{or} \ x = \frac{-13 - 17}{12}$$

$$x = \frac{1}{3} \qquad x = -\frac{5}{2}$$

The solution set is $\left\{ -\frac{5}{2}, \ \frac{1}{3} \right\}$.

49. $x^2 + 2x + 3 = 3x^2 + 7x - 3$

       $-2x^2 - 5x + 6 = 0$ Standard form.

        $2x^2 + 5x - 6 = 0$ Multiply both sides by -1.

       $a = 2, \ b = 5, \ c = -6$

$$x = \frac{-b \pm \sqrt{b^2 - 4ac}}{2a}$$

$$x = \frac{-(5) \pm \sqrt{(5)^2 - 4(2)(-6)}}{2(2)}$$ Substitute.

$$x = \frac{-5 \pm \sqrt{25 + 48}}{4}$$ Simplify.

$$x = \frac{-5 \pm \sqrt{73}}{4}$$

The solution set is $\left\{\dfrac{-5 \pm \sqrt{73}}{4}\right\}$.

53. $ax^2 - 3abx + 2b^2 = 0$ Standard form.

       $a = a, \ b = -3ab, \ c = 2b^2$ Identify a, b, and c.

$$x = \frac{-b \pm \sqrt{b^2 - 4ac}}{2a}$$ Quadratic formula.

$$x = \frac{-(-3ab) \pm \sqrt{(-3ab)^2 - 4(a)(2b^2)}}{2(a)}$$

$$x = \frac{3ab \pm \sqrt{9a^2 b^2 - 8ab^2}}{2a}$$

$$x = \frac{3ab \pm \sqrt{b^2(9a^2 - 8a)}}{2a}$$ Simplify the radical.

$$x = \frac{3ab \pm b\sqrt{9a^2 - 8a}}{2a}$$

The solution set is $\left\{\dfrac{3ab \pm b\sqrt{9a^2 - 8a}}{2a}\right\}$.

57. $2x^2 + \sqrt{2}x - 12 = 0$

$a = 2$, $b = \sqrt{2}$, $c = -12$

$$x = \frac{-b \pm \sqrt{b^2 - 4ac}}{2a}$$

$$x = \frac{-\sqrt{2} \pm \sqrt{(\sqrt{2})^2 - 4(2)(-12)}}{2(2)}$$

$$x = \frac{-\sqrt{2} \pm \sqrt{2 + 96}}{4}$$

$$x = \frac{-\sqrt{2} \pm \sqrt{98}}{4}$$

$$x = \frac{-\sqrt{2} \pm 7\sqrt{2}}{4}$$

$$x = \frac{-\sqrt{2} + 7\sqrt{2}}{4} \text{ or } x = \frac{-\sqrt{2} - 7\sqrt{2}}{4}$$

$$x = \frac{6\sqrt{2}}{4} \text{ or } \qquad x = \frac{-8\sqrt{2}}{4}$$

$$x = \frac{3\sqrt{2}}{2} \text{ or } \qquad x = -2\sqrt{2}$$

The solution set is $\left\{-2\sqrt{2}, \; \frac{3\sqrt{2}}{2}\right\}$.

61. $6x^2 + 7ix + 3 = 0$
$a = 6$, $b = 7i$, $c = 3$

$$x = \frac{-7i \pm \sqrt{(7i)^2 - 4(6)(3)}}{2(6)}$$

$$x = \frac{-7i \pm \sqrt{-49 - 72}}{12} \qquad\qquad i^2 = -1.$$

$$x = \frac{-7i \pm \sqrt{-121}}{12}$$

$$x = \frac{-7i \pm 11i}{12}$$

$$x = \frac{-7i + 11i}{12} \text{ or } x = \frac{-7i - 11i}{12}$$

$$x = \frac{1}{3}i \text{ or } \qquad x = -\frac{3}{2}i$$

The solution set is $\left\{-\frac{3}{2}i, \; \frac{1}{3}i\right\}$.

65. Elna wants to put her $1000 into an account that will be worth $1210 in two years. What rate of interest must she look for? (*Hint*: Use the formula in Example 7, with A = 1210 and I = 1000.)

Formula:
$$A = I(1 + r)^2$$

Substitute:
$$1210 = 1000(1 + r)^2$$

Solve:
$$1210 = 1000(1 + 2r + r^2)$$
$$1210 = 1000 + 2000r + 1000r^2$$
$$1000r^2 + 2000r - 210 = 0$$
$$100r^2 + 200r - 21 = 0 \qquad \text{Divide both sides by 10.}$$

$$x = \frac{-b \pm \sqrt{b^2 - 4ac}}{2a} \qquad \text{Use a = 100, b = 200,} \\ \text{c = -21.}$$

$$x = \frac{-200 \pm \sqrt{(200)^2 - 4(100)(-21)}}{2(100)}$$

$$x = \frac{-200 \pm \sqrt{40000 + 8400}}{200}$$

$$x = \frac{-200 \pm \sqrt{48400}}{200}$$

$$x = \frac{-200 \pm 220}{200}$$

$$x = \frac{-200 + 220}{200} \text{ or } x = \frac{-200 - 220}{200}$$

$$x = \frac{1}{10} \text{ or } 0.1 \text{ or } \quad x = -2\frac{1}{10} \text{ or } -2.1$$

Reject -2.1 as an answer since rate of interest is not negative.

Answer:
The interest rate is 0.1 or 10%.

69. The product of four more than a number and seven more than the same number is 108. What is the number?

Simpler word form:
A product equals 108.

Select variable:
Let x represent the number. Then four more than the number is represented by x + 4, and x + 7 represents seven more than the number.

Translate to algebra:
$(x + 4)(x + 7) = 108$

Solve:
$x^2 + 11x + 28 = 108$
$x^2 + 11x - 80 = 0$

$$x = \frac{-11 \pm \sqrt{11^2 - 4(1)(-80)}}{2(1)}$$ Quadratic formula.

$$x = \frac{-11 \pm \sqrt{441}}{2}$$

$$x = \frac{-11 \pm 21}{2}$$

$$x = \frac{-11 + 21}{2} \text{ or } x = \frac{-11 - 21}{2}$$

$$x = \frac{10}{2} \text{ or } \qquad x = \frac{-32}{2}$$

$$x = 5 \text{ or } \qquad x = -16$$

Answer:
The number is either 5 or -16.

CHALLENGE EXERCISES

73. Using the formula below, how far will an arrow travel if the initial velocity is 144 feet per second and it reaches a maximum height of 162 feet?

    Formula:
    $$y = \frac{-32x^2}{v^2} + x$$

    Substitution:
    $$162 = \frac{-32x^2}{144^2} + x$$ Replace v with 144 and y with 162.

Solve:

$$(144)^2(162) = -32x^2 + (144)^2x$$

Multiply both sides by 144^2.

$$32x^2 - 144^2x + 144^2(162) = 0$$

Standard form.

$$x^2 - 648x + 104976 = 0$$

Divide both sides by 32 and simplify.

$$x = \frac{648 \pm \sqrt{648^2 - 4(104976)}}{2}$$

Quadratic formula.

$$x = \frac{648 \pm \sqrt{0}}{2}$$

$$x = 324$$

Answer:
The arrow will travel 324 feet.

MAINTAIN YOUR SKILLS

Simplify using only positive exponents:

77. $\left(p^{-1/2}q^{2/3}\right)^{3/4}\left(pq^4\right)^{-1/4}$

$$= p^{(-1/2)(3/4)}q^{(2/3)(3/4)}p^{-1/4}q^{(4)(-1/4)}$$

Multiply exponents.

$$= p^{-3/8}q^{1/2}p^{-1/4}q^{-1}$$

$$= \left(p^{-3/8}\right)\left(p^{-1/4}\right)\left(q^{1/2}q^{-1}\right)$$

Regroup.

$$= p^{-5/8}q^{-1/2}$$

Add exponents.

$$= \frac{1}{p^{5/8}q^{1/2}}$$

Definition of negative exponents.

EXERCISES 5.3   PROPERTIES OF QUADRATIC EQUATIONS

A

Find the value of the discriminant and describe the roots of each of the following:

1. $x^2 + 10x + 16 = 0$
   $a = 1, b = 10, c = 16$ Identify a, b, and c.
   $b^2 - 4ac = (10)^2 - 4(1)(16)$ Substitute into the formula
   $= 100 - 64$ for the discriminant.
   $= 36$ The discriminant is positive, so the equation has two unequal real roots.

   The discriminant is 36, and the equation has two unequal real roots.

5. $x^2 + 16x + 64 = 0$
   $a = 1, b = 16, c = 64$ Identify a, b, and c.
   $b^2 - 4ac = (16)^2 - 4(1)(64)$ Substitute into the formula for the discriminant.
   $= 256 - 256$
   $= 0$

   The value of the discriminant is zero, and the equation has two equal real roots.

9. $x^2 = -x + 11$
   $x^2 + x - 11 = 0$ Standard form.
   $a = 1, b = 1, c = -11$ Identify a, b, and c.
   $b^2 - 4ac = (1)^2 - 4(1)(-11)$ Substitute into the formula for the discriminant.
   $= 1 + 44$
   $= 45$

   The discriminant is 45, and the equation has two unequal real roots.

Write a quadratic equation in standard form that has the following roots:

13. $x = 7$ or $x = -5$
    $x - 7 = 0$ or $x + 5 = 0$ Rewrite each equation so the right side is zero.
    $(x - 7)(x + 5) = 0$ Zero-product law.
    $x^2 - 2x - 35 = 0$ Write in standard form.

Solve and check using the sum and the product of roots:

17. $x^2 + 8x - 16 = 0$
     $a = 1, \ b = 8, \ c = -16$ Solve using the quadratic formula.

$$x = \frac{-8 \pm \sqrt{8^2 - 4(1)(-16)}}{2(1)}$$ Substitute in the formula.

$$x = \frac{-8 \pm \sqrt{128}}{2}$$

$$x = \frac{-8 \pm 8\sqrt{2}}{2}$$

$$x = -4 \pm 4\sqrt{2}$$

<u>Check:</u>

$$r_1 + r_2 = -\frac{8}{1}$$

$$(-4 + 4\sqrt{2}) + (-4 - 4\sqrt{2}) = -8$$
$$-8 + 0 = -8$$
$$-8 = -8$$

$$(r_1)(r_2) = -\frac{16}{1}$$

$$(-4 + 4\sqrt{2})(-4 - 4\sqrt{2}) = -16$$
$$16 - 32 = -16$$
$$-16 = -16$$

The solution set is $\{-4 \pm 4\sqrt{2}\}$.

B

Find the value of the discriminant and describe the roots of each of the following:

21. $3x^2 - 8x + 11 = 0$
     $a = 3, \ b = -8, \ c = 11$ Identify a, b, and c.
     $b^2 - 4ac = (-8)^2 - 4(3)(11)$ Substitute.
                $= 64 - 132$
                $= -68$

The discriminant is negative, and the equation has two conjugate complex roots.

25.
$$7x^2 - 2x = -1$$
$$7x^2 - 2x + 1 = 0 \qquad \text{Standard form.}$$
$$a = 7, \ b = -2, \ c = 1$$
$$b^2 - 4ac = (-2)^2 - 4(7)(1) \qquad \text{Substitute.}$$
$$= 4 - 28$$
$$= -24$$

The discriminant is negative, so the equation has two conjugate complex roots.

29.
$$7x^2 - 2x = x + 14$$
$$7x^2 - 3x - 14 = 0 \qquad \text{Standard form.}$$
$$a = 7, \ b = -3, \ c = -14 \qquad \text{Identify a, b, and c.}$$
$$b^2 - 4ac = (-3)^2 - 4(7)(-14) \qquad \text{Substitute into the formula for the discriminant.}$$
$$= 9 + 392$$
$$= 401$$

The discriminant is 401, and the equation has two unequal real roots.

Write a quadratic equation in standard form that has the following roots:

33.
$$x = 1 + \sqrt{5} \text{ or } \qquad x = 1 - \sqrt{5}$$
$$x - 1 - \sqrt{5} = 0 \text{ or } x - 1 + \sqrt{5} = 0 \qquad \text{Rewrite each equation so that the right side is zero.}$$

$$(x - 1 - \sqrt{5})(x - 1 + \sqrt{5}) = 0 \qquad \text{Zero-product property.}$$

$$[(x - 1) - \sqrt{5}][(x - 1) + \sqrt{5}] = 0 \qquad \text{Write factors on left side as conjugates.}$$

$$[(x - 1)^2 - (\sqrt{5})^2] = 0 \qquad \text{Product of conjugates.}$$
$$x^2 - 2x + 1 - 5 = 0 \qquad \text{Multiply.}$$
$$x^2 - 2x - 4 = 0 \qquad \text{Standard form.}$$

Solve and check using the sum and the product of roots:

37. $5x^2 - 10x + 6 = 0$
     $a = 5, \quad b = -10, \quad c = 6$

$$x = \frac{10 \pm \sqrt{10^2 - 4(5)(6)}}{2(5)}$$

$$x = \frac{10 \pm \sqrt{-20}}{10}$$

$$x = \frac{10 \pm 2i\sqrt{5}}{10}$$

$$x = 1 \pm \frac{\sqrt{5}}{5}i$$

Check:

$$r_1 + r_2 = -\left(\frac{-10}{5}\right)$$

$$1 + \frac{\sqrt{5}}{5}i + 1 - \frac{\sqrt{5}}{5}i = 2$$

$$1 + 1 = 2$$

$$2 = 2$$

$$(r_1)(r_2) = \frac{6}{5}$$

$$\left(1 + \frac{\sqrt{5}}{5}i\right)\left(1 - \frac{\sqrt{5}}{5}i\right) = \frac{6}{5}$$

$$1 - \frac{5}{25}i^2 = \frac{6}{5}$$

$$1 + \frac{5}{25} = \frac{6}{5} \qquad\qquad i^2 = -1$$

$$\frac{30}{25} = \frac{6}{5}$$

$$\frac{6}{5} = \frac{6}{5}$$

The solution set is $\left\{1 \pm \frac{\sqrt{5}}{5}i\right\}$.

c

Find the value of the discriminant and describe the roots of each of the following:

41.
$$7x^2 + 4x = 3x - 10$$
$$7x^2 + x + 10 = 0$$ Rewrite in standard form.
$$a = 7, \ b = 1, \ c = 10$$ Identify a, b, and c.
$$b^2 - 4ac = (1) - 4(7)(10)$$ Substitute into the formula for the discriminant.

$$= 1 - 280$$
$$= -279$$

The discriminant is -279 so the equation has two conjugate complex roots.

45.
$$2x^2 - x + 2 = -2x^2 + 11x - 7$$
$$4x^2 - 12x + 9 = 0$$ Rewrite in standard form.
$$a = 4, \ b = -12, \ c = 9$$ Identify a, b, and c.
$$b^2 - 4ac = (-12)^2 - 4(4)(9)$$ Substitute.
$$= 144 - 144$$
$$= 0$$

The discriminant is zero, so the equation has two equal real roots.

49.
$$(3x - 1)^2 - (x + 5)^2 = 6$$
$$9x^2 - 6x + 1 - (x^2 + 10x + 25) = 6$$ Multiply and simplify.
$$9x^2 - 6x + 1 - x^2 - 10x - 25 = 6$$
$$8x^2 - 16x - 24 = 6$$
$$8x^2 - 16x - 30 = 0$$ Rewrite in standard form.
$$a = 8, \ b = -16, \ c = -30$$ Identify a, b, and c.
$$b^2 - 4ac = (-16)^2 - 4(8)(-30)$$ Substitute.
$$= 256 + 960$$
$$= 1216$$

The discriminant is 1216, so the equation has two unequal real roots.

Write an equation in standard form that has the following roots:

53. $x = \dfrac{2 \pm \sqrt{5}}{3}$

$x = \dfrac{2 + \sqrt{5}}{3}$ or $x = \dfrac{2 - \sqrt{5}}{3}$

$3x = 2 + \sqrt{5}$ or $3x = 2 - \sqrt{5}$ Multiply both sides of each equation by 3 to clear the fraction.

$3x - 2 - \sqrt{5} = 0$ or $3x - 2 + \sqrt{5} = 0$ Write each equation with the right side equal to zero.

$(3x - 2 - \sqrt{5})(3 - 2 + \sqrt{5}) = 0$ Zero-product property.

$[(3x - 2) - \sqrt{5}][(3x - 2) + \sqrt{5}] = 0$ Write the left side using conjugate factors.

$(3x - 2)^2 - (\sqrt{5})^2 = 0$ Product of conjugates.

$9x^2 - 12x + 4 - 5 = 0$ Multiply.

$9x^2 - 12x - 1 = 0$ Standard form.

57. $x = -3 \pm 2i\sqrt{5}$

$x = -3 + 2i\sqrt{5}$ or $x = -3 - 2i\sqrt{5}$

$x + 3 - 2i\sqrt{5} = 0$ or $x + 3 + 2i\sqrt{5} = 0$

$[(x + 3) - 2i\sqrt{5}][(x + 3) + 2i\sqrt{5}] = 0$

$(x + 3)^2 - (2i\sqrt{5})^2 = 0$

$x^2 + 6x + 9 - 4i^2(5) = 0$

$x^2 + 6x + 29 = 0$ $i^2 = -1$

D

61. The number of units (N) produced each day at WHAT-CO, Inc. is related to the number of employees. The relationship can be expressed by the quadratic equation $N = x^2 - 20x + 121$, where the number of employees is the larger of the two values of x found by solving the equation. What are the number of units for which the equation is valid (has real solutions)? Find the minimum number of employees for which the equation is valid.

Formula:
$N = x^2 - 20x + 121$
$0 = x^2 - 20x + (121 - N)$ Subtract N from both sides.

$a = 1$, $b = -20$, $c = 121 - N$
$(-20)^2 - 4(1)(121 - N) \geq 0$ To have real solutions,

$$400 - 484 + 4N \geq 0$$ $b^2 - 4ac$ must be positive,

$$4N \geq 84$$ or equal to zero.

$$N \geq 21$$

$N = x^2 - 20x + 121$ Substitute $N = 21$.
$21 = x^2 - 20x + 121$
$0 = x^2 - 20x + 100$ Subtract 21 from both sides.

$0 = (x - 10)^2$ Factor.
$x - 10 = 0$ or $x - 10 = 0$ Zero-product property.
$x = 10$ A double root.

So the numbers for which the equation has real solutions are the numbers greater than or equal to 21. The minimum number of employees (at $x = 21$) is 10.

65. Using the general roots of a quadratic equation,

$x = \dfrac{-b \pm \sqrt{b^2 - 4ac}}{2a}$, show that the product of the roots is $\dfrac{c}{a}$.

$$\left(\frac{-b + \sqrt{b^2 - 4ac}}{2a} \right) \left(\frac{-b - \sqrt{b^2 - 4ac}}{2a} \right)$$

$$= \frac{(-b + \sqrt{b^2 - 4ac})(-b - \sqrt{b^2 - 4ac})}{4a^2}$$

$$= \frac{b^2 - (b^2 - 4ac)}{4a^2}$$

$$= \frac{b^2 - b^2 + 4ac}{4a^2}$$

$$= \frac{4ac}{4a^2}$$

$$= \frac{c}{a}$$

69. Given the equation $x^2 + bx + 48 = 0$, for what values of b will the equation have one real solution?

$x^2 + bx + 48 = 0$
$a = 1$, $b = b$, $c = 48$

If $b^2 - 4ac = 0$, then there is one real solution.

$b^2 - 4(1)(48) = 0$

Set $b^2 - 4ac$ equal to zero.

$b^2 - 192 = 0$
$b^2 = 192$

$b = \pm\sqrt{192}$

$192 = (64)(3)$

$b = \pm 8\sqrt{3}$

The equation will have one real solution when $b = 8\sqrt{3}$ or $b = -8\sqrt{3}$.

CHALLENGE EXERCISES

Find the value of b in the following:

73. $6bx^2 + 13x + 6b = 0$ if the sum of the roots is $-\frac{13}{6}$.
$a = 6b$, $b = 13$, $c = 6b$

$$x = \frac{-13 \pm \sqrt{(13)^2 - 4(6b)(6b)}}{2(6b)}$$

Solve using the quadratic formula.

$$x = \frac{-13 \pm \sqrt{169 - 144b^2}}{12b}$$

$$r_1 + r_2 = -\frac{13}{6}$$

Form the sum of the roots and set equal to $\frac{13}{6}$.

$$\left(\frac{-13 + \sqrt{169 - 144b^2}}{12b}\right) + \left(\frac{-13 - \sqrt{169 - 144b^2}}{12b}\right) = -\frac{13}{6}$$

$$\frac{(-13 + \sqrt{169 - 144b^2}) + (-13 - \sqrt{169 - 144b^2})}{12b} = -\frac{13}{6}$$

The common denominator is 12b.

$$\frac{-26}{12b} = -\frac{13}{6}$$

Simplify.

$$-156b = -156$$

Cross-multiply.

$$b = 1$$

Write in radical form:

77. $\left(a^2 - 3\right)^{3/5} = \sqrt[5]{(a^2 - 3)^3}$

Simplify. Assume that variables represent real numbers.

81. $4\sqrt{256a^2 b^4} = 4 \cdot 16 \cdot a \cdot b^2 = 64ab^2$

EXERCISES 5.4 ABSOLUTE VALUE EQUATIONS

A

Solve:

1. $|x| = 3$
   $x = \pm 3$

   3 and -3 are the two numbers that are three units from zero.

   The solution set is $\{\pm 3\}$.

5. $|x - 4| = 5$
   $x - 4 = 5$ or $x - 4 = -5$
   $x = 9$ $x = -1$

   If $|x| = c$, $c \geq 0$, then $x = c$ or $x = -c$.

   The solution set is $\{-1, 9\}$.

9. $|x + 5| = 0$
   $x + 5 = 0$
   $x = -5$

   The solution set is $\{-5\}$.

13. $|x - 9| = 20$
    $x - 9 = 20$ or $x - 9 = -20$
    $x = 29$ $x = -11$

    The solution set is $\{-11, 29\}$.

17. $|x - 5| = 12$
    $x - 5 = 12$ or $x - 5 = -12$
    $x = 17$ or $x = -7$

    The solution set is $\{-7, 17\}$.

B

21. $|x + 3| - 8 = 5$

$\qquad |x + 3| = 13$

Isolate the absolute value on one side of the equation.

$x + 3 = 13$ or $x + 3 = -13$

$\qquad x = 10 \qquad\qquad x = -16$

The solution set is $\{-16, 10\}$.

25. $|3x + 6| = 9$

$\qquad 3x + 6 = 9$ or $3x + 6 = -9$

$\qquad\qquad 3x = 3 \qquad\qquad 3x = -15$

$\qquad\qquad\quad x = 1 \qquad\qquad\quad x = -5$

The solution set is $\{-5, 1\}$.

29. $-2|2x - 14| = 40$

$\qquad |2x - 14| = -20$

Multiply both sides by the reciprocal of -2.
An absolute value cannot be negative.

No solution.

The solution set is empty, $\emptyset$.

33. $6 - |2x + 7| = -4$

$\qquad |2x + 7| = 10$

Isolate the absolute value on one side of the equation.

$2x + 7 = 10$ or $2x + 7 = -10$

$\qquad 2x = 3 \qquad\qquad 2x = -17$

$\qquad\quad x = \dfrac{3}{2} \qquad\qquad x = \dfrac{-17}{2}$

The solution set is $\left\{-\dfrac{17}{2}, \dfrac{3}{2}\right\}$.

37. $5|2x - 6| - 9 = 31$

$\qquad 5|2x - 6| = 40$

Isolate the absolute value on the left.

$\qquad |2x - 6| = 8$

$2x - 6 = 8$ or $2x - 6 = -8$

$\qquad 2x = 14$ or $\qquad 2x = -2$

$\qquad\quad x = 7$ or $\qquad\quad x = -1$

The solution set is $\{-1, 7\}$.

C

41. $|x| = 2x - 3$

$2x - 3 \geq 0$

$2x \geq 3$

$x \geq \dfrac{3}{2}$

$x = 2x - 3$ or $x = -(2x - 3)$

$-x = -3$ $x = -2x + 3$

$x = 3$ $3x = 3$

                $x = 1$

Restrict the variable so that $|x|$ will not be negative so $2x - 3 \geq 0$.

$x = cx + d$ or $x = -(cx + d)$.

Reject $x = 1$ since it is a condition that $x \geq \dfrac{3}{2}$.

Check:

$|x| = 2x - 3$

$|3| = 2(3) - 3$

$3 = 6 - 3$

$3 = 3$

Substitute $x = 3$.

The statement is true.

The solution set is $\{3\}$.

45. $-3|x - 5| = 9 - 3x$

$-3|x - 5| = -3(x - 3)$

$|x - 5| = x - 3$

$x - 5 = x - 3$ or $x - 5 = -(x - 3)$

$-5 = -3$ $x - 5 = -x + 3$

Contradiction $2x = 8$

                $x = 4$

The solution set is $\{4\}$.

Factor on the right.

Divide both sides by the common factor, -3.

If $|x| = c$, then $x = c$ or $x = -c$.

49. $|2x + 3| = 6 - x$

$6 - x \geq 0$

$-x \geq -6$

$x \leq 6$

$2x + 3 = 6 - x$ or $2x + 3 = -(6 - x)$

$3x = 3$ $2x + 3 = -6 + x$

$x = 1$ $x = -9$

Restrict the variable so that $|2x + 3|$ cannot be negative, so $6 - x \geq 0$.

<u>Check:</u>
$$|2x + 3| = 6 - x \qquad \text{Substitute } x = 1.$$

$$|2(1) + 3| = 6 - 1$$
$$|5| = 5$$
$$5 = 5$$

$$|2(-9) + 3| = 6 - (-9) \qquad \text{Substitute } x = -9.$$

$$|-18 + 3| = 6 + 9$$
$$|-15| = 15$$
$$15 = 15$$

The solution set is $\{-9, 1\}$.

53. $|7x - 2| - 13 = 3x$
   $|7x - 2| = 3x + 13$ Isolate the absolute value on one side of the equation.

$7x - 2 = 3x + 13$ or $7x - 2 = -(3x + 13)$
$\qquad 4x = 15 \qquad\qquad 7x - 2 = -3x - 13$
$\qquad\quad x = \dfrac{15}{4} \qquad\qquad\quad 10x = -11$
$$x = \dfrac{-11}{10}$$

The solution set is $\left\{- \dfrac{11}{10}, \dfrac{15}{4}\right\}$.

57. $|4x + 5| = |3x + 9|$ For this equation to be true, the expressions $4x + 5$ and $3x + 9$ must be either equal or opposites.

$4x + 5 = 3x + 9$ or $4x + 5 = -(3x + 9)$ Solve the
$\qquad x = 4$ or $\qquad 4x + 5 = -3x - 9$ related equations.
$$7x = -14$$
$$x = -2$$

The solution set is $\{-2, 4\}$.

D

61. The weekly cost (C) of producing crowbars at the Old Crow Manufacturing Company is given by $C = 10000 + 2.5x$ where x is the number of crowbars produced. Find the maximum and minimum number of crowbars if the weekly costs are held at $25,000 plus or minus $3000.

Simpler word form:
$25000 minus weekly cost $= \pm 3000$.

In absolute value form:
$|\$25000 - \text{weekly cost}| = 3000$

Select a variable:
Let x represent the number of crobars.

Translate to algebra:
$|25000 - (10000 + 2.5x)| = 3000$ The formula for cost is given in the problem: $10000 + 2.5x$.

Solve:
$|15000 - 2.5x| = 3000$
$15000 - 2.5x = 3000$ or $15000 - 2.5x = -3000$
$\qquad -2.5x = -12000 \qquad\qquad -2.5x = -18000$
$\qquad\quad x = 4800 \qquad\qquad\qquad x = 7200$
The maximum number of crowbars is 7200, and the minimum number is 4800.

STATE YOUR UNDERSTANDING

65. Explain why sometimes $|x| = x$ and other times $|x| = -x$.

$|x|$ is always positive. But x can be either positive or negative. Therefore, for $x \geq 0$, $|x| = x$, and for $x < 0$, $|x| = -x$.

CHALLENGE EXERCISES

69. $|x + 5| = -(x + 5)$ Since $|x + 5|$ is nonnegative by definition, then $-(x + 5)$ is also nonnegative.
                                   Write the related equation and solve.

$\quad -(x + 5) \geq 0$

$\quad\quad -x - 5 \geq 0$

$\quad\quad\quad -x \geq 5$

$\quad\quad\quad\quad x \leq -5$ Divide both sides by -1, reversing the direction of the inequality.

MAINTAIN YOUR SKILLS

Find the LCD of each of the following sets of rational expressions:

73. $\dfrac{7}{x^2 + 6x + 8}$, $\dfrac{7}{x^2 - 2x - 8}$, $\dfrac{7}{x^2 - 8x + 16}$

$\dfrac{7}{(x + 2)(x + 4)}$, $\dfrac{7}{(x - 4)(x + 2)}$, $\dfrac{7}{(x - 4)(x - 4)}$ Factor each denominator.

The LCD is $(x + 2)(x + 4)(x - 4)^2$.

203

77. $\dfrac{4x}{x^2 + 8x + 15}$, $\dfrac{5x}{2x^2 + 7x - 15}$, $\dfrac{x + 2}{2x^2 + 3x - 9}$

$\dfrac{4x}{(x + 5)(x + 3)}$, $\dfrac{5x}{(2x - 3)(x + 5)}$, $\dfrac{x + 2}{(2x - 3)(x + 3)}$

The LCD is $(x + 5)(x + 3)(2x - 3)$.

EXERCISES 5.5   EQUATIONS IN QUADATIC FORM

A

Solve:

1. $\dfrac{x^2 + 1}{x + 3} = \dfrac{11}{x + 3}$

$x \neq -3$ Restrict the variable.
$x^2 + 1 = 11$ Multiply both sides by the
                      LCM, $x + 3$.
$x^2 = 10$ Solve for x.

$x = \pm\sqrt{10}$ Neither of these values is
                      restricted.

The solution set is $\{\pm\sqrt{10}\}$.

5. $\dfrac{2}{x} = x - 1$

$x \neq 0$ Restrict the variable.
$2 = x(x - 1)$ Multiply both sides by the
                               LCM, x.

$2 = x^2 - x$ Write in standard form.
$x^2 - x - 2 = 0$ Factor to solve.
$(x + 1)(x - 2) = 0$ Neither of these values is
$x = -1$ or $x = 2$ restricted.

The solution set is $\{-1, 2\}$.

9. $\dfrac{12}{y + 6} = y - 5$

$y \neq -6$ Restrict the variable.
$12 = (y + 6)(y - 5)$ Multiply both sides by the
                               LCM, $y + 6$.

$12 = y^2 + y - 30$ Write in standard form.
$y^2 + y - 42 = 0$ Factor and solve.
$(y + 7)(y - 6) = 0$ Neither value is
$y = -7$ or $y = 6$ restricted.

The solution set is $\{-7, 6\}$.

13.
$$x^4 - 16 = 0$$
$$(x^2 - 4)(x^2 + 4) = 0 \qquad \text{Factor.}$$
$$(x - 2)(x + 2)(x^2 + 4) = 0$$
$$x - 2 = 0 \text{ or } x + 2 = 0 \text{ or } x^2 + 4 = 0 \qquad \text{Zero-product}$$
$$x = 2 \text{ or } \qquad x = -2 \text{ or } \qquad x^2 = -4 \qquad \text{property.}$$
$$x = \pm 2i$$

The solution set is $\{\pm 2, \pm 2i\}$.

17. $x^4 - 22x^2 = -120$

Let $u = x^2$

$$u^2 - 22u = -120 \qquad \text{Substitute.}$$
$$u^2 - 22u + 120 = 0$$
$$(u - 10)(u - 12) = 0 \qquad \text{Factor}$$
$$u - 10 = 0 \text{ or } u - 12 = 0 \qquad \text{Solve.}$$
$$u = 10 \text{ or } \qquad u = 12$$
$$x^2 = 10 \text{ or } \qquad x^2 = 12 \qquad \text{Replace } u \text{ by } x^2 \text{ and solve.}$$
$$x = \pm\sqrt{10} \text{ or } \qquad x = \pm 2\sqrt{3}$$

The solution set is $\{\pm 2\sqrt{3}, \pm 10\}$.

21. $16x^4 - 8x^2 = -1$

Let $u = x^2$

$$16u^2 - 8u + 1 = 0 \qquad \text{Substitute.}$$
$$(4u - 1)(4u - 1) = 0$$
$$(4u - 1)^2 = 0$$
$$4u - 1 = 0$$
$$4u = 1$$
$$u = \frac{1}{4}$$
$$x^2 = \frac{1}{4} \qquad \text{Replace } y \text{ by } x^2 \text{ and solve.}$$
$$x = \pm\sqrt{\frac{1}{4}}$$
$$x = \pm\frac{1}{2}$$

The solution set is $\left\{\pm\frac{1}{2}\right\}$.

B

25. $\dfrac{3}{x + 2} + \dfrac{9x}{1} = \dfrac{6x}{x + 2}$

                $x \neq -2$ Restrict the variable.

       $3 + 9x(x + 2) = 6x$ Multiply both sides by the
       $3 + 9x^2 + 18x = 6x$ LCD of the fractions.
        $3x^2 + 4x + 1 = 0$ Divide both sides by 3.
     $(3x + 1)(x + 1) = 0$ Factor.

     $x = -\dfrac{1}{3}$ or $x = -1$ Solve.

The solution set is $\left\{-1, -\dfrac{1}{3}\right\}$.

29. $\dfrac{1}{x + 2} + \dfrac{1}{x - 2} = \dfrac{x^2 - 8}{x^2 - 4}$

               $x \neq \pm 2$ Restrict the variable.
    $x - 2 + x + 2 = x^2 - 8$ Multiply by the LCD on
                                both sides. The LCD is
                                $x^2 - 4$ or $(x - 2)(x + 2)$.
       $x^2 - 2x - 8 = 0$ Standard form.
    $(x - 4)(x + 2) = 0$ Factor.
    $x = 4$ or $x = -2$ Solve. $x = -2$ is a
                                restricted value so reject
                                it.

The solution set is $\{4\}$.

33. $x - 3 = \dfrac{1}{x} + \dfrac{x^2 + 2}{x - 2}$

          $x \neq 0, 2$ Restrict the
                                variable.
$(x - 3)x(x - 2) = x - 2 + x^3(x^2 + 2)$ Multiply both sides
   $x^3 - 5x^2 + 6x = x - 2 + x^3 + 2x$ by $x(x - 2)$.
       $-5x^2 + 6x = 3x - 2$
     $5x^2 - 3x - 2 = 0$ Standard form.
   $(5x + 2)(x - 1) = 0$ Factor and solve.

   $x = -\dfrac{2}{5}$ or $x = 1$

The solution set is $\left\{-\dfrac{2}{5}, 1\right\}$.

37. $(x^2 - 1)^2 - 4(x^2 - 1) + 3 = 0$

Let $u - x^2 - 1$
$$u^2 - 4u + 3 = 0$$ 　　　　　　　　Substitute u for $x^2 - 1$.
$$(u - 3)(u - 1) = 0$$ 　　　　　　Factor.
$$u = 3 \text{ or } \quad u = 1$$ 　　　　Solve.
$$x^2 - 1 = 3 \text{ or } x^2 - 1 = 1$$ 　　Substitute $x^2 - 1$ for u,
$$x^2 = 4 \qquad x^2 = 2$$ 　　　and continue solving for
$$x = \pm\sqrt{4} \qquad x = \pm\sqrt{2}$$ 　x.
$$x = \pm 2$$

The solution set is $\{\pm\sqrt{2},\ \pm 2\}$.

41. $8(x^2 - 2)^2 - 10(x^2 - 2) - 63 = 0$
Let $u = x^2 - 2$
$$8u^2 - 10u - 63 = 0$$
$$(4u + 9)(2u - 7) = 0$$
$$4u + 9 = 0 \text{ or } \quad 2u - 7 = 0$$
$$y = -\frac{9}{4} \text{ or } \qquad u = \frac{7}{2}$$
$$x^2 - 2 = -\frac{9}{4} \text{ or } x^2 - 2 = \frac{7}{2}$$
$$4x^2 - 8 = -9 \qquad\qquad x^2 = \frac{7}{2} + 2 = \frac{7}{2} + \frac{4}{2}$$
$$4x^2 = -1$$
$$x^2 = -\frac{1}{4} \qquad\qquad x^2 = \frac{11}{2}$$
$$x = \pm\sqrt{-\frac{1}{4}} \qquad\quad x = \pm\sqrt{\frac{11}{2}} \cdot \frac{\sqrt{2}}{\sqrt{2}}$$
$$x = \pm \frac{1}{2}\sqrt{-1} \qquad\quad x = \pm \frac{\sqrt{22}}{2}$$
$$x = \pm \frac{1}{2}i$$

The solution set is $\left\{\pm \frac{\sqrt{22}}{2},\ \pm \frac{1}{2}i\right\}$.

45. $(x^2 - 10x)^2 = -11(x^2 - 10x) + 210$
Let $u = x^2 - 10x$

$$u^2 = -11u + 210 \qquad \text{Substitute } u \text{ for } x^2 - 10x.$$

$$u^2 + 11u - 210 = 0$$
$$(u + 21)(u - 10) = 0$$
$$u = -21 \text{ or } u = 10 \qquad \text{Solve for } u.$$
$$x^2 - 10x = -21 \text{ or} \qquad x^2 - 10x = 10 \qquad \text{Substitute and solve for } x.$$

$$x^2 - 10x + 21 = 0 \qquad x^2 - 10x - 10 = 0 \qquad \text{Since } x^2 - 10x - 10$$
$$(x - 3)(x - 7) = 0 \qquad\qquad\qquad\qquad \text{cannot be fac-}$$
$$x = 3 \text{ or } x = 7 \qquad\qquad\qquad\qquad \text{tored, solve}$$

for x by using the quadratic formula, where $a = 1$, $b = -10$, and $c = -10$.

$$x = \frac{-b \pm \sqrt{b^2 - 4ac}}{2a}$$

$$= \frac{10 \pm \sqrt{100 - 4(1)(-10)}}{2}$$

$$= \frac{10 + \sqrt{140}}{2}$$

$$= \frac{10 \pm \sqrt{4 \cdot 35}}{2}$$

$$= \frac{10 \pm 2\sqrt{35}}{2} = \frac{2(5 \pm \sqrt{35})}{2}$$

$$= 5 \pm \sqrt{35}$$

The solution set is $\{3, 7, 5 \pm\sqrt{35}\}$.

c

49. $$\frac{1}{x^2 - 7x + 12} - \frac{x}{x^2 - 2x - 8} = \frac{2}{x^2 - x - 6}$$

$$\frac{1}{(x - 3)(x - 4)} - \frac{x}{(x - 4)(x + 2)} = \frac{2}{(x - 3)(x + 2)}$$

Factor each denominator.

$$x + 2 - x(x - 3) = 2(x - 4)$$

Multiply both sides by the LCD: $(x-3)(x-4)(x+2)$.

$$x + 2 - x^2 + 3x = 2x - 8$$
$$x^2 - 2x - 10 = 0$$

208

$$x = \frac{2 \pm \sqrt{(-2)^2 - 4(-10)}}{2}$$

Solve using the quadratic formula.

$$x = \frac{2 \pm 2\sqrt{11}}{2}$$

$$x = 1 \pm \sqrt{11}$$

The solution set is $\{1 \pm \sqrt{11}\}$.

53. $$\frac{1}{2x^2 + 7x - 4} - \frac{x - 1}{2x^2 + 5x - 3} = \frac{5}{x^2 + 7x + 12}$$

$$\frac{1}{(2x - 1)(x + 4)} - \frac{x - 1}{(2x - 1)(x + 3)} = \frac{5}{(x + 4)(x + 3)}$$

The LCD is $(2x-1)(x+3)(x+4)$.

$$x + 3 - (x - 1)(x + 4) = 5(2x - 1)$$
$$x + 3 - (x^2 + 3x - 4) = 10-x - 5$$
$$x + 3 - x^2 - 2x + 7 = 10x - 5$$
$$x^2 + 12x - 12 = 0$$

$$x = \frac{-12 \pm \sqrt{12^2 - 4(-12)}}{2}$$

$$x = \frac{-12 \pm \sqrt{192}}{2}$$

$$x = \frac{-12 \pm 8\sqrt{3}}{2}$$

$$x = -6 \pm 4\sqrt{3}$$

The solution set is $\{-6 \pm 4\sqrt{3}\}$.

57. $x^{-2} + 11x^{-1} + 30 = 0$
Let $u = x^{-1}$.

$u^2 + 11u + 30 = 0$ Substitute.
$(u + 5)(u + 6) = 0$ Factor and solve.
$u = -5$ or $u = -6$
$x^{-1} = 5$ or $x^{-1} = -6$ Replace u with x^{-1}.
$\frac{1}{x} = -5$ or $\frac{1}{x} = -6$
$x = -\frac{1}{5}$ or $x = -\frac{1}{6}$

The solution set is $\left\{-\frac{1}{6}, -\frac{1}{5}\right\}$.

61. $x - 4\sqrt{x} - 45 = 0$

Let $u = \sqrt{x}$.

$u^2 - 4u - 45 = 0$	Substitute.
$(u - 9)(u + 5) = 0$	Factor and solve.
$u = 9$ or $u = -5$	
$\sqrt{x} = 9$ or $\sqrt{x} = -5$	Replace u with $\sqrt{x}$ and
$x = 81$	solve. Reject $\sqrt{x} = -5$, since the left side is defined as positive and cannot equal a negative quantity.

The solution set is {81}.

65. $x^{-4} + 2x^{-2} - 24 = 0$
Let $u = x^{-2}$.

$u^2 + 2u - 24 = 0$	Substitute.
$(u + 6)(u - 4) = 0$	Factor and solve.
$u = -6$ or $u = 4$	
$x^{-2} = -6$ or $x^{-2} = 4$	Replace u with x^{-2}.

$$\frac{1}{x^2} = -6 \text{ or } \frac{1}{x^2} = 4$$

$$x^2 = -\frac{1}{6} \text{ or } x^2 = \frac{1}{4}$$

$$x = \pm\sqrt{-\frac{1}{6}} \text{ or } x = \pm\frac{1}{2}$$

$x = \pm\dfrac{1}{\sqrt{6}}i$	Rationalize the denominator.

$$x = \pm\frac{\sqrt{6}}{6}i$$

The solution set is $\left\{\pm\dfrac{1}{2}, \pm\dfrac{\sqrt{6}}{6}i\right\}$.

69. $x^{-4} + 5x^{-2} + 4 = 0$
Let $u = x^{-2}$.

$$u^2 + 5u + 4 = 0 \qquad \text{Substitute.}$$
$$(u + 1)(u + 4) = 0 \qquad \text{Factor and solve.}$$
$$u = -1 \quad \text{or} \quad u = -4$$
$$x^{-2} = -1 \quad \text{or} \quad x^{-2} = -4 \qquad \text{Replace } u \text{ with } x^{-2} \text{ and solve.}$$

$$\frac{1}{x^2} = -1 \quad \text{or} \quad \frac{1}{x^2} = -4$$

$$x^2 = -1 \quad \text{or} \quad x^2 = -\frac{1}{4}$$

$$x = \pm\sqrt{-1} \quad \text{or} \quad x = \pm\sqrt{-\frac{1}{4}}$$

$$x = \pm i \quad \text{or} \quad x = \pm\frac{1}{2}i$$

The solution set is $\left\{\pm\frac{1}{2}i, \pm i\right\}$.

D

45. Jane and Sally each earned $600 last month. Sally is paid one dollar an hour more than Jane. If together they worked a total of 220 hours, how much is each paid per hour?

Select a variable:
Let x represent the pay per hour that Jane receives. So x + 1 represents Sally's pay per hour.

Make a chart to organize the information.

	Pay per hour	Earnings	Hours worked (Earnings/Pay per hour)
Jane	x	600	$\dfrac{600}{x}$
Sally	x + 1	600	$\dfrac{600}{x + 1}$

Simpler word form:
$$\begin{pmatrix}\text{Hours that}\\ \text{Jane works}\end{pmatrix} + \begin{pmatrix}\text{Hours that}\\ \text{Sally works}\end{pmatrix} = 220$$

211

Translate to algebra:

$$\frac{600}{x} + \frac{600}{x + 1} = 220$$

$600(x + 1) + 600(x) = 220(x)(x + 1)$ Multiply by the LCM.

$600x + 600 + 600x = 220x^2 + 220x$

$220x^2 - 980x - 600 = 0$ Simplify.

$11x^2 - 49x - 30 = 0$ Divide by 20.

$(11x + 6)(x - 5) = 0$ Factor.

$x = -\frac{6}{11}$ or $x = 5$ Reject $x = -\frac{6}{11}$ since x represents Jane's pay.

Check:

	Pay per hour	Earnings	Hours worked (Earnings/Pay per hour
Jane	5	600	$\frac{600}{5} = 120$
Sally	5+1=6	600	$\frac{600}{6} = 100$

Check in the original information chart. The total hours worked is 220.

Answer:

Jane earns $5 per hour and Sally earns $6 per hour.

STATE YOUR UNDERSTANDING

77. What does it mean for an equation to be in quadratic form?

When the equation can be transformed to one in which the variable is squared, then it is said to be quadratic in form. Usually, a substitution is made using the letter u so that the resulting equation looks like: $au^2 + bu + c = 0$, $a \neq 0$.

MAINTAIN YOUR SKILLS

Multiply and simplify:

81. $(6\sqrt{14})(3\sqrt{21}) = (6 \cdot 3)(\sqrt{14} \cdot \sqrt{21})$

$= 18(\sqrt{2 \cdot 7 \cdot 3 \cdot 7})$

$= 18(\sqrt{49} \cdot \sqrt{6})$

$= 126\sqrt{6}$

85. $\sqrt{6}(\sqrt{30} - \sqrt{54}) = \sqrt{6 \cdot 6 \cdot 5} - \sqrt{6 \cdot 6 \cdot 9}$
$= \sqrt{36} \cdot \sqrt{5} - \sqrt{36} \cdot \sqrt{9}$
$= 6\sqrt{5} - 18$

EXERCISES 5.10   ABSOLUTE VALUE INEQUALITIES

A

Solve and graph the solution:

1. $|x| < 4$
   $x < 4$ and $x > -4$ Write the absolute value
   or   inequality in its
   $-4 < x < 4$ equivalent form.

   The graph of the solution is:

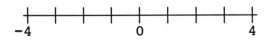

   -4   0   4

Solve and write the solution in set builder notation:

5. $|2x| \leq 10$
    $-10 \leq 2x \leq 10$ Rewrite an equivalent
                                           compound inequality.
     $-5 \leq x \leq 5$

   The solution set is $\{x | -5 \leq x \leq 5\}$.

9. $|2x| - 5 < 3$
      $|2x| < 8$ Add 5 to both sides.
    $-8 < 2x < 8$ Rewrite an equivalent
    $-4 < x < 4$ compound inequality.

   The solution set is $\{x | -4 < x < 4\}$.

Solve and write the solution in interval notation:

13. $|x + 2| - 3 < 4$
       $|x + 2| < 7$ Simplify by adding 3 to
                                           both sides.
    $x + 2 < 7$ and $x + 2 > -7$ Write an equivalent form
                                           for the inequality.
      $x < 5$ and $x > -9$ Solve each inequality.

   The solution is $(-9, 5)$.

17. $|x + 3| > -3$
x can be any real number.

Since absolute value is always non-negative, the left side, being non-negative, is always greater than any negative quantity.

The solution is $(-\infty, \infty)$.

B

21. $|2x + 3| \leq 5$

$2x + 3 \leq 5$ and $2x + 3 \geq -5$ Write an equivalent form.

$2x \leq 2$ $2x \geq -8$ Solve each inequality.

$x \leq 1$ and $x \geq -4$

The solution is $[-4, 1]$.

25. $|3x + 5| < 2$
$3x + 5 < 2$ and $3x + 5 > -2$ Write an equivalent form.
$3x < -3$ $3x > -7$ Solve.
$x < -1$ $x > -\dfrac{7}{3}$

or

$-\dfrac{7}{3} < x < -1$

The solution is $\left(-\dfrac{7}{3}, -1\right)$.

29. $|2x + 4| > x - 5$
$2x + 4 > x - 5$ or $2x + 4 < -(x - 5)$ Write an equivalent form for the inequality.

$x > -9$ or $2x + 4 < -x + 5$ Solve.
$3x < 1$
$x < \dfrac{1}{3}$ Since either one or the other condition is always satisfied by any number, the solution set includes all real numbers.

The solution set is $(-\infty, \infty)$.

Solve and write the solution in set builder notation:

33. $|4x - 7| \geq 3x - 12$

   $4x - 7 \geq 3x - 12$ or $4x - 7 \leq - (3x - 12)$ Write the inequality in its equivalent form.

   $x \geq -5$ $4x - 7 \leq -3x + 12$

                                  $7x \leq 19$

   $x \geq -5$ or $x \leq \frac{19}{7}$ All real numbers make either the one or the other condition true, and so the solution set includes all real numbers.

   The solution is $\{x \mid x \in \mathbb{R}\}$.

37. $|4 - x| \geq 4$

   $4 - x \geq 4$ or $4 - x \leq -4$ Write an equivalent form.

   $-x \geq 0$ $-x \leq -8$

   $x \leq 0$ or $x \geq 8$

   The solution is $\{x \mid x \leq 0\} \cup \{x \mid x \geq 8\}$.

C

41. $|12 - 4x| - 8 \leq 12$

   $|12 - 4x| \leq 20$ Simplify first by adding 8 to both sides.

   $12 - 4x \leq 20$ and $12 - 4x \geq -20$ Write an equivalent form, and solve.

   $-4x \leq 8$ $-4x \geq -32$

   $x \geq -2$ and $x \leq 8$
   or $-2 \leq x \leq 8$

   The solution is $\{x \mid -2 \leq x \leq 8\}$.

45. $|2x + 1| > -x$
   $2x + 1 > -x$ or $2x + 1 < -(-x)$ Write an equivalent form.
   $3x > -1$ $2x + 1 < x$
   $x > -\frac{1}{3}$ or $x < -1$
   The solution is $\{x \mid x < -1\} \cup \left\{x \mid x > -\frac{1}{3}\right\}$.

49. $\left|\dfrac{2x - 3}{3}\right| < \dfrac{1}{3}$

   $-\dfrac{1}{3} < \dfrac{2x - 3}{3} < \dfrac{1}{3}$ Write an equivalent form.

   $-1 < 2x - 3 < 1$

   $2 < 2x < 4$

   $1 < x < 2$

   The solution set is $\{x \mid 1 < x < 2\}$.

53. $\left|\dfrac{1}{2}x - 1\right| > 2$

   $\dfrac{1}{2}x - 1 > 2$ or $\dfrac{1}{2}x - 1 < -2$ Write the inequality in its equivalent form.

   $\dfrac{1}{2}x > 3$ or $\dfrac{1}{2}x < -1$

   $x > 6$ or $x < -2$

   The solution set is $\{x \mid x > 6\} \cup \{x \mid x < -2\}$.

57. $|4x + 5| > x$

   $4x + 5 > x$ or $4x + 5 < -x$ Write an equivalent form.

   $3x > -5$ or $5x < -5$

   $x > -\dfrac{5}{3}$ or $x < -1$

   The solution set is $\{x \mid x < -1\} \cup \left\{x \mid x > -\dfrac{5}{3}\right\}$.

D

In 61-72 write each as an absolute value inequality and solve:

61. The distance from a number and 4 is less than 12.

       Use x to represent the number.

   $|x - 4| < 12$ The geometric meaning of absolute value is the distance between the number, x, and 4. Equivalent form.

   $x - 4 < 12$ and $x - 4 > -12$

      $x < 16$ and $x > -8$

   or

   $-8 < x < 16$

65. The distance from a number and 18 is at least 8.

$|x - 18|$

        The geometric meaning of absolute value is the distance from x to 18.

$|x - 18| \geq 8$

        "At least 8" is equivalent to "greater than or equal to 8."

$x - 18 \geq 8$ or $x - 18 \leq -8$

    $x \geq 26$ or $\qquad x \leq 10$

69. The storage of food in a home freezer is best accomplished if the temperature in the freezer is maintained at 6°F plus or minus 8°. Find the range of permissible temperatures using absolute value inequalities.

$|T - 6| \leq 8$

        The difference in the absolute value of the temperature in the freezer and 6°F must be less than or equal to 8°. Solve.

$T - 6 \leq 8$ and $T - 6 \geq -8$

    $T \leq 14$ and $\qquad T \geq -2$
or
$-2 \leq t \leq 14$

Thus, the inside temperatures should be between -2°F and 14°F.

73. Write $-5 \leq x + 7 \leq 5$ as an absolute value inequality.

$|x + 7| \leq 5$

        Rewrite the compound inequality as an equivalent absolute value.

77. Write $-8 \leq 2x \leq 2$ as an absolute value inequality.

$-5 \leq 2x + 3 \leq 5$

    $|2x + 3| \leq 5$

        Add 3 to all parts of the compound inequality so that the left side and right side will be opposites. Then rewrite as an absolute value.

CHALLENGE EXERCISES

Solve:

81. $|x + a| < b,\ b > 0$

    $-b < x + a < b$ Write an equivalent form.
    $-a - b < x < b - a$

    The solution is $\{x | -a - b < x < b - a\}$.

85. $|x - a| \geq 0$

    $x - a \geq 0$ or $x - a \leq 0$
         $x \geq a$ or $x \leq a$

    The solution is $\{x | x \epsilon \mathbb{R}\}$.

MAINTAIN YOUR SKILLS

Factor completely:

89. $56a - 32b + 21ab - 12b^2$ No common monomial
                                        factors.

    $(56a - 32b) + (21ab - 12b^2)$ Group in pairs.
    $8(7a - 4b) + 3b(7a - 4b)$ Factor each pair.
    $(8 + 3b)(7a - 4b)$
    or
    $(3b + 8)(7a - 4b)$

93. $y^8 - 81 = (y^4)^2 - (9)^2$ Difference of squares.
           $= (y^4 - 9)(y^4 + 9)$
           $= (y^2 - 3)(y^2 + 3)(y^4 + 9)$ Again, difference of
                                            squares.

A

Solve and graph. Write the solution in interval notation:

1. $(x + 2)(x - 5) > 0$

   The critical numbers are -2 and 5, shown below with vertical
   lines.

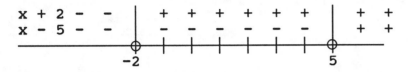

   Since $(x + 2)(x - 5) > 0$, the solution contains those
   numbers for which the factors are both negative and both
   positive.

   The solution is $(-\infty, -2) \cup (5, \infty)$.

5. $\dfrac{(5x + 3)}{(2x - 1)} \geq 0$ Standard form.

   $5x + 3 = 0 \qquad 2x - 1 = 0$

   $\qquad x = -\dfrac{3}{5} \qquad\quad x = \dfrac{1}{2}$

   The critical values are $-\dfrac{3}{5}$ and $\dfrac{1}{2}$.

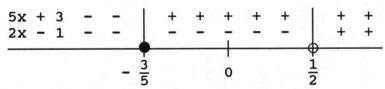

   Exclude the value $\dfrac{1}{2}$, since division by zero is
   undefined.

   Since the quotient is greater than or equal to 0, the
   solutions will be those numbers for which the factors are
   both negative, and those numbers for which the factors are
   both positive, including the critical value $-\dfrac{3}{5}$.

   The solution is $\left(-\infty, -\dfrac{3}{5}\right] \cup \left(\dfrac{1}{2}, \infty\right)$.

9. $-5x(x - 3)(x + 2) \geq 0$

The critical numbers are 0, 3, and -2.

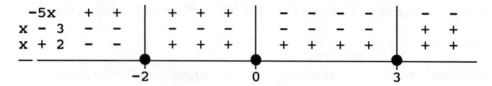

The solution is $(-\infty, -2] \cup [0, 3]$.

13. $\dfrac{2x(2x + 3)}{(x - 5)} \geq 0$

The critical numbers are   Set each factor equal to
$0, -\dfrac{3}{2},$ and 5   zero to determine
                                 critical numbers.

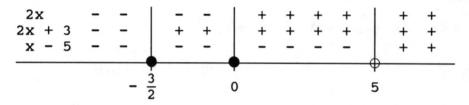

The solution is $\left[-\dfrac{3}{2}, 0\right] \cup (5, \infty)$.

17. $x(x + 2)(x - 3) < 0$

The critical numbers are 0, -2, and 3.

x	$-$ $-$		$-$ $-$		$+$ $+$		$+$ $+$
x + 2	$-$ $-$		$+$ $+$		$+$ $+$		$+$ $+$
x - 3	$-$ $-$		$-$ $-$		$-$ $-$		$+$ $+$

```
               -2        0        3
```

The solution is $(-\infty, -2) \cup (0, 3)$.

B

21. $3x^2 + 5x + 4 > x^2 - 2x - 2$

        $2x^2 + 7x + 6 > 0$ Standard form.

     $(2x + 3)(x + 2) > 0$ Factor, and identify the
                                          critical numbers.

The critical numbers are $-\dfrac{3}{2}$ and -2.

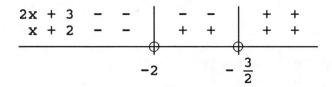

```
2x + 3   -   -  |  -   -  |  +   +
 x + 2   -   -  |  +   +  |  +   +
_____⊕_____⊕_____
                -2        -3/2
```

The solution is $(-\infty, -2) \cup (-\dfrac{3}{2}, \infty)$.

25. $\dfrac{2}{x} + 5 \geq 3$

       $\dfrac{2}{x} + 2 \geq 0$ Simplify. Subtract 3 from
                                       both sides.

     $\dfrac{2 + 2x}{x} \geq 0$ Write the left side with a
                                       common denominator, x.

     $\dfrac{2(1 + x)}{x} \geq 0$

      $\dfrac{1 + x}{x} \geq 0$ Divide both sides by 2.

The critical numbers are   Set denominator and
0 and -1.   numerator equal to zero to
                                       find the critical numbers.

```
1 + x   -   -  |  +   +  |  +   +
  x     -   -  |  -   -  |  +   +
_____●_____⊕_____
              -1         0
```

The solution is $(-\infty, -1] \cup (0, \infty)$.

221

29. $x^2 + 5x + 10 < -5x - 15$

$x^2 + 10x + 25 < 0$ Standard form.

$(x + 5)^2 < 0$ Factor, to identtify critical numbers.

The critical number is -5.

```
x + 5   -   -  |  +   +
x + 5   -   -  |  +   +
_____⊕_____
              -5
```

The solution contains numbers for which the factors have opposite signs (to produce a negative product).
So the solution is the empty set, $\emptyset$.

33. $8x^2 + 20x + 15 \lessgtr -8x^2 \quad 20x - 10$

$16x^2 + 40x + 25 \lessgtr 0$ Standard form.

$(4x + 5)^2 \lessgtr 0$

The critical number is $-\dfrac{5}{4}$.

```
4x + 25   -   -  |  +   +
4x + 25   -   -  |  +   +
_____●_____
              -5
              --
               4
```

No numbers in either of the intervals have factors that are opposite in sign, so no negative products are produced.
However, the critical number satisfies the equality.
The solution is $\left\{-\dfrac{5}{4}\right\}$.

37. $$\dfrac{-4}{x + 5} > 4$$

$$\dfrac{-4}{x + 5} - 4 > 0$$ Write the inequality in standard form.

$$\dfrac{-4}{x + 5} - \dfrac{4(x + 5)}{x + 5} > 0$$

$$\dfrac{-4 - 4x - 20}{x + 5} > 0$$

$$\dfrac{-4x - 24}{x + 5} > 0$$ Standard form.

The critical numbers are -5 and -6.

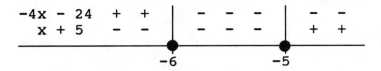

The solution set is (-6, -5).

C

41. $\dfrac{1}{x^2 + 1} < -1$

$x^2 + 1 = 0$ Set $x^2 + 1$ equal to zero
to find critical numbers.

$x^2 = -1$ Contradiction. x^2 is
nonnegative. No critical
values.
Both numerator and
denominator are positive,
so quotient is
nonnegative.

The solution set is $\emptyset$.

45. $\dfrac{1}{x - 1} > \dfrac{3}{x - 2}$

$\dfrac{1}{x - 1} - \dfrac{3}{x - 2} > 0$

$\dfrac{1(x - 2) - 3(x - 1)}{(x - 1)(x - 2)} > 0$

$\dfrac{x - 2 - 3x + 3}{(x - 1)(x - 2)} > 0$

$\dfrac{-2x + 1}{(x - 1)(x - 2)} > 0$ Standard form.

The critical numbers are $\frac{1}{2}$, 1, and 2.

```
-2x + 1   + +  |   -  |   - -  |   - -
  x - 1   - -  |   -  |   + +  |   + +
  x - 2   - -  |   -  |   - -  |   + +
        ⊕     ⊕      ⊕
        1/2    1      2
```

The solution set is $\left(-\infty, \frac{1}{2}\right) \cup (1, 2)$.

49.

$$\frac{1}{x - 5} \geq \frac{1}{x + 6}$$

$$\frac{1}{x - 5} - \frac{1}{x + 6} \geq 0$$

$$\frac{1(x + 6) - 1(x - 5)}{(x - 5)(x + 6)} \geq 0$$

$$\frac{x + 6 - x + 5}{(x - 5)(x + 6)} \geq 0$$

$$\frac{11}{(x - 5)(x + 6)} \geq 0 \qquad \text{Standard form.}$$

The critical numbers are -6 and 5.

```
  11      +  +  |  +  +  +  +  |  +  +
x - 5     -  -  |  -  -  -  -  |  +  +
x + 6     -  -  |  +  +  +  +  |  +  +
       ─────────⊕──────┼──────⊕─────────
              -6       0       5
```

The solution set is $(-\infty, -6) \cup (5, \infty)$.

53.

$$\frac{x - 5}{(x - 1)(x + 4)} \leq 2$$

$$\frac{x - 5}{(x - 1)(x + 4)} - 2 \leq 0$$

$$\frac{x - 5 - 2(x - 1)(x + 4)}{(x - 1)(x + 4)} \leq 0$$

$$\frac{x - 5 - 2(x^2 + 3x - 4)}{(x - 1)(x + 4)} \leq 0$$

$$\frac{x - 5 - 2x^2 - 6x + 8}{(x - 1)(x + 4)} \leq 0$$

$$\frac{-2x^2 - 5x + 3}{(x - 1)(x + 4)} \leq 0 \qquad \text{Standard form.}$$

$$\frac{(-2x + 1)(x + 3)}{(x - 1)(x + 4)} \leq 0 \qquad \begin{array}{l}\text{Factor to identify} \\ \text{critical numbers.}\end{array}$$

The critical numbers are -4, -3, $\frac{1}{2}$, and 1.

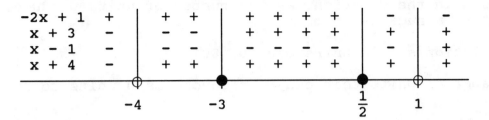

The solution set is $(-\infty, -4) \cup \left[-3, \frac{1}{2}\right] \cup (1, \infty)$.

57. $\dfrac{5}{x^2} + 4 \geq 0$

$\dfrac{5 + 4x^2}{x^2} \geq 0$ Numerator is nonnegative; denominator is non-negative. Therefore the quotient will be either zero (if numerator is zero) or positive).

The solution set is $(-\infty, \infty)$.   All values of x satisfy the conditions.

D

61. In an electronics factory, the cost (C) of producing x units of an item is given by $C = 8000 + 5x^2 - 350x$. If the cost is to be held at \$12,000 or below, what is the range of units that can be produced?

$C = 8000 + 5x^2 - 350x$
or
$8000 + 5x^2 - 350x = C$ Symmetric property.
$8000 + 5x^2 - 350x \leq 12000$ The cost, C, is less than or equal to 12,000.

$5x^2 - 350x - 4000 \leq 0$ Standard form.
$5(x^2 - 70x - 800) \leq 0$ Factor.
$5(x - 80)(x + 10) \leq 0$

The critical numbers are -10 and 80.

```
x - 80   -  |  -  |  -  -  -  |  +  +
x + 10   -  |  +  |  +  +  +  |  +  +
         ───●─────●───────────●──────
           -10    0          80
```

Although numbers between -10 and 0 result in factors whose products are below 12,000, they are nevertheless not included in the solution, as the number of units produced cannot be a negative number.

The solution set is therefore [0, 80].

The range of units that can be produced is 0 units to 80 units.

STATE YOUR UNDERSTANDING

65. How are critical numbers determined?

Critical numbers are values of the variable for which the left side is equal to zero (when the inequality is written in standard form). Determine critical values by factoring the left side.   Then set each factor equal to zero and solve.

CHALLENGE EXERCISES

Solve:

69. $(x - 1)(x + 1)^2 < 0$

The critical numbers are 1 and -1.

```
x - 3   -   -   |   -   -   |   +   +
x + 1   -   -   |   +   +   |   +   +
x + 1   -   -   |   +   +   |   +   +
_____⊕_____⊕_____
               -1          1
```

The solution set is $(-\infty, -1) \cup (-1, 1)$.
Or in set builder notation: $\{x \mid x < -1\} \cup \{x \mid -1 < x < 1\}$.

MAINTAIN YOUR SKILLS

Simplify:

73. $(-27)^{2/3} = \left(\sqrt[3]{-27}\right)^2$
$$= (-3)^2$$
$$= 9$$

226

77. $$\frac{5}{1 + \sqrt{2}} = \frac{5}{1 + \sqrt{2}} \cdot \frac{1 - \sqrt{2}}{1 - \sqrt{2}}$$

$$= \frac{5 - 5\sqrt{2}}{1 - 2}$$

$$= \frac{5 - 5\sqrt{2}}{-1}$$

or $-5 + 5\sqrt{2}$

CHAPTER 6

LINEAR EQUATIONS AND GRAPHING

EXERCISES 6.1   LINEAR EQUATIONS IN TWO VARIABLES

A

Find the ordered pairs that are solutions of the following equations that have the given values of x or y.

1. $2x + y = 5$

x	y
0	5

$$2(0) + y = 5$$ Substitute 0 for x in the
$$0 + y = 5$$ given equation. Solve for
$$y = 5$$ y.

5. $-2x + y = 4$

x	y
0	4

$$-2(0) + y = 4$$ Substitute 0 for x in the
$$0 + y = 4$$ given equation and solve
$$y = 4$$ for y.

9. $-3x + 2y = 6$

x	y
-4	-3

$$-3x + 2(-3) = 6$$ Substitute -3 for y in
$$-3x - 6 = 6$$ the given equation.
$$x = -4$$ Solve for x.

13. $5x - 2y = 10$

x	y
0	-5

$$5x - 2(-5) = 10$$ Substitute -5 for y in
$$5x + 10 = 10$$ the given equation, and
$$x = 0$$ solve for x.

Draw the graph of each equation:

17. y = x - 2

x	y
-2	-4
0	-2
2	0

Find a minimum of three points by assigning values to x and solving for y.

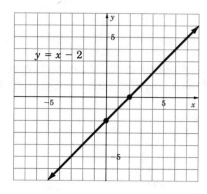

Plot each of the points on the coordinate system and draw the straight line that passes through the points.

B

Find the ordered pairs that are solutions of the following equations that have the given values of x and y:

0.5x - 0.2y = 0.75.

21.

x	y
0.1	

        0.5x - 0.2y = 0.75
    0.5(0.1) - 0.2y = 0.75   Substitute 0.1 for x.
        0.05 - 0.2y = 0.75
              -0.2y = 0.7
                  y = -3.5

The ordered pair with x = 0.1 is (0.1, -3.5).

$\frac{1}{3}x + \frac{3}{5}y = 1$

25.

x	y
$\frac{1}{2}$	

$\frac{1}{3}x + \frac{3}{5}y = 1$

$\frac{1}{3}\left(\frac{1}{2}\right) + \frac{3}{5}y = 1$ Substitute $\frac{1}{2}$ for x.

$\frac{1}{6} + \frac{3}{5}y = 1$

$\frac{3}{5}y = \frac{5}{6}$ $1 - \frac{1}{6} = \frac{6}{6} - \frac{1}{6} = \frac{5}{6}$

$y = \frac{25}{18}$ $\frac{5}{6} \div \frac{3}{5} = \frac{5}{6} \cdot \frac{5}{3} = \frac{25}{18}$

The ordered pair with $x = \frac{1}{2}$ is $\left(\frac{1}{2}, \frac{25}{18}\right)$.

29.   y = -2

x	y
-2	-2
0	-2
1	-2

Find three points.
Regardless of the value of
x, the value of y is -2.

Plot the points and draw
the line.

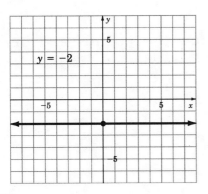

33. $2x + y = 3$

x	y
-2	7
0	3
3	-3

Assign values to x, and solve for y.

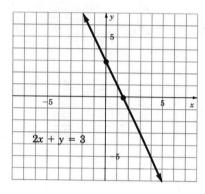

$2x + y = 3$

Plot the points and draw the line.

37. $7y - 6x = 42$

x	y
0	6
-7	0
-3	$\dfrac{24}{7}$

Substitute assigned values for x, and solve for y.

$7y - 6x = 42$

Plot the points and draw the line.

41. x - 5y = -6

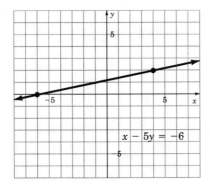

x	y
-6	0
0	$\frac{6}{5}$
4	2

Find three points.

Plot the points and draw
the line.

45. 2x - 9y = 18

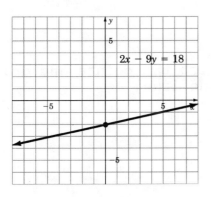

x	y
0	$- 2$
3	$- \frac{4}{3}$
-3	$- \frac{24}{9}$

Find a minimum of three
points.

Plot the points and draw
the line.

49. 8x - y = 4

x	y
0	-4
$\frac{1}{2}$	0
1	4

Find three points.

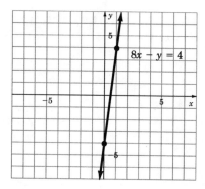

Plot each point and draw
the line.

C

Draw the graph of each equation:

53. $y = \frac{3}{8}x + 6$

x	y
0	6
8	9
-8	3

Find three points.

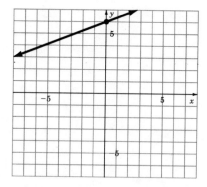

Plot the points and draw
the line.

57. $\frac{1}{7}x + \frac{1}{4}y + 1 = 0$

x	y
-7	0
0	- 4
3	- $\frac{40}{7}$

Find a minimum of three points.

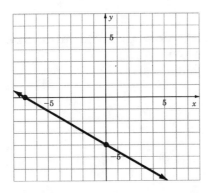

Draw the line connecting the points.

In Exercise 61, one unit on the x-axis equals 5, and one unit on the y-axis equals 10.

61. 25x - y = 125

x	y
5	0
7	50
3	-50

Find a minimum of three points.

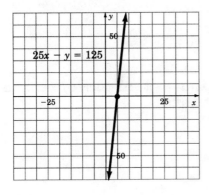

Plot each point and draw the line.

235

In exercise 65, let one unit on the x-axis equal 50 and one unit on the y-axis equal 25.

65. $x - y = 200$

x	y
75	-125
200	0
275	75

Find three points.

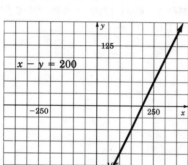

Plot the points and draw the graph.

D

69. A furniture refinishing outlet pays its employees $40 per day plus $10 per unit refinished. Write an equation to express an employee's daily wage (w) in terms of the units finished (x). Draw a graph for $x \geq 0$.

$w = 10x + 40$ Let x represent the number of units finished. Then 10x represents the amount paid for x units. The base pay is 40, so add 40.

x	y
0	40
1	50
2	60

Find a minimum of three points.

236

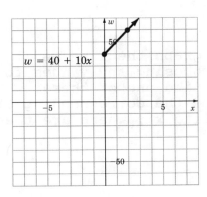

$w = 40 + 10x$

Since $x \geq 0$, do not extend
the line past the y-axis
where $x = 0$.

73. The height, H in feet, of an object dropped from a 100 ft
    high building after a time (t, in seconds) can be
    approximated by H = -40t + 100. Using H as the vertical
    axis and t as the horizontal axis, graph this relationship.

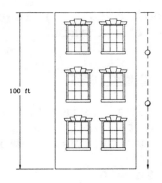

100 ft

H = -40t + 100

t	H
0	100
1	60
2	20
2.5	0

Make a table of values.

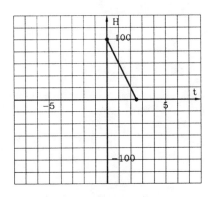

Plot the points and draw
the graph. Each unit on
the y-axis is 20.

Since the height, H, is
greater than or equal to
zero, do not graph points
beyond x = 2.5.

77. Given the equation 2x - 3y = 6, to graph the line we
     generally find three solutions by choosing any values for
     either variable. If, however, you find one of these
     solutions by letting x = 0 (and find the corresponding y)
     and another by letting y = 0 (and find x), where are these
     points on your graph?

     2x - 3y = 6

     2(0) - 3y = 6   x = 0
          -3y = 6
            y = -2

     2x - 3(0) = 6   y = 0
           2x = 6
            x = 3

     The point (0, -2) is on the y-axis, while the point (3, 0)
     is on the x-axis.

CHALLENGE EXERCISES

81. This time, using "m" as the vertical axis and "n" as the
     horizontal axis, graph 3m + 2n = 6 and compare it to the two
     previous graphs. Is it the same as either?

     3m + 2n = 6

n	m
0	2
3	0

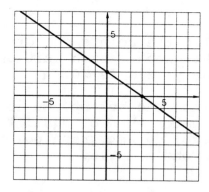

     Plot the points.

     This graph 3m + 2n = 6 is the same as 2m + 3n = 6 (when m
     was the horizontal axis and n was the vertical axis).

Solve by factoring:

85. $$4x^2 - 225 = 0$$
$$(2x - 15)(2x + 15) = 0 \qquad \text{Difference of squares}$$
$$2x - 15 = 0 \text{ or } 2x + 15 = 0$$
$$x = \frac{15}{2} \text{ or } \qquad x = -\frac{15}{2}$$

The solution set is $\left\{\pm \frac{15}{2}\right\}$.

89. $$4(3 + 4x) = 3x^2$$
$$3x^2 - 4(3 + 4x) = 0$$
$$3x^2 - 12 - 16x = 0$$
$$3x^2 - 16x - 12 = 0 \qquad \text{Standard form.}$$
$$(3x + 2)(x - 6) = 0$$
$$3x + 2 = 0 \text{ or } x - 6 = 0$$
$$x = -\frac{2}{3} \text{ or } \quad x = 6$$

The solution set is $\left\{-\frac{2}{3}, 6\right\}$.

EXERCISES 6.2 SLOPE, INTERCEPTS, AND THE DISTANCE FORMULA

A

Write the coordinates of the x- and y-intercepts:

1. $3x - 7y = 21$

y-intercept	x-intercept	
$x = 0$	$y = 0$	
$3(0) - 7y = 21$	$3x - 7(0) = 21$	The intercepts occur at
$y = -3$	$x = 7$	$x = 0$ and $y = 0$.
$(0, -3)$	$(7, 0)$	Write the intercepts as ordered pairs.

The x-intercept is $(7, 0)$, and the y-intercept is $(0, -3)$.

5. Is the graph of the line that passes through $(5, 8)$ and $(3, 10)$ parallel to the line that passes through $(9, 6)$ and $(4, 12)$?

$$m_1 = \frac{8 - 10}{5 - 3} = \frac{-2}{2} = -1$$

$$m_2 = \frac{6 - 12}{9 - 4} = \frac{-6}{5} = -\frac{6}{5}$$

The slopes are not equal, so the lines are not parallel.

Find the slope of the line joining the following points:

9. (4, 0), (0, -2)

$$m = \frac{y_2 - y_1}{x_2 - x_1} = \frac{-2 - 0}{0 - 4}$$

Formula for slope. Let $(x_2, y_2) = (0, -2)$ and $(x_1, y_1) = (4, 0)$.

$$= \frac{-2}{-4} = \frac{1}{2}$$

13. (1, -1), (-1, 1)

$$m = \frac{1 - (-1)}{-1 - 1} = \frac{1 + 1}{-2} = \frac{2}{-2} = -1$$

Let $(x_2, y_2) = (-1, 1)$ and $(x_1, y_1) = (1, -1)$.

Find the distance between the following pair of points:

17. (0, -2), (3, 0)

$$d = \sqrt{(x_2 - x_1)^2 + (y_2 - y_1)^2}$$

Distance formula.

$$= \sqrt{(3 - 0)^2 + (0 - -2)^2}$$

Let $(x_2, y_2) = (3, 0)$ and $(x_1, y_1) = (0, -2)$

$$= \sqrt{3^2 + 2^2}$$

$$= \sqrt{9 + 4}$$

$$= \sqrt{13}$$

The distance is $\sqrt{13}$.

B

Write the coordinates of the x- and y-intercepts:

17. 2x - 3y = 11

$y = 0$	$x = 0$	Substitute $y = 0$, and
$2x - 3(0) = 11$	$2(0) - 3y = 11$	solve for x. Then
$x = \frac{11}{2}$	$y = -\frac{11}{3}$	substitute $x = 0$, and solve for y.

The x-intercept is $\left(\frac{11}{2}, 0\right)$, and the y-intercept is $\left(0, -\frac{11}{3}\right)$.

25. $0.4x + 0.3y = 1.2$

       $y = 0$ Substitute $y = 0$, and

       $0.4x + 0.3(0) = 1.2$ solve for x.

                $0.4x = 1.2$

                     $x = 3$

       $x = 0$ Then substitute $x = 0$,

       $0.4(0) + 0.3y = 1.2$ and solve for y.

                $0.3y = 1.2$

                     $y = 4$

The x-intercept is (3, 0), and the y-intercept is (0, 4).

Find the slope of the line joining the following points and find the distance between the points:

29. (3, -7), (8, -4)

$$m = \frac{-7 - (-4)}{3 - 8} = \frac{-7 + 4}{-5} = \frac{3}{5}$$

$$d = \sqrt{(8 - 3)^2 + [(-4) - (-7)]^2}$$

$$= \sqrt{25 + 9}$$

$$= \sqrt{34}$$

The slope is $\frac{3}{5}$ and the distance is $\sqrt{34}$.

33. (-2, 3), (-2, -3)

$$m = \frac{3 - (-3)}{-2 - (-2)} = \frac{6}{0} \qquad \text{Division by zero is undefined.}$$

$$d = \sqrt{[(-2) - (-2)]^2 + [3 - (-3)]^2}$$

$$= \sqrt{0^2 + 6^2}$$

$$= \sqrt{36}$$

$$= 6$$

The slope is undefined, and the distance is 6.

Are the line segments joining the following pairs of points perpendicular?

37. (3, 5) and (-4, 3)
     (4, 1) and (-3, 3)

$$m_1 = \frac{3 - 5}{-4 - 3} = \frac{-2}{-7} = \frac{2}{7}$$

$$m_2 = \frac{1 - 3}{4 - (-3)} = \frac{-2}{7} = -\frac{2}{7}$$

No, since $\left(\frac{2}{7}\right)\left(-\frac{2}{7}\right) \neq -1$.

C

Write the coordinates of the x- and y-intercepts:

41. $\frac{2}{3}x + 6 = \frac{1}{2}y$

$y = 0$

$\frac{2}{3}x + 6 = \frac{1}{2}(0)$

$\frac{2}{3}x + 6 = 0$

$\frac{2}{3}x = -6$

$x = -6\left(\frac{3}{2}\right)$

$x = -9$

$x = 0$

$\frac{2}{3}(0) + 6 = \frac{1}{2}y$

$6 = \frac{1}{2}y$

$2(6) = y$

$12 = y$

The x-intercept is (-9, 0) and the y-intercept is (0, 12).

Find the slope of the line joining the following pairs of points and find the distance:

45. $\left(\frac{1}{2}, 2\right)$, $\left(3, \frac{1}{3}\right)$

$$m = \frac{\frac{1}{3} - 2}{3 - \frac{1}{2}} = \frac{-\frac{5}{3}}{\frac{5}{2}} = -\frac{2}{3}$$

$$d = \sqrt{\left(3 - \frac{1}{2}\right)^2 + \left(\frac{1}{3} - 2\right)^2}$$

$$= \sqrt{\left(\frac{5}{2}\right)^2 + \left(-\frac{5}{3}\right)^2}$$

$$= \sqrt{\frac{25}{4} + \frac{25}{9}}$$

$$= \sqrt{\frac{225 + 100}{36}}$$

$$= \frac{\sqrt{325}}{6}$$

$$= \frac{5\sqrt{13}}{6} \text{ or } \frac{5}{6}\sqrt{13}$$

The slope is $-\frac{2}{3}$ and the distance is $\frac{5}{6}\sqrt{13}$.

49. $(0.39, 0.25), (-0.21, 0.40)$

$$m = \frac{0.40 - 0.25}{-0.21 - 0.39} = \frac{0.15}{-0.6} = -0.25$$

$$d = \sqrt{[0.39 - (-0.21)]^2 + (0.25 - 0.40)^2}$$

$$= \sqrt{0.36 + 0.0225}$$

$$= \sqrt{0.3825}$$

$$= \sqrt{\frac{3825}{10000}}$$

Rewrite the decimal as a fraction in order to simplify.

$$= \frac{\sqrt{25} \cdot \sqrt{153}}{100} = \frac{5\sqrt{153}}{100}$$

$$= \frac{\sqrt{153}}{20}$$

The slope is -0.25 or $-\frac{1}{4}$, and the distance is $\frac{\sqrt{153}}{20}$.

53. Determine if the line 2y + x = 3 is parallel or perpendicular to the line y = 2x + 1.

2y + x = 3   y = 2x + 1

x	y
0	$\frac{3}{2}$
1	1

x	y
0	1
1	3

Find two points on the graph of each equation. note that the y-intercepts are different.

$$m = \frac{\frac{3}{2} - 1}{0 - 1} = \frac{\frac{1}{2}}{-1} \qquad m = \frac{1 - 3}{0 - 1} = \frac{-2}{-1}$$

$$m = -\frac{1}{2} \qquad\qquad m = 2$$

Compute each slope.

$$\left(-\frac{1}{2}\right)(2) = -1$$

Find the product of the slopes.

Since the product of the slopes is -1, the lines are perpendicular.

Are the line segments joining the following pairs of points parallel, perpendicular, or neither?

57. (17, -42) and (-18, 21)
     (-67, 113) and (133, -247)

$$m_1 = \frac{21 - (-42)}{-18 - 17} = -\frac{63}{35} = -\frac{9}{5}$$

$$m_2 = \frac{-247 - 113}{133 - (-67)} = \frac{-360}{200} = -\frac{9}{5}$$

Since $m_1 = m_2$, the segments are parallel.

D

61. The enrollment in geography classes is projected on a straight line. If the enrollment in the first year is 50, represented by (1, 50), and the enrollment in the second year is 60, represented by (2, 60), find the rate of growth. Give the enrollment in the seventh year.

(1, 50), (2, 60)

$$m = \frac{60 - 50}{2 - 1} = 10$$

The rate of growth is represented by the value of the slope.

(1, 50), (7, x)

$$\frac{x - 50}{7 - 1} = 10$$

Let x represent the enrollment in the seventh year. Use the definition for m and set it equal to 10. Solve for x.

$$\frac{x - 50}{6} = 10$$

$$x - 50 = 60$$

$$x = 110$$

The rate of growth is 10, and the seventh-year enrollment is 110.

65. Show that the points (2, -3), (8, -3), and (5, 0) are vertices of an isosceles triangle (two equal sides).

Compute the length of each side by using the distance formula.

(2, -3), (8, -3)

$$d_1 = \sqrt{(8 - 2)^2 + [-3 - (-3)]^2}$$

$$= \sqrt{36 + 0}$$
$$= 6$$

(2, -3), (5, 0)

$$d_2 = \sqrt{(5 - 2)^2 + [0 - (-3)]^2}$$

$$= \sqrt{9 + 9}$$
$$= 3\sqrt{2}$$

(8, -3), (5, 0)

$$d_3 = \sqrt{(8 - 5)^2 + (-3 - 0)^2}$$

$$= \sqrt{9 + 9}$$
$$= 3\sqrt{2}$$

The sides have lengths $3\sqrt{2}$, $3\sqrt{2}$, and 6 and therefore are the vertices of an isosceles triangle.

69. The points A (2, 1), B (5, 1), C (5, 4), and D (2, 4) are vertices of a square. Show that the diagonals of the square are perpendicular.

The vertices of the diagonals are the pair (2, 4) and (5, 1) and the pair (2, 1) and (5, 4). Find the slope of each.

(2, 4), (5, 1)

$$m_1 = \frac{4 - 1}{2 - 5} = \frac{3}{-3} = -1$$

(2, 1), (5, 4)

$$m_2 = \frac{4 - 1}{5 - 2} = \frac{3}{3} = 1$$

Since $m_1 \cdot m_2 = -1$, the diagonals are perpendicular

STATE YOUR UNDERSTANDING

73. The slope formula can be given in two forms:

$$\frac{y_2 - y_1}{x_2 - x_1} \quad \text{or} \quad \frac{y_1 - y_2}{x_1 - x_2}$$

Why, mathematically, are these forms equivalent?

The first form

$$\frac{y_2 - y_1}{x_2 - x_1} \quad \text{can be written:} \quad \frac{-(y_1 - y_2)}{-(x_1 - x_2)}$$

But the form

$$\frac{-(y_1 - y_2)}{-(x_1 - x_2)} = \frac{y_1 - y_2}{x_1 - x_2}$$ since a negative value divided by a negative value is positive.

Therefore, the two forms are equivalent.

77. A college observes that in 1985, the average score on the verbal section of the SAT is 500, represented by the ordered pair (0, 500). Four years later, in 1989, the average score was 485, represented by (4, 485). Using the scores (S) as the dependent variable and the year (t) as the independent variable, locate the two points on a graph and connect them with a straight line. If the trend continues, what will the average score be in 1993?

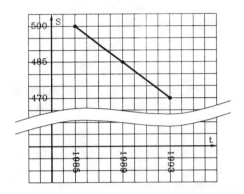

Label the horizontal axis t with each unit equal to 1. Label the vertical axis S with each unit equal to 5.

Locate the points (0, 500) and (4, 485). Draw a line through the points and intersecting the vertical line corresponding to 1993.

The point with t = 8 intersects the horizontal line that represents 500 - 30 or 470. So if the trend continues, the average score for 1993 will be 470.

MAINTAIN YOUR SKILLS

Simplify. Assume that all variables represent positive numbers:

81. $\sqrt[3]{160a^7 b^{10}} = \sqrt[3]{8a^6 b^9} \cdot \sqrt[3]{20ab}$

$$= 2a^2 b^3 \sqrt[3]{20ab}$$

247

Solve by square roots:

85. $y^2 = 54$

$$y = \pm\sqrt{54} = \pm\sqrt{9} \cdot \sqrt{6}$$
$$= \pm 3\sqrt{6}$$

The solution set is $\{\pm 3\sqrt{6}\}$.

EXERCISES 6.3 DIFFERENT FORMS OF THE LINEAR EQUATION

A

Write the equation of a line in standard form, given its slope and a point:

1.   (3, 6), m = 4

$y - y_1 = m(x - x_1)$	Use the point-slope form of the equation.
$y - 6 = 4(x - 3)$	Substitute.
$y - 6 = 4x - 12$	
$6 = 4x - y$	
or $4x - y = 6$	Standard form.

5.   (-7, 8), m = $-\dfrac{1}{3}$

$y - y_1 = m(x - x_1)$	Point-slope form.
$y - 8 = -\dfrac{1}{3}[x - (-7)]$	Substitute.
$y - 8 = -\dfrac{1}{3}(x + 7)$	Simplify.
$3y - 24 = -x - 7$	
$x + 3y = 17$	Standard form.

Find the slope and the y-intercept of the following:

9. $3x - 7y = 35$

$-7y = -3x + 35$	Solve for y.
$y = \dfrac{3}{7}x - 5$	Slope-intercept form.

The slope is $\dfrac{3}{7}$ and the y-intercept is (0, -5).

Find the slope and the y-intercept of the graphs of the following equations. Draw the graph using the slope and the y-intercept.

13. y = -x + 3   Slope-intercept form.
     m = -1 and the y-intercept is (0,3).

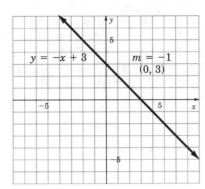

Plot the intercept. Move one unit right and one unit down. Draw the line passing through the points.

17. y = - $\frac{2}{3}$x + 5   Slope-intercept form.

     m = - $\frac{2}{3}$ and the y-intercept is (0,5).

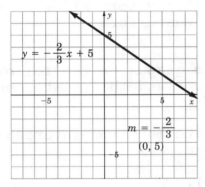

Plot the intercept. Move three units right and two units down.

Write the equation of a line in standard form, given its slope and a point on the line:

21. (-11, 14), m = -3

     y - y₁ = m(x - x₁)   Point-slope form.
     y - 14 = -3[x - (-11)]   Substitute.
     y - 14 = -3(x + 11)
     y - 14 = -3x - 33
     3x + y = -19   Standard form.

249

B

Write the equation of the line in standard form, given two points on the line:

25. (6, -5), (8, -3)

$$m = \frac{-3 - (-5)}{8 - 6} = \frac{2}{2} = 1$$

$$y - y_1 = m(x - x_1)$$

$$y - (-5) = 1(x - 6)$$

Since the slope is not given, first calculate m. Now use point-slope form with either given point. Substitute the point (6, -5).

$$y + 5 = x - 6$$
$$11 = x - y$$
or $x - y = 11$

Standard form.

29. (-11, 12), (3, -5)

$$m = \frac{-5 - 12}{3 - (-11)} = \frac{-17}{14} = -\frac{17}{14}$$

First calculate m.

$$y - (-5) = -\frac{17}{14}(x - 3)$$

Next, substitute either point using the point-slope form.
Here the point (3, -5) is substituted.

$$y + 5 = -\frac{17}{14}(x - 3)$$

$$14y + 70 = -17x + 51$$
$$17x + 14y = -19$$

Multiply both sides by 14.
Standard form.

Find the slope and the y-intercept of the graphs of the following equations. Draw the graphs using the slope and the y-intercepts:

33. 4x - y = 7
$$y = 4x - 7$$

Write in slope-intercept form.

m = 4 and the y-intercept is (0, -7).

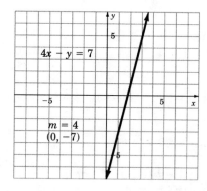

Plot the y-intercept. Move one unit right and four units up.

250

37. $2x - 5y - 20 = 0$
$$-5y = -2x + 20$$
$$y = \frac{2}{5}x - 4$$

$m = \frac{2}{5}$ and the y-intercept is (0, -4).

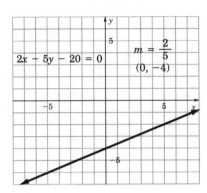

Plot the y-intercept. Move five units right and two units up.

Write the equation of a line in slope-intercept form, given its slope and a point on the line:

41. (-34, -55), $m = -\frac{7}{12}$

$$y - y_1 = m(x - x_1)$$ Point-slope form.

$$y - (-55) = -\frac{7}{12}[x - (-34)]$$

$$y + 55 = -\frac{7}{12}(x + 34)$$

$$y = -\frac{7}{12}x - \frac{238}{12} - 55$$

$$y = -\frac{7}{12}x - \frac{898}{12}$$

$$y = -\frac{7}{12}x - \frac{449}{6}$$ Slope-intercept form.

C

Write the equation of the line in standard form given two points on the line:

45. $\left(\frac{1}{2}, 4\right), \left(-\frac{2}{3}, 4\right)$

$$m = \frac{4 - 4}{\frac{1}{2} - \left(-\frac{2}{3}\right)} = 0$$ Calculate m.

$$y - 4 = 0\left(x - \frac{1}{2}\right)$$ Substitute $\left(\frac{1}{2}, 4\right)$ using the point-slope formula.

$$y - 4 = 0$$
$$y = 4$$

49. $\left(\frac{1}{3}, \frac{3}{4}\right), \left(\frac{1}{2}, \frac{2}{3}\right)$

$$m = \frac{\frac{2}{3} - \frac{3}{4}}{\frac{1}{2} - \frac{1}{3}} = \frac{-\frac{1}{12}}{\frac{1}{6}}$$

$$= -\frac{1}{2}$$

$$y - \frac{3}{4} = -\frac{1}{2}\left(x - \frac{1}{3}\right)$$ Using the point-slope form substitute either point. Here the point $\left(\frac{1}{3}, \frac{3}{4}\right)$ is substituted.

$$12y - 9 = -6x + 2$$ Multiply both sides by 12.
$$6x + 12y = 11$$ Simplify.

Write the following equations in standard form and slope-intercept form.

53. $3(x - 1) - 5 = 2(y - 1) + 7$
$$3x - 3 - 5 = 2y - 2 + 7$$ Simplify.
$$3x - 8 = 2y + 5$$
$$3x - 2y = 13$$ Write in standard form.

$$-2y = -3x + 13$$ Now write in slope-intercept form by
$$y = \frac{3}{2}x - \frac{13}{2}$$ solving for y.

57. $4(x + 1) - 6 = 2(y + 4) + 5$
     $4x + 4 - 6 = 2y + 8 + 5$ Simplify.
          $4x - 2 = 2y + 13$
          $4x - 2y = 15$ Write in standard form.

               $-2y = -4x + 15$ Now write in
                 $y = 2x - \dfrac{15}{2}$ slope-intercept form.

D

61. The Uptown Corp. expects its profit to increase $1500 per
     year. If the profit in year three was $25,000, write the
     equation of the profit line. What will be the profit in
     year 12?

                                        Since the profit is $1500
                                        per year, the slope is
                                        1500.

     Formula:
     $y - y_1 = m(x - x_1)$ Given the point
                                        $(x_1, y_1) = (3, 25000)$ and
                                        the slope, m = 1500, use
                                        the point-slope form to
                                        find the equation.

     Substitute:
     $y - 25000 = 1500(x - 3)$

     Solve:
        $y - 25000 = 1500x - 4500$
     $1500x - 4500 = y - 25000$
         $1500x - y = -20500$ Standard form.

     $1500(12) - y = -20500$ Evaluate the equation when
               $-y = -20500 - 18000$ x = 12.
                $y = 38500$ Profit in year 12.

     The standard form of the profit line is 1500x - y = -20500,
     and the profit in year 12 will be $38,500.

65. Write the equation of a line that passes through the point
    (-2, 5) and is parallel to the line whose equation is
    $2x - 3y = 5$.

    $2x - 3y = 4$ Parallel lines have the
                                                 same slope. Write the
                                                 given equation in the
        $-3y = -2x + 4$ slope-intercept form to
                                                     find the slope.

$$y = \frac{2}{3}x - \frac{4}{3}$$

So $m = \frac{2}{3}$

$$y - y_1 = m(x - x_2)$$ Use the given point and the
                                                   slope to find the equation.

$$y - 5 = \frac{2}{3}[(x - (-2)]$$

$$y - 5 = \frac{2}{3}(x + 2)$$ Simplify.

$3y - 15 = 2x + 4$ Multiply both sides by 3.

       $-19 = 2x - 3y$

or $2x - 3y = -19$

69. Write the equation of the line that passes through the point
    (2, -5) and is parallel to $y + 3 = 0$.

$y + 3 = 0$

    $y = -3$ Solve for y.

    $y = 0x - 3$ Slope-intercept form.

So $m = 0$.

    $y - y_1 = m(x - x_1)$

$y - (-5) = 0(x - 2)$ Substitute the given point

    $y + 5 = 0$ (2, -5) for (x_1, y_1) and

        $y = -5$ replace m with 0.

73. The voltage (V) in a circuit is 40 volts when the resistance (R) is 8 ohms. The voltage is 80 volts when the resistance is 12 ohms. Using V as the dependent variable and the resistance (R) as the independent variable, write the equation of the voltage in terms of the resistance. What would the voltage be when R = 20 ohms?

(8, 40)

(12, 80)

Write ordered pairs where the first coordinate represents R and the second coordinate represents V.

$$m = \frac{80 - 40}{12 - 8} = \frac{40}{4} = 10$$

Calculate the slope.

$$y - y_1 = m(x - x_1)$$

Use the point-slope formula to generate an equation.

$$y - 40 = 10(x - 8)$$

Use (8, 40) as (x_1, y_1).

$$y - 40 = 10x - 80$$
$$y = 10x - 40$$
$$V = 10R - 40$$

Slope intercept form. Substitute R for x, and V for y.

$$V = 10(20) - 40$$
$$V = 200 - 40$$
$$V = 160$$

R = 20.

The equation of the voltage in terms of resistance is V = 10R - 40. When R = 20 ohms, V = 160 volts.

CHALLENGE EXERCISES

77. The midpoint of a line segment connecting two points $P_1(x_1, y_1)$ and $P_2(x_2, y_2)$ is given by the formula $\left(\dfrac{x_1 + x_2}{2}, \dfrac{y_1 + y_2}{2}\right)$. Find the midpoint of the line segment joining (-3, 6) and (5, 2).

$$\left(\frac{x_1 + x_2}{2}, \frac{y_1 + y_2}{2}\right)$$

$$\left(\frac{-3 + 5}{2}, \frac{6 + 2}{2}\right)$$

Use (-3, 6) as (x_1, y_1) and (5, 2) as (x_2, y_2).

$$\left(\frac{2}{2}, \frac{8}{2}\right)$$

(1, 4)

The midpoint is (1, 4).

Perform the indicated operations and reduce if possible:

81. $\dfrac{2a^2 + 9a - 35}{18a^2 - 21a - 4} \cdot \dfrac{12a^2 - 13a - 4}{2a^2 + 19a + 35} \cdot \dfrac{12a^2 + 32a + 5}{8a^2 - 22a + 5}$

$= \dfrac{(2a - 5)(a + 7)}{(3a - 4)(6a + 1)} \cdot \dfrac{(4a + 1)(3a - 4)}{(2a + 5)(a + 7)} \cdot \dfrac{(6a + 1)(2a + 5)}{(4a - 1)(2a - 5)}$

$= \dfrac{4a + 1}{4a - 1}$

Simplify:

85. $\dfrac{x}{\dfrac{1}{3} + \dfrac{1}{y} + \dfrac{1}{6y}} = \dfrac{x}{\dfrac{2y + 6 + 1}{6y}}$

$= \dfrac{x}{\dfrac{2y + 7}{6y}}$

$= x\left(\dfrac{6y}{2y + 7}\right)$

$= \dfrac{6xy}{2y + 7}$

256

A

Draw the graph of each inequality:

1. y > x + 1
    y = x + 1

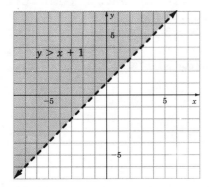

Draw the graph of the corresponding equality.

x	y
0	1
-1	0
1	2

The graph of the line is broken since equality is not included. Test the point (0, 0). 0 > 0 + 1 False.
The graph is the half plane not containing the origin.

5. y ≤ -x + 5
    y = -x + 5

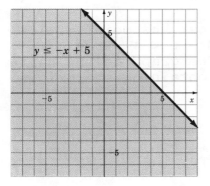

Draw the graph of the equality.

x	y
0	5
5	0
2	3

The graph of the line is solid since the relation includes equality. Test the point (0, 0).
0 ≤ 0 + 5 True.
The graph is the half plane containing the origin.

9. $x > -3$
    $x = -3$

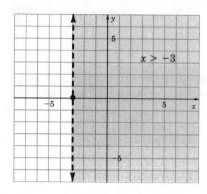

Vertical line with
x-intercept at (-3, 0).
Draw the graph of the
equality.

The graph of the line is
broken since equality is
not included. Test the
point (0, 0).
0 > -3 True.
The graph is the half-plane
containing the origin.

B

13. $x - y < 4$
     $x - y = 4$

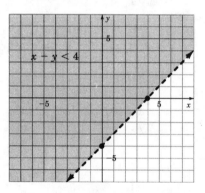

Draw the graph of the
corresponding equality.

x	y
0	-4
4	0
2	-2

Test the point (0, 0).
0 - 0 < 4 True.
The graph is the half-plane
containing the origin.

258

17. 3x - 5y < 10
     3x - 5y = 10

Draw the graph of the
corresponding equality.

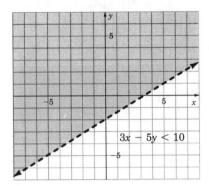

x	y
0	-2
$\frac{10}{3}$	0
-5	-5

The graph of the line is
broken since equality is
not included. Test the
point (0, 0).
3(0) - 5(0) < 10
          0 < 10   True.
The graph is the half-plane
containing the origin.

C

21. 2x - 7y $\leq$ 14
     2x - 7y = 14

Draw the graph of the
equality.

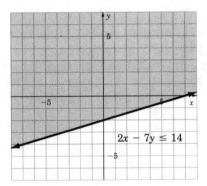

x	y
0	-2
7	0
3	$-\frac{8}{7}$

The graph of the line is
solid since equality is
included in the relaion.
Test the point (0, 0).
0 - 0 $\leq$ 14   True.
The graph of the inequality
is the half-plane
containing the origin.

259

25. $\frac{1}{2}y + \frac{1}{2} - x \leq 0$

$\frac{1}{2}y + \frac{1}{2} - x = 0$

Draw the graph of the equality.

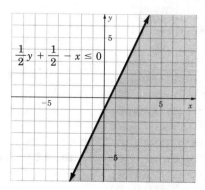

x	y
0	-1
$\frac{1}{2}$	0
2	3

The graph of the line is solid. Test the point

(0, 0). $0 + \frac{1}{2} - 0 \leq 0$

False.
The graph of the inequality is the half-plane not containing the origin.

Draw the graph of the union of the solution sets of the given inequalities:

29. $x + y \leq 1$, $x - y \leq 1$

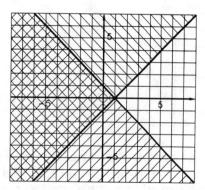

Draw the graph of each inequality separately. The graph of the union of the two inequalities is the entire shaded area.

33. $\frac{1}{3}x - \frac{1}{4}y \leq 1$, $\frac{1}{3}x + \frac{1}{4}y \geq 1$

   $4x - 3y \leq 12$, $4x + 3y \geq 12$

   $y \leq \frac{4}{3}x - 4$, $y \geq -\frac{4}{3}x + 4$

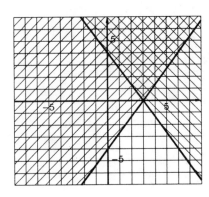

Write each inequality in slope-intercept form for easier graphing. Draw the graph of each inequality separately. The graph of the union of the two inequalities is the entire shaded area.

Draw the graph of the intersection of the solution sets of the given inequalities:

37. $3x + y \leq 12$, $2x + 5y \geq 10$

   or $y \leq -3x + 12$ $5y \geq -2x + 10$

                              or $y \geq -\frac{2}{5}x + 2$

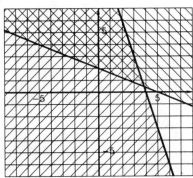

Draw the graph of each of the inequalities. The graph of the intersection is the region common to both inequalities.

261

41. $x > y + 1$, $x < y - 3$
    or $y < x - 1$ or $y > x + 3$

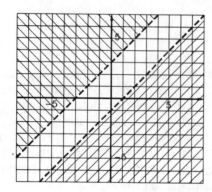

Draw the graph of each of the inequalities. Since the inequalities do not overlap, there is no intersection.

45. $2x - 3y \geq 6$, $3x + 2y \leq 6$, $x - y \leq 6$
    or $y \leq \frac{2}{3}x - 2$, or $y \leq -\frac{3}{2}x + 3$, or $y \geq x - 6$

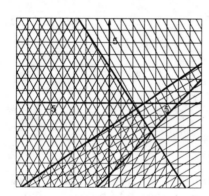

Draw the graph of each of the inequalities. The graph of the intersection is the region common to all three.

D

49. The Miniature Collection Company manufactures tea sets at a cost of $8 per set and horse and buggy sets at a cost of $5 per set. If manufacturing costs for the week cannot exceed $1000, what are the possible combinations of tea sets (y) and horse and buggy sets (x) that can be made? Graph the solution. (Note: Negative values to represent sets have no meaning and are disregarded.)

Simpler word form:
$$\begin{pmatrix} \text{Cost of} \\ \text{tea sets} \end{pmatrix} + \begin{pmatrix} \text{Cost of} \\ \text{buggy sets} \end{pmatrix} \leq 1000$$

Translate to algebra:
5x + 8y ≤ 1000

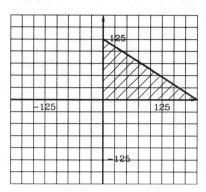

Let y represent the number of tea sets, so 8y represents the cost of the tea sets. Let x represent the number of buggy sets, so 5x represents the cost of the buggy sets.

Draw the graph of 8x + 5y < 1000.

53.   Given the same costs to manufacture the miniature sets in Exercise 49, the maufacturer wants to make at least 50 tea sets. Graph the solution of the original inequality with this new added condition.

    5x + 8y ≤ 1000
         y ≥ 50

From Exercise 49.
The variable y represents the number of tea sets.

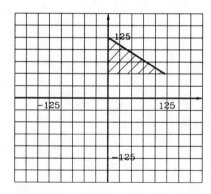

Draw the graph of each of the inequalities. The shaded triangle represents the solution.

57.   In Exercise 55, suppose Sue does not want to invest more than $2,000 in stocks because of the market at this time. Graph the original inequality with this added condition.

    x + 2y ≤ 10,000

From Exercise 55.

        x ≤ 2000

The variable x represents the amount invested in stocks.

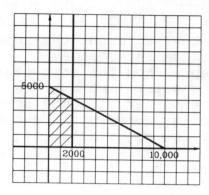

Draw the graph of both inequalities. The shaded region represents the solution.

61. If the nutritionist in Exercise 60 varies the amount of iron and calcium to be x mg of iron and 3y mg of calcium, rewrite the inequality and graph the solution.

$$1000 < x + 3y \leq 2000$$

Let x represent the amount of iron and y represent the amount of calcium.

Draw the graph of both inequalities. The shaded region represents the solution.

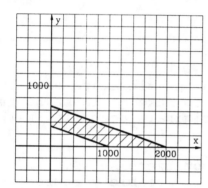

<u>MAINTAIN YOUR SKILLS</u>

Perform the indicated operations and reduce if possible:

65. $\dfrac{x^2 - 15x + 54}{2x^2 + 13x + 21} \div \dfrac{x^2 - x - 72}{2x^2 - 3x - 35}$

$= \dfrac{(x - 9)(x - 6)}{(2x + 7)(x + 3)} \cdot \dfrac{(2x + 7)(x - 5)}{(x - 9)(x + 8)}$

$= \dfrac{(x - 6)(x - 5)}{(x + 3)(x + 8)}$

264

Divide and write the answer in standard form:

69. $\dfrac{8}{3 + i} = \dfrac{8}{3 + i} \cdot \dfrac{3 - i}{3 - i}$

$\qquad = \dfrac{24 - 8i}{9 - i^2}$

$\qquad = \dfrac{24 - 8i}{10} \qquad\qquad\qquad i^2 = -1$

$\qquad = \dfrac{24}{10} - \dfrac{8}{10}i$

$\qquad = \dfrac{12}{5} - \dfrac{4}{5}i \qquad\qquad\qquad$ Standard form.

EXERCISES 6.5 VARIATION

A

1. If x varies directly as y, and if x = 6 when y = 15, find y when x = 30.

$k = \dfrac{x}{y}$
$\qquad\qquad\qquad\qquad\qquad$ This is an example of direct variation so the quotient $\dfrac{x}{y}$ is used.

So x = ky

First case:
6 = 15k
$\qquad\qquad\qquad\qquad\qquad$ Replace x with 6 and y with 15 to find k.

$k = \dfrac{6}{15} = \dfrac{2}{5}$

Second case:
$30 = \dfrac{2}{5}y$
$\qquad\qquad\qquad\qquad\qquad$ Substitute $k = \dfrac{2}{5}$ and x = 30 in the formula in Step 1.

y = 75

5. If x varies inversely as y, and if x = 24 when y = 6, find x when y = 15.

    $k = xy$ This is an example of indirect variation, so the product xy is used.

    First case:
    $k = (24)(6) = 144$ Replace x with 24 and y with 6 to find k.

    Second case:
    $144 = 15x$ Substitute k = 144 and y = 15 in the formula at Step 1.

      $x = 9.6$

9. If z varies jointly as x and y, and if z = 36 when x = 4 and y = 9, find z if x = 28 and y = 7.

    $k = \dfrac{z}{xy}$ Since this is an example of joint variation, use the quotient $\dfrac{z}{xy}$.

    First case:
    $k = \dfrac{36}{4 \cdot 9} = 1$ Replace z with 36, x with 4 and y with 9 and solve for k.

    Second case:
    $1 = \dfrac{z}{(28)(7)}$ Substitute x = 28, y = 7, and k = 1 in the formula in Step 1.
    $z = 196$ Solve for z.

B

13. The pressure P per square inch in water varies directly as the depth d. If P = 5.77 when d = 13, find P when d = 32. (To the nearest tenth.)

$k = \dfrac{P}{d}$	This is an example of direct variation so use the quotient, $\dfrac{P}{d}$.

First case:

$k = \dfrac{5.77}{13}$

Replace P with 5.77 and d with 13.

Second case:

$\dfrac{5.77}{13} = \dfrac{P}{32}$

Substitute $\dfrac{5.77}{13}$ for k and 32 for d.

$P = \dfrac{5.77}{13}(32) \approx 14.2$

Round to the nearest tenth.

So if the distance is 32, the pressure is 14.2.

17. The number of amperes varies directly as the number of watts. For a reading of 550 watts, the number of amperes is 5. What is the number of amperes when the reading of watts is 1600? (To the nearest tenth.)

Let a represent amperes, and w represent watts.

$k = \dfrac{w}{a}$

This is an example of direct variation.

First case:

$k = \dfrac{550}{5} = 110$

To find k, the constant of variation, replace a with 5 and w with 550.

Second case:

$110 = \dfrac{1600}{a}$

$110a = 1600$

$a = 14.5$

Substitute k with 110 and w with 1600.
Solve.
Round to the nearest tenth.

So when the number of watts is 1600, the number of amperes is 14.5.

c

21. At PPL Express the cost of shipping goods (C) varies jointly as the distance shipped (d) and the weight (w). If it costs $338 to ship 6.5 tons of goods 8 miles, how much will it cost to ship 24 tons of goods 1535 miles? (Round to the nearest dollar.)

$k = \dfrac{C}{dw}$

This is an example of joint variation.

First case:

$k = \dfrac{338}{8(6.5)} = 6.5$

Solve for k by replacing C with 338, w with 6.5, and d with 8.

Second case:

$6.5 = \dfrac{C}{24(1535)}$

Substitute d with 1535, w with 24, and k with 6.5.

$C = (6.5)(24)(1535) = 239,460$

Solve for C.

It costs $239,460 to ship 24 tons a distance of 1535 miles.

25. The volume of a box with a fixed depth varies jointly as the length and width of the bottom of the box. If the volume of the box is 1152 cm³ when the dimensions of the bottom are 16 cm × 12 cm, find the volume when the bottom dimensions are 15 cm × 7 cm.

Let V represent the volume, ℓ the length and w the width. k represents the constant of variation:

$k = \dfrac{V}{\ell w}$

This is an example of joint variation.

First case:

$k = \dfrac{1152}{(16)(12)} = 6$

Replace the variables and solve for k.

Second case:

$6 = \dfrac{V}{(15)(7)}$
$V = 630$

Replace the variables and solve for V.

When the dimensions of the bottom of the box are 15 cm × 7 cm, the volume is 630 cm³.

29. If the pressure (P) of a gas varies directly with the absolute temperature (T) in degrees Kelvin and inversely with the volume (V), write the equation of the relationship using the constant k. If P = 15 lb/in^2 when the volume V = 33 in^2 and the absolute temperature is 330°K, what is the value of k (the constant of proportionality)? What will be the value of P when V = 25in^3 and T = 350°K?

$$\frac{PV}{T} = k$$

P varies directly with T and inversely with V.

First case:
$$\frac{15(33)}{330} = k$$
$$1.5 = k$$

Solve for k, using the values in the first case.

Second case:
$$\frac{P(25)}{350} = 1.5$$

Use the value of k from the first case to solve for P.

$$25P = 525$$
$$P = 21$$

The value of P is 21 lb/in^2 when V = 25 in^3 and T = 350°K.

CHALLENGE EXERCISES

33. Using the equation in Exercise 32, find the value of y when r = 49 and t = 4.

$$\frac{yt^2}{\sqrt{r}} = k, \text{ with } k = 36$$

From Exercise 32.

$$\frac{4^2 y}{\sqrt{49}} = 36$$

Substitute k - 36, r = 49 and t = 4 and solve for y.

$$16y = 7(36)$$
$$16y = 252$$
$$y = 15.75$$

$\sqrt{49} = 7$

MAINTAIN YOUR SKILLS

Simplify using only positive exponents:

37. $(w^{-1}y^{-1})^{-1} (wy)^{-1}$

$= (wy)\left(\frac{1}{wy}\right) = 1$

or $(w^{-1}y^{-1})^{-1} (wy)^{-1}$

$$\frac{1}{w^{-1}y^{-1}} \cdot w^{-1}y^{-1} = 1$$

Write in scientific notation:

41.   7,843,000,000 = 7.843 × 10^9

CHAPTER 7

SYSTEMS OF LINEAR EQUATIONS

EXERCISES 7.1 SOLVING SYSTEMS BY GRAPHING AND SUBSTITUTION

A

Solve by graphing:

1. $\begin{cases} x + y = -4 \\ x - y = 8 \end{cases}$

Write each equation in slope-intercept form to determine the type of system.

$x + y = -4$ $x - y = 8$

    $y = -x - 4$ $y = x - 8$

$m = -1$ $m = 1$

y-intercept (0, -4) y-intercept (0, -8)

The slopes are not the same, so the system has one solution. Use the slope and y-intercept of each line to draw the graphs.

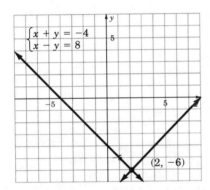

The solution is {(2, -6)}.

The lines intersect at (2, -6).

5. $\begin{cases} y = x - 4 \\ y = -x - 2 \end{cases}$

$y = x - 4 \qquad y = -x - 2$
$m = 1 \qquad\qquad\qquad m = -1$
y-intercept $(0, -4)$ y-intercept $(0, -2)$

The slopes are not the same, so the system has one solution. Draw the graphs.

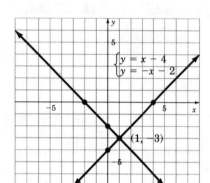

The solution set is $\{(1, -3)\}$.   The lines intersect at $(1, -3)$.

Solve the following systems by substitution:

9. $\begin{cases} y = x + 2 & (1) \\ x + y = 10 & (2) \end{cases}$

$x + y = 10 \qquad (2)$
$x + (x + 2) = 10$

Equation (1) is solved for y, so substitute $x + 2$ for y in equation (2). Solve for x.

$2x + 2 = 10$
$2x = 8$
$x = 4$

$y = x + 2 \qquad\qquad (1)$
$y = 4 + 2$
$y = 6$

To find y, substitute $x = 4$ in equation (1). Solve for y.

The solution set is $\{(4, 6)\}$.

272

13. $\begin{cases} x + y = 4 & (1) \\ 3x - 2y = 7 & (2) \end{cases}$

$$x + y = 4 \qquad (1) \qquad \text{Solve equation (1) for x.}$$

$$x = -y + 4$$

$$3x - 2y = 7 \qquad (2) \qquad \text{Substitute } -y + 4 \text{ for x in}$$
$$3(-y + 4) - 2y = 7 \qquad\qquad \text{equation (2).}$$
$$-3y + 12 - 2y = 7$$
$$-5y = -5$$
$$y = 1$$

$$x + y = 4 \qquad (1) \qquad \text{Substitute } y = 1 \text{ in}$$
$$x + 1 = 4 \qquad\qquad \text{equation (1).}$$
$$x = 3$$
The solution set is $\{(3, 1)\}$.

17. $\begin{cases} y = 2x - 3 & (1) \\ x = 2y - 9 & (2) \end{cases}$

$$x = 2y - 9 \qquad (2) \qquad \text{Substitute } 2x - 3 \text{ for y in}$$
$$x = 2(2x - 3) - 9 \qquad\qquad \text{equation (2).}$$
$$x = 4x - 6 - 9$$
$$-3x = -15$$
$$x = 5$$

$$y = 2x - 3 \qquad (1) \qquad \text{Substitute } x = 5 \text{ in}$$
$$y = 2(5) - 3 \qquad\qquad \text{equation (1).}$$
$$y = 10 - 3$$
$$y = 7$$
The solution set is $\{(5, 7)\}$.

B

21. $\begin{cases} 4x - y = 16 & (1) \\ 3x + 2y = 1 & (2) \end{cases}$

$$4x - y = 16 \qquad (1) \qquad \text{Solve for y in}$$
$$4x - 16 = y \qquad\qquad \text{equation (1).}$$

$$3x - 2y = 1 \qquad (2) \qquad \text{Substitute } 4x - 16 \text{ for y}$$
$$3x + 2(4x - 16) = 1 \qquad\qquad \text{in equation (2).}$$
$$3x + 8x - 32 = 1$$
$$11x = 33$$
$$x = 3$$

$$4x - y = 16 \qquad (1) \qquad \text{Substitute } x = 3 \text{ in}$$
$$4(3) - y = 16 \qquad\qquad \text{equation (1).}$$
$$12 - y = 16$$
$$12 - 16 = y$$
$$-4 = y$$
The solution set is $\{(3, -4)\}$.

273

25.
$$\begin{cases} x = \frac{1}{2}y - 3 & (1) \\ 5x - 2y = -17 & (2) \end{cases}$$

$$2x = y - 6$$
$$\begin{cases} 2x - y = -6 & (1') \\ 5x - 2y = -17 & (2) \end{cases}$$

Rewrite the first equation, clearing the fraction. Relabel it (1').

$$y = 2x + 6$$

Solve equation (1') for y.

$$5x - 2(2x + 6) = -17$$
$$5x - 4x - 12 = -17$$
$$x - 12 = -17$$
$$x = -5$$

Substitute $2x + 6$ for y in (2) and solve for x.

$$y = 2(-5) + 6$$
$$= -10 + 6$$
$$= -4$$

To find y, substitute $x = -5$ in $y = 2x + 6$.

The solution set is $\{(-5, -4)\}$.

29.
$$\begin{cases} 6x + y = -5 & (1) \\ 3x + 5y = 2 & (2) \end{cases}$$

$$y = -6x - 5$$

Solve equation (1) for y.

$$3x + 5(-6x - 5) = 2$$
$$3x - 30x - 25 = 2$$
$$-27x - 25 = 2$$
$$-27x = 27$$
$$x = -1$$

Substitute $-6x - 5$ for y in equation (2).

$$y = -6(-1) - 5$$
$$= 6 - 5$$
$$= 1$$

Substitute -1 for x in $y = -6x - 5$ to find y.

The solution set is $\{(-1, 1)\}$.

33.
$$\begin{cases} x = 2y + 6 & (1) \\ 3x - 4y = 12 & (2) \end{cases}$$

$$3(2y + 6) - 4y = 12$$
$$6y + 18 - 4y = 12$$
$$2x + 18 = 12$$
$$2y = -6$$
$$y = -3$$

Substitute $2y + 6$ for x in equation (2) and solve for y.

$$x = 2(-3) + 6$$
$$= -6 + 6$$
$$= 0$$

To find x, substitute -3 for y in equation (1).

The solution set is $\{(0, -3)\}$.

37. $\begin{cases} x + 3y = 5 & (1) \\ 3x - 5y = 12 & (2) \end{cases}$

$x + 3y = 5$ (1)   Solve equation (1) for x.
$x = -3y + 5$

$3x - 5y = 12$ (2)   Substitute $-3y + 5$ for x
$3(-3y + 5) - 5y = 12$ in equation (2).
$-9y + 15 - 5y = 12$
$-14y = -3$
$y = \dfrac{3}{14}$

$x + 3y = 5$ (1)   Substitute $\dfrac{3}{14}$ for y in
$x + 3\left(\dfrac{3}{14}\right) = 5$ equation (1).
$14x + 9 = 70$ Solve for x.
$14x = 61$
$x = \dfrac{61}{14}$

The solution set is $\left\{\left(\dfrac{61}{14}, \dfrac{3}{14}\right)\right\}$.

C

41. $\begin{cases} 8x - 4y = 11 & (1) \\ x + 3y = -3 & (2) \end{cases}$

$x = -3y - 3$ Solve equation (2) for x.

$8(-3y - 3) - 4y = 11$ Substitute $-3y - 3$ for x
$-24y - 24 - 4y = 11$ in equation (1) and solve
$-28y = 35$ for y.
$y = -\dfrac{35}{28}$
$y = -\dfrac{5}{4}$

$x + 3\left(-\dfrac{5}{4}\right) = -3$ Substitute $-\dfrac{5}{4}$ for y in
$x - \dfrac{15}{4} = -3$ equation (2) and solve
       for x.
$x = -3 + \dfrac{15}{4}$
$x = -\dfrac{12}{4} + \dfrac{15}{4}$
$x = \dfrac{3}{4}$

The solution set is $\left\{\left(\dfrac{3}{4}, -\dfrac{5}{4}\right)\right\}$.

45. $\begin{cases} x - 2y = -5 & (1) \\ 6x + 3y = -10 & (2) \end{cases}$

$x = 2y - 5$ — Solve equation (1) for x.

$6(2y - 5) + 3y = -10$
$12y - 30 + 3y = -10$
$15y = 20$
$y = \dfrac{4}{3}$ — Substitute $2y - 5$ for x in equation (2) and solve for y.

$x = 2\left(\dfrac{4}{3}\right) - 5$
$= \dfrac{8}{3} - \dfrac{15}{3}$
$= -\dfrac{7}{3}$ — Substitute $\dfrac{4}{3}$ for y in $x = 2y - 5$.

The solution set is $\left\{ \left(-\dfrac{7}{3}, \dfrac{4}{3} \right) \right\}$.

49. $\begin{cases} x - 9y = -5 & (1) \\ -x - 3y = 1 & (2) \end{cases}$

$x = 9y - 5$ — Solve equation (1) for x.

$-(9y - 5) - 3y = 1$
$-9y + 5 - 3y = 1$
$-12y = -4$
$y = \dfrac{1}{3}$ — Substitute $9y - 5$ for x in equation (2) and solve for y.

$x = 9\left(\dfrac{1}{3}\right) - 5$
$= 3 - 5$
$= -2$ — Substitute $\dfrac{1}{3}$ for y in $x = 9y - 5$.

The solution set is $\left\{ \left(-2, \dfrac{1}{3} \right) \right\}$.

53. $\begin{cases} x - 2y = -1 & (1) \\ 4x + 4y = 5 & (2) \end{cases}$

$x = 2y - 1$ — Solve equation (1) for x.

$4(2y - 1) + 4y = 5$
$8y - 4 + 4y = 5$
$12y - 4 = 5$
$12y = 9$
$y = \dfrac{3}{4}$ — Substitute $2y - 1$ for x in equation (2). Solve for y.

$$x = 2\left(\frac{3}{4}\right) - 1$$

$$= \frac{6}{4} - \frac{4}{4}$$

$$= \frac{2}{4}$$

$$= \frac{1}{2}$$

Substitute $\frac{3}{4}$ for y in

$x = 2y - 1.$

The solution set is $\left\{\left(\frac{1}{2}, \frac{3}{4}\right)\right\}.$

57.
$$\begin{cases} x + \frac{3}{4}y = 6 & (1) \\ \frac{2}{3}x + y = 6 & (2) \end{cases}$$

$$\begin{array}{ll} 4x + 3y = 24 & (1') \\ 2x + 3y = 18 & (2') \end{array}$$

Clear the fractions in both equations. Multiply equation (1) by 4, and relabel it (1'). Multiply equation (2) by 3 and relabel it (2').

$$2x = -3y + 18$$

$$x = -\frac{3}{2}y + 9$$

Solve equation (2') for x.

$$4\left(-\frac{3}{2}y + 9\right) + 3y = 24$$

$$-6 + 3y + 3y = 24$$

$$-3y = -12$$

$$y = 4$$

Substitute $-\frac{3}{2}y + 9$ for x in equation (1'). Solve for y.

$$x = -\frac{3}{2}(4) + 9$$

$$= -6 + 9$$

$$= 3$$

Substitute 4 for y in

$x = -\frac{3}{2}y + 9.$

The solution set is $\{(3, 4)\}.$

D

61. A total of $40,000 is invested, part at 18% and part at 12%. The annual return is $5220. How much is invested at each rate?

Simpler word form:

$$\begin{pmatrix} \text{Amount of money} \\ \text{invested at 18\%} \end{pmatrix} + \begin{pmatrix} \text{Amount of money} \\ \text{invested at 12\%} \end{pmatrix} = \$40,000$$

$$\begin{pmatrix} \text{Interest at} \\ 18\% \end{pmatrix} + \begin{pmatrix} \text{Interest at} \\ 12\% \end{pmatrix} = \$5220.$$

Translate to algebra:

$x + \quad y = 40000 \qquad (1)$ $0.18x + 0.12y = 5220 \qquad (2)$	Let x represent amount of money invested at 18%, and y represent amount of money invested at 12%.
$x + y = 40000 \qquad (1)$ $18x + 12y = 522000 \qquad (2')$	Clear the decimals in equation (2) by multiplying the equation by 100.
$y = -x + 40000$	Solve equation (1) for y.
$18x + 12(-x + 40000) = 522000$ $18x - 12x + 480000 = 522000$ $6x = 42000$ $x = 7000$	Substitute $-x + 40000$ for y in equation (2′). Solve for x.
$7000 + y = 40000$ $y = 33000$	Substitute 7000 for x in equation (1).

So $7000 is invested at 18% and $33,000 is invested at 12%.

65. The local Farmer's Feed Store in Ames, Iowa, has two grades of rabbit food. One grade sells for $0.82 per pound and the other for $0.62 per pound. How many pounds of each grade must be used to form 1000 pounds of a mixture that will sell for $0.75 per pound?

Simpler word form:

$$\begin{pmatrix} \text{Amount of} \\ \text{one grade} \\ \text{rabbit food} \end{pmatrix} + \begin{pmatrix} \text{Amount of} \\ \text{second grade} \\ \text{rabbit food} \end{pmatrix} = 1000 \text{ pounds}$$

$$\$0.82 \begin{pmatrix} \text{Pounds} \\ \text{of} \\ \text{first grade} \end{pmatrix} + \$0.62 \begin{pmatrix} \text{Pounds} \\ \text{of} \\ \text{second grade} \end{pmatrix} = \$0.75 \begin{pmatrix} 1000 \text{ lbs} \\ \text{of the} \\ \text{mixture} \end{pmatrix}$$

Translate to algebra:

$$x + \quad y = 1000 \qquad (1)$$
$$0.82x + 0.62y = 0.75(1000) \qquad (2)$$

278

$$y = -x + 1000 \qquad \text{Solve equation (1) for y.}$$

$$82x + 62y = 75000 \qquad (2') \quad \begin{array}{l}\text{Clear the decimals in}\\ \text{equation (2).}\end{array}$$

$$\begin{array}{rl} 82x + 62(-x + 1000) &= 75000 \\ 82x - 62x + 62000 &= 75000 \\ 20x &= 13000 \\ x &= 650 \end{array} \qquad \begin{array}{l}\text{Substitute } -x + 1000 \text{ for}\\ \text{y in equation } (2'). \quad \text{Solve}\\ \text{for x.}\end{array}$$

$$\begin{array}{rl} y &= -(650) + 1000 \\ &= 350 \end{array} \qquad \begin{array}{l}\text{Substitute 650 for x}\\ \text{in } y = -x + 1000.\end{array}$$

To form 1000 pounds of a mixture, 650 pounds of the grade selling at $0.82 a pound and 350 pounds of the grade selling at $0.62 a pound must be used.

69. The current through two circuits on a television set is given by the equations $2I_1 + 3I_2 = 8$ and $3I_1 + I_2 = 5$. Using I_2 as the dependent variable and I_1 as the independent, graph both and determine the common solution. Would the common solution be different if you reversed the axes?

$$\begin{cases} 2I_1 + 3I_2 = 8 & (1) \\ 3I_1 + I_2 = 5 & (2) \end{cases}$$

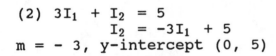

(1) $2I_1 + 3I_2 = 8$ Write each equation in
$$I_2 = -\frac{2}{3}I_1 + \frac{8}{3} \qquad \text{in slope-intercept form.}$$

$m = -\frac{2}{3}$, y-intercept $\left(0, \frac{8}{3}\right)$

(2) $3I_1 + I_2 = 5$
$$I_2 = -3I_1 + 5$$
$m = -3$, y-intercept $(0, 5)$

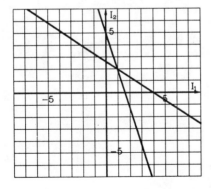

If you reversed the axes, the ordered pair representing the solution would also be reversed. The solution would be the same, however. $I_1 = 1$, $I_2 = 2$.

73.   How do you know that a system is independent and consistent
       when you solve the system using substitution?

       The system is independent and consistent if the variables
       are not eliminated using substitution.

CHALLENGE EXERCISES

77.   Solve $\begin{cases} 2x - y = 4 \\ y + 3 = 0 \end{cases}$ both graphically and by substitution.

       Graphically:
       2x - y = 4   (1)
        y + 3 = 0   (2)

       2x - y = 4   Write each equation in
              y = -2x + 4   slope-intercept form.
       m = -2, y-intercept (0, 4)

       y + 3 = 0
           y = -3
       m = 0, y-intercept (0, -3)

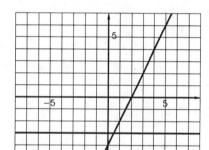

       The solution is $\left\{ \left(\frac{1}{2}, -3 \right) \right\}$.

Substitution:

$$2x - y = 4 \qquad (1)$$
$$y + 3 = 0 \qquad (2)$$
$$y = -3$$

Solve equation (2) for y.

$$2x - (-3) = 4$$
$$2x + 3 = 4$$
$$2x = 1$$
$$x = \frac{1}{2}$$
$$y = -3$$

Substitute -3 for y in equation (1).

Since there is no variable x in equation (2), no further substitution is needed.

The solution set is $\left\{ \left(\frac{1}{2}, -3 \right) \right\}$.

MAINTAIN YOUR SKILLS

Solve:

81.
$$x^4 - 26x^2 + 25 = 0$$
$$(x^2 - 25)(x^2 - 1) = 0$$
$$(x + 5)(x - 5)(x + 1)(x - 1) = 0 \quad \text{Factor completely.}$$
$$x + 5 = 0 \text{ or } x - 5 = 0 \text{ or } x + 1 = 0 \text{ or } x - 1 = 0$$
$$x = -5 \text{ or } \qquad x = 5 \text{ or } \qquad x = -1 \text{ or } \qquad x = 1$$

The solution set is {±1, ±5}.

Draw the graph of:

85. $2x + 3y = 18$

x	y
0	6
9	0
3	4

Make a table of values and find the intercepts.

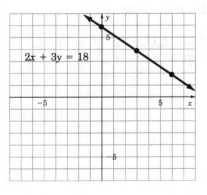

Plot each of the points. Draw the line that passes through the points.

A

Solve using linear combinations:

1. $\begin{cases} x + y = 5 & (1) \\ x - y = 3 & (2) \end{cases}$

$$\begin{array}{ll} x + y = 5 & (1) \\ \underline{x - y = 3} & (2) \\ 2x \quad\quad = 8 \\ \quad\quad x = 4 \end{array}$$ Add equations (1) and (2).

    Solve for x.

$$\begin{array}{l} 4 + y = 5 \\ \quad\quad y = 1 \end{array}$$ Substitute x = 4 in
equation (1).

The solution set is {(4, 1)}.

5. $\begin{cases} 4x - 3y = 8 & (1) \\ -4x + 3y = 7 & (2) \end{cases}$

$$\begin{array}{l} 4x - 3y = 8 \\ \underline{-4x + 3y = 7} \\ \quad\quad\quad 0 = 15 \end{array}$$ Add.

    Both variables are
eliminated but the result
is a contradiction. The
system is independent and
inconsistent.

There are no solutions.

9. $\begin{cases} x + 3y = 6 & (1) \\ x - y = 2 & (2) \end{cases}$

$$\begin{array}{ll} -1(x + 3y) = -1 \cdot 6 \\ \quad -x - 3y \;\; = -6 & (3) \end{array}$$ Multiply equation (1) by
-1 so that the
coefficients of the x
terms are opposites.

$$\begin{array}{ll} -x - 3y = -6 & (3) \\ \underline{x - y = 2} & (2) \\ \quad\quad -4y = -4 \\ \quad\quad\quad\; y = 1 \end{array}$$ Solve for y.

$$\begin{array}{l} x + 3 \cdot 1 = 6 \\ \quad\quad\quad x = 3 \end{array}$$ Substitute y = 1 in
equation (1).

The solution set is {(3, 1)}.

13. $\begin{cases} x + 4y = 8 & (1) \\ -2x + 5y = 23 & (2) \end{cases}$

$\quad\quad 2x + 8y = 16 \quad\quad (3)$ Multiply equation (1) by 2 so that the coefficients of the x term are opposites.

$$\begin{aligned} 2x + 8y &= 16 \\ \underline{-2x + 5y} &= \underline{23} \\ 13y &= 39 \\ y &= 3 \end{aligned}$$

Add (3) and (2).

$$\begin{aligned} x + 4(3) &= 8 \\ x + 12 &= 8 \\ x &= -4 \end{aligned}$$

Substitute 3 for y in equation (1).

The solution set is {(-4, 3)}.

17. $\begin{cases} 4x - 5y = 15 & (1) \\ x + y = -3 & (2) \end{cases}$

$\quad\quad 5x + 5y = -15 \quad\quad (3)$ Multiply equation (2) by 5.

$$\begin{aligned} 4x - 5y &= 15 \\ \underline{5x + 5y} &= \underline{-15} \\ 9x &= 0 \\ x &= 0 \end{aligned}$$

Add equations (1) and (3).

$$\begin{aligned} 4(0) - 5y &= 15 \\ -5y &= 15 \\ y &= -3 \end{aligned}$$

Substitute 0 for x in equation (1).

The solution set is {(0, -3)}.

B

21. $\begin{cases} 3x + y = 17 & (1) \\ 4x + y = 22 & (2) \end{cases}$

$\quad\quad -4x - y = -22 \quad\quad (3)$ Multiply equation (2) by -1.

$$\begin{aligned} 3x + y &= 17 \\ \underline{-4x - y} &= \underline{-22} \\ -x &= -5 \\ x &= 5 \end{aligned}$$

Add equations (1) and (3).

$$\begin{aligned} 3(5) + y &= 17 \\ 15 + y &= 17 \\ y &= 2 \end{aligned}$$

Substitute 5 for x in equation (1).

The solution set is {(5, 2)}.

25. $\begin{cases} 3x + 2y = -4 & (1) \\ 4x + 3y = -5 & (2) \end{cases}$

$\begin{array}{ll} -12x - 8y = 16 & (3) \\ \underline{12x + 9y = -15} & (4) \\ \quad\quad\quad y = 1 \end{array}$ Multiply equation (1) by -4 and equation (2) by 3.

$\begin{array}{l} 3x + 2(1) = -4 \\ \quad 3x + 2 = -4 \\ \quad\quad 3x = -6 \\ \quad\quad\; x = -2 \end{array}$ Substitute 1 for y in equation (1).

The solution set is {(-2, 1)}.

29. $\begin{cases} 2x - 7y = -1 & (1) \\ 6x - 21y = -3 & (2) \end{cases}$

$\begin{array}{ll} -6x + 21y = 3 & (3) \\ \underline{6x - 21y = -3} & (2) \\ \quad\quad\quad 0 = 0 \end{array}$ Multiply equation (1) by -3. Add equations (3) and (2). Identity. The system is dependent.

The solution set is {(x, y)|2x - 7y = -1}.

33. $\begin{cases} 2x + 9y = -31 & (1) \\ 5x - 4y = 55 & (2) \end{cases}$

$\begin{array}{ll} -10x - 45y = 155 & (3) \\ \underline{10x - 8y = 110} & (4) \\ \quad\quad -53y = 265 \\ \quad\quad\quad\; y = -5 \end{array}$ Multiply equation (1) by -5, and equation (2) by 2.

$\begin{array}{l} 2x + 9(-5) = -31 \\ \quad 2x - 45 = -31 \\ \quad\quad\; 2x = 14 \\ \quad\quad\;\; x = 7 \end{array}$ Substitute -5 for y in equation (1).

The solution set is {(7, -5)}.

37. $\begin{cases} \dfrac{2}{3}x - y = \dfrac{13}{3} & (1) \\[2mm] \dfrac{1}{6}x + y = -\dfrac{8}{3} & (2) \end{cases}$

$\begin{array}{ll} 2x - 3y = 13 & (3) \\ x + 6y = -16 & (4) \end{array}$ Clear the fractions in both equations. Multiply (1) by 3 and (2) by 6.

$\begin{array}{l} 4x - 6y = 26 \\ \underline{x + 6y = -16} \\ 5x \quad\quad = 10 \\ \quad\; x = 2 \end{array}$ Now multiply equation (1) by 2, and add it to equation (4).

$$\frac{1}{6}(2) + y = -\frac{8}{3}$$

$$\frac{1}{3} + y = -\frac{8}{3}$$

$$y = -\frac{9}{3}$$

$$y = -3$$

The solution set is $\{(2, -3)\}$.

Substitute 2 for x in equation (2).

41. $\begin{cases} 20x + 25y = -7 & (1) \\ 10x - 15y = 13 & (2) \end{cases}$

$$\begin{array}{ll} -20x + 30y = -26 & (3) \\ \underline{20x + 25y = -7} & (1) \\ \quad\quad\quad 55y = -33 \\ \quad\quad\quad\; y = -\frac{3}{5} \end{array}$$

Multiply equation (2) by -2. Add to equation (1).

$$20x + 25\left(-\frac{3}{5}\right) = -7$$

$$20x - 15 = -7$$

$$20x = 8$$

$$x = \frac{2}{5}$$

Substitute $-\frac{3}{5}$ for y in equation (1).

The solution set is $\left\{\left(\frac{2}{5}, -\frac{3}{5}\right)\right\}$.

45. $\begin{cases} 3x + 3y = -2 & (1) \\ \;\; x - \;\; y = -1 & (2) \end{cases}$

$$\begin{array}{ll} -3x + 3y = 3 & (3) \\ \underline{3x + 3y = -2} & \\ \quad\quad 6y = 1 \\ \quad\quad\; y = \frac{1}{6} \end{array}$$

Multiply equation (2) by -3. Add to equation (1).

$$3x + 3\left(\frac{1}{6}\right) = -2$$

$$3x + \frac{1}{2} = -2$$

$$6x + 1 = -4$$

$$6x = -5$$

$$x = -\frac{5}{6}$$

Substitute $\frac{1}{6}$ for y in equation (1).

Clear the fraction; multiply by 2.

The solution set is $\left\{\left(-\frac{5}{6}, \frac{1}{6}\right)\right\}$.

49. $\begin{cases} 4x - 6y = 5 & (1) \\ 10x - 8y = 9 & (2) \end{cases}$

$\begin{array}{ll} -20x + 30y = -25 & (3) \\ \underline{20x - 16y = 18} & (4) \\ \quad\quad\,\, 14y = -7 \\ \quad\quad\quad\, y = -\dfrac{1}{2} \end{array}$

Multiply equation (1) by -5, and equation (2) by 2.

$4x - 6\left(-\dfrac{1}{2}\right) = 5$
$\quad\quad 4x + 3 = 5$
$\quad\quad\quad 4x = 2$
$\quad\quad\quad\,\, x = \dfrac{1}{2}$

Substitute $-\dfrac{1}{2}$ for y in equation (1).

The solution set is $\left\{\left(\dfrac{1}{2},\, -\dfrac{1}{2}\right)\right\}$.

53. $\begin{cases} \dfrac{9}{4}x - \dfrac{15}{4}y = 2 & (1) \\[2mm] \dfrac{15}{8}x - \dfrac{9}{8}y = 1 & (2) \end{cases}$

$\begin{array}{ll} 9x - 15y = 8 & (3) \\ 15x - 9y = 8 & (4) \end{array}$

Clear fractions in both equations. Multiply equation (1) by 4, and equation (2) by 8.

$\begin{array}{ll} -45x + 75y = -40 & (5) \\ \underline{45x - 27y = 24} \\ \quad\quad\,\, 48y = -16 \\ \quad\quad\quad\, y = -\dfrac{1}{3} \end{array}$

Multiply equation (3) by -5, and equation (4) by 3.

$\dfrac{9}{4}x - \dfrac{15}{4}\left(-\dfrac{1}{3}\right) = 2$
$\quad\quad \dfrac{9}{4}x + \dfrac{5}{4} = 2$
$\quad\quad 9x + 5 = 8$
$\quad\quad\quad 9x = 3$
$\quad\quad\quad\,\, x = \dfrac{1}{3}$

Substitute $-\dfrac{1}{3}$ for y in equation (1).

Multiply both sides by 4 to clear the fractions.

The solution set is $\left\{\left(\dfrac{1}{3},\, -\dfrac{1}{3}\right)\right\}$.

57. $\begin{cases} 0.8x + 1.5y = 0.16 & (1) \\ 1.4x - 0.5y = -2.22 & (2) \end{cases}$

$\begin{array}{lll} 80x + 150y = 16 & (3) & \text{Multiply both equations} \\ 140x - 50y = -222 & (4) & \text{by 100 to clear the} \\ & & \text{decimals.} \end{array}$

$\begin{array}{ll} 420x - 150y = -666 \qquad (5) & \text{Multiply equation (4) by} \\ \underline{80x + 150y = 16} & \text{3, and add it to equation} \\ 500x = -650 & \text{(3).} \\ x = -1.3 \end{array}$

$\begin{array}{ll} 0.8(-1.3) + 1.5y = 0.16 & \text{Substitute } -1.3 \text{ for } x \\ -1.04 + 1.5y = 0.16 & \text{in equation (1).} \\ 1.5y = 1.2 \\ y = 0.8 \end{array}$

The solution set is $\{(-1.3, 0.8)\}$.

D

61. How many cubic centimeters (cc) of a 50% saline solution must be mixed with how many cc of a 10% saline solution to yield 80 cc of a 25% saline solution?

Simpler word form:

$\left(\begin{array}{c} \text{cc of 10\%} \\ \text{saline solution} \end{array}\right) + \left(\begin{array}{c} \text{cc of 50\%} \\ \text{saline solution} \end{array}\right) = \begin{array}{c} \text{80 cc of 25\%} \\ \text{saline solution} \end{array}$

$\left(\begin{array}{c} \text{Salt in} \\ \text{10\% mixture} \end{array}\right) + \left(\begin{array}{c} \text{Salt in} \\ \text{50\% mixture} \end{array}\right) = \begin{array}{c} \text{Salt in} \\ \text{25\% mixture} \end{array}$

Select variable:
Let x represent the number of cubic centimeters of the 10% saline solution and y represent the number of cubic centimeters of the 50% saline solution.

Translate to algebra:
$\begin{cases} x + y = 80 & (1) \\ 0.10x + 0.50y = 0.25(80) & (2) \end{cases}$

$\begin{array}{ll} x + y = 80 & \text{Multiply equation (2) by} \\ \underline{-x - 5y = -200} & -10 \text{ to clear the decimals} \\ -4y = -120 & \text{and make the coefficients} \\ y = 30 & \text{of } x \text{ opposite.} \end{array}$

$\begin{array}{ll} x + 30 = 80 & \text{Substitute 30 for } y \text{ in} \\ x = 50 & \text{equation (1).} \end{array}$

To get the 25% solution, 30 cc of the 50% solution and 50cc of the 10% solution should be mixed together.

287

65. The WOW I Feel Good health food store is to fill a diet for a client that calls for two kinds of food. The first contains 40% of nutrient A and 16% of nutrient B. The second contains 35% of nutrient A and 20% of nutrient B. If the diet calls for exactly 4 ounces of nutrient A and 2 ounces of nutrient B, how many ounces of each food should be used?

Simpler word form:

$$\left(\begin{array}{c}\text{Amount of nutrient A} \\ \text{in first food}\end{array}\right) + \left(\begin{array}{c}\text{Amount of nutrient A} \\ \text{in second food}\end{array}\right) = \begin{array}{c}\text{4 oz of} \\ \text{nutrient A} \\ \text{in mixture}\end{array}$$

$$\left(\begin{array}{c}\text{Amount of nutrient B} \\ \text{in first food}\end{array}\right) + \left(\begin{array}{c}\text{Amount of nutrient B} \\ \text{in second food}\end{array}\right) = \begin{array}{c}\text{2 oz of} \\ \text{nutrient B} \\ \text{in mixture}\end{array}$$

Select variables:
Let x represent the number of ounces needed of first food and y represent the number of ounces needed of second food.

Translate to algebra:
$0.40x + 0.35y = 4$ (1)
$0.16x + 0.20y = 2$ (2)

$40x + 35y = 400$ (3)   Clear the decimals.
$16x + 20y = 200$ (4)

$$\begin{array}{r} 160x + 140y = 1600 \\ -112x - 140y = -1400 \\ \hline 48x \quad\quad\quad = 200 \\ x = \dfrac{25}{6} \text{ or } 4\dfrac{1}{6} \end{array}$$

Multiply equation (3) by 4, and equation (4) by -7 to make the coefficients of y opposites.

$$0.40\left(\frac{25}{6}\right) + 0.35y = 4$$
$$40\left(\frac{25}{6}\right) + 35y = 400$$
$$1000 + 210y = 2400$$
$$210y = 1400$$
$$y = \frac{20}{3} \text{ or } 6\frac{2}{3}$$

Substitute $\frac{25}{6}$ for x in equation (1) and solve for y.
Multiply the equation by 6 to clear the fraction.

To get 4 ounces of nutrient A and 2 ounces of nutrient B, $4\frac{1}{6}$ ounces of the first food (40% of nutrient A and 16% of nutrient B) and $6\frac{2}{3}$ ounces of the second food (35% of nutrient A and 20% of nutrient B) should be mixed together.

69. In construction, one concrete mixture contains four times as much gravel as cement. If the total volume needed is 48 cubic yards, how much of each ingredient is needed?

Simpler word form:

$$\begin{pmatrix} \text{Volume} \\ \text{of} \\ \text{gravel} \end{pmatrix} + \begin{pmatrix} \text{Volume} \\ \text{of} \\ \text{cement} \end{pmatrix} = \begin{matrix} \text{Total volume} \\ \text{of mixture} \end{matrix}$$

$$4\begin{pmatrix} \text{Volume} \\ \text{of} \\ \text{gravel} \end{pmatrix} = \begin{pmatrix} \text{Volume} \\ \text{of} \\ \text{cement} \end{pmatrix}$$

Select variables:
Let x represent the volume of gravel needed, and y represent the volume of cement that is needed.

Translate to algebra:
$$\begin{cases} x + y = 48 & (1) \\ \quad 4y = x & (2) \end{cases}$$

$$\begin{array}{ll} x + y = 48 & (1) \\ \underline{-x + 4y = 0} & (2) \\ \quad 5y = 48 \\ \qquad y = \dfrac{48}{5} \text{ or } 9\frac{3}{5} \end{array}$$

Rewrite equation (2) in a form equivalent to equation (1). Add.

$$x + 9\frac{3}{5} = 48$$
$$x = 48 - 9\frac{3}{5}$$
$$x = 38\frac{2}{5}$$

Substitute $9\frac{3}{5}$ for y in equation (1).

For a total volume of 48 cubic yards, $38\frac{2}{5}$ cubic yards of gravel and $9\frac{3}{5}$ cubic yards of cement must be used.

73. Given that the system $\begin{cases} 4x - 2y = 5 \\ 6x - 3y = 7 \end{cases}$ is independent and consistent, what is the "ratio" of the coefficients of x in the two equations? What is the ratio of the coefficients of y in the two equations? Conclusion? What is the ratio of the constants?

The ratio of the coefficients of x is $\frac{4}{6}$ or $\frac{2}{3}$.

The ratio of the coefficients of y is $\frac{2}{3}$.

If the ratios of the coefficients of both variables are equal, the system may be independent and inconsistent, or dependent and consistent.

The ratio of the constants is $\frac{5}{7} \neq \frac{2}{3}$.

So if the ratios of the coefficients of both variables are equal to each other but not equal to the ratio of the constants, the system is independent and inconsistent.

CHALLENGE EXERCISES

77. Replacing $\frac{1}{u}$ by x and $\frac{1}{v}$ by y, solve the following system for u and v.

$$\begin{cases} \dfrac{1}{u} + \dfrac{1}{v} = 4 \quad\quad (1) \\[3mm] \dfrac{1}{u} - \dfrac{1}{v} = 2 \quad\quad (2) \end{cases}$$

$x + y = 4$ (3)		Replace $\frac{1}{u}$ with x, and
$\underline{x - y = 2}$ (4)		$\frac{1}{v}$ with y.
$2x \quad\quad = 6$		Add.
$\quad\quad x = 3$		
$3 + y = 4$		Substitute 3 for x in
$\quad\quad y = 1$		equation (3).

$$3 + \frac{1}{v} = 4$$

$$\frac{1}{v} = 1$$

$$v = 1$$

$$\frac{1}{u} + \frac{1}{1} = 4$$

$$\frac{1}{u} + 1 = 4$$

$$\frac{1}{u} = 3$$

$$1 = 3u$$

$$\frac{1}{3} = u$$

Now replace $\frac{1}{u}$ with the value of x, 3, in equation (1).

Replace v with 1 in equation (1).

Multiply both sides by u to clear the fraction.

The solution set is $\left\{ \left(\frac{1}{3}, 1 \right) \right\}$.

MAINTAIN YOUR SKILLS

81. Write the coordinates of the x- and y-intercepts of the graph $x = \frac{3}{5}y + \frac{1}{5}$.

x-intercept, y = 0

$$x = \frac{3}{5}(0) + \frac{1}{5}$$

$$x = \frac{1}{5}$$

x-intercept: $\left(\frac{1}{5}, 0 \right)$

y-intercept, x = 0

$$0 = \frac{3}{5}y + \frac{1}{5}$$

$$0 = 3y + 1$$

$$-1 = 3y$$

$$-\frac{1}{3} = y$$

y-intercept: $\left(0, -\frac{1}{3} \right)$

85. Is a line with slope $\frac{11}{2}$ perpendicular to the line containing the points (7, 8) and (5, -3)?

$m = \dfrac{y_2 - y_1}{x_2 - x_1}$

$\quad = \dfrac{-3 - 8}{5 - 7}$

$\quad = \dfrac{-11}{-2}$

$\quad = \dfrac{11}{2}$

$\dfrac{11}{2} \cdot \dfrac{11}{2} \neq -1$

First find the slope of the line containing the two points.
Let (7, 8) be (x_1, y_1) and (5, -3) be (x_2, y_2).

Two lines are perpendicular if the product of their slopes is -1.

Therefore, these lines are not perpendicular.

EXERCISES 7.3   SOLVING SYSTEMS IN THREE VARIABLES BY LINEAR COMBINATIONS

A

Solve using linear combinations:

1. $\begin{cases} x + y + z = 4 & (1) \\ x - y + 2z = 8 & (2) \\ x + y - z = 3 & (3) \end{cases}$

$\begin{array}{ll} x + y + z = 4 & (1) \\ \underline{x - y + 2z = 8} & (2) \\ 2x \quad\quad + 3z = 12 & (4) \end{array}$

The y term is eliminated if equations (1) and (2) are added.

$\begin{array}{ll} x - y + 2z = 8 & (2) \\ \underline{2x + y - z = 3} & (3) \\ 3x \quad\quad + z = 11 & (5) \end{array}$

Again eliminate the y term by adding equations (2) and (3).

$\begin{array}{l} -3(3x + z) = -3 \cdot 11 \\ -9x - 3z = -33 \\ \underline{2x + 3z = 12} \\ -7x \quad\quad = -21 \\ \quad\quad x = 3 \end{array}$

Multiply equation (5) by -3 and add to equation (4) to eliminate z.

$\begin{array}{l} 2 \cdot 3 + 3x = 12 \\ \quad\quad 3z = 6 \\ \quad\quad z = 2 \end{array}$

Substitute x = 3 in equation (4) to find z.

$\begin{array}{l} 3 + y + 2 = 4 \\ \quad\quad\quad y = -1 \end{array}$

Substitute x = 3 and z = 2 in equation (1) to find y.

The solution set is {(3, -1, 2)}.

5.
$$\begin{cases} x - y + z = 2 & (1) \\ 2x + y - 2z = -9 & (2) \\ 3x + y + 3z = -6 & (3) \end{cases}$$

$$\begin{array}{l} x - y + z = 2 \\ \underline{2x + y - 2z = -9} \\ 3x - z = -7 \quad (4) \end{array}$$
Add equations (1) and (2) to eliminate y.

$$\begin{array}{l} x - y + z = 2 \\ \underline{3x + y + 3z = -6} \\ 4x + 4z = -4 \quad (5) \end{array}$$
Add equations (1) and (3) which will also eliminate y.

$$\begin{array}{l} 3x - z = -7 \quad (4) \\ 4x - 4z = -4 \quad (5) \end{array}$$
A system of two equations in two variables.

$$\begin{array}{l} 12x - 4z = -28 \\ \underline{4x + 4z = -4} \\ 16x = -32 \\ x = -2 \end{array}$$
Multiply equation (4) by 4, and add it to equation (5).

$$\begin{array}{l} 3(-2) - z = -7 \\ -6 - z = -7 \\ -z = -1 \\ z = 1 \end{array}$$
Substitute -2 for x in equation (4) and solve for z.

$$\begin{array}{l} -2 - y + 1 = 2 \\ -y - 1 = 2 \\ -y = 3 \\ y = -3 \end{array}$$
Substitute -2 for x and 1 for z in equation (1) and solve for y.

The solution set is $\{(-2, -3, 1)\}$.

9.
$$\begin{cases} x + 3y - 2z = 5 & (1) \\ -x + y + 3z = 20 & (2) \\ x - y - z = -10 & (3) \end{cases}$$

$$\begin{array}{l} x + 3y - 2z = 5 \\ \underline{-x + y + 3z = 20} \\ 4y + z = 25 \quad (4) \end{array}$$
Add equations (1) and (2) to eliminate x.

$$\begin{array}{l} -x + y + 3z = 20 \\ \underline{x - y - z = -10} \\ 2z = 10 \\ z = 5 \quad (5) \end{array}$$
Add equations (2) and (3) to again eliminate x.

$$\begin{array}{l} 4y + 5 = 25 \\ 5y = 20 \\ y = 5 \end{array}$$
Since both x and y were eliminated, substitute 5 for z in equation (4) and solve for y.

$$x + 3(5) - 2(5) = 5$$
$$x + 15 - 10 = 5$$
$$x + 5 = 5$$
$$x = 0$$

Substitute 5 for y and 5 for z in equation (1) and solve for x.

The solution set is {(0, 5, 5)}.

B

13.
$$\begin{cases} 3x + 4y + 2z = -13 & (1) \\ 2x - y + z = -1 & (2) \\ x - 2y + 3z = 0 & (3) \end{cases}$$

$$\begin{array}{r} 8x - 4y + 4z = -4 \\ \underline{3x + 4y + 2z = -13} \\ 11x \qquad + 6z = -17 \end{array} \quad (4)$$

Multiply equation (2) by 4 and add to equation (1) to eliminate y.

$$\begin{array}{r} -4x + 2y - 2z = 2 \\ \underline{x - 2y + 3z = 0} \\ -3x \qquad + z = 2 \end{array} \quad (5)$$

Multiply equation (2) by -2 and add to equation (3) to eliminate y.

$$11x + 6z = -17 \quad (4)$$
$$-3x + z = 2 \quad (5)$$

A new system in two variables.

$$\begin{array}{r} 18x - 6z = -12 \\ \underline{11x + 6z = -17} \\ 29x \qquad = -29 \\ x = -1 \end{array}$$

Multiply equation (5) by -6 and eliminate z by adding the result to equation (4).

$$11(-1) + 6z = -17$$
$$6z = -6$$
$$z = -1$$

Replace -1 for x in equation (4) and solve for z.

$$3(-1) + 4y + 2(-1) = -13$$
$$4y - 5 = -13$$
$$4y = -8$$
$$y = -2$$

Replace -1 for z and -1 for x in equation (1) and solve for y.

The solution set is {(-1, -2, -1)}.

17.
$$\begin{cases} 2x - 3y + 5z = -3 & (1) \\ 2x + 5y + 6z = 28 & (2) \\ 3x - 2y + z = 12 & (3) \end{cases}$$

$$\begin{array}{r} -18x + 12y - 6z = -72 \\ \underline{2x + 5y + 6z = 28} \\ -16x + 17y \qquad = -44 \end{array} \quad (4)$$

Multiply equation (3) by -6 and add to equation (2) to eliminate z.

$$\begin{array}{r} -15x + 10y - 5z = -60 \\ \underline{2x - 3y + 5z = -3} \\ -13x + 7y \qquad = -63 \end{array} \quad (5)$$

Multiply equation (3) by -5 and add to equation (1) to eliminate z.

$$-16x + 17y = -44 \quad (4)$$
$$-13x + 7y = -63 \quad (5)$$

A new system in two variables.

$$112x - 119y = 308$$
$$\underline{-221x + 119y = -1071}$$
$$-109x \qquad = -763$$
$$x = 7$$

Multiply equation (4) by -7 and equation (5) by 17. Eliminate y.

$$-16(7) + 17y = -44$$
$$-112 + 17y = -44$$
$$17y = 68$$
$$y = 4$$

Replace 7 for x in equation (4) and solve for y.

$$2(7) - 3(4) + 5z = -3$$
$$14 - 12 + 5z = -3$$
$$5z = -5$$
$$z = -1$$

Replace 7 for x and 4 for y in equation (1).

The solution set is $\{(7, 4, -1)\}$.

21. $\begin{cases} 9x + 3y + 2z = 3 & (1) \\ 4x + 2y + 3z = 9 & (2) \\ 3x + 5y + 4z = 19 & (3) \end{cases}$

$$-27x - 9y - 6z = -9$$
$$\underline{8x + 4y + 6z = 18}$$
$$-19x - 5y \qquad = 9 \quad (4)$$

Multiply equation (1) by -3. Multiply equation (3) by 2. Add.

$$-16x - 8y - 12z = -36$$
$$\underline{9x + 15y + 12z = 57}$$
$$-7x + 7y \qquad = 21 \quad (5)$$
$$-x + y \qquad = 3 \quad (6)$$

Multiply equation (2) by -4. Multiply equation (3) by 3. Add. Simplify equation (5) by dividing by 7.

$$-5x + 5y = 15$$
$$\underline{-19x - 5y = 9}$$
$$-24x \qquad = 24$$
$$x = -1$$

Now multiply equation (6) by 5 and add to equation (4).

$$19 - 5y = 9$$
$$-5y = -10$$
$$y = 2$$

Substitute x = -1 in equation (4) to find y.

$$-9 + 6 + 2z = 3$$
$$2z = 6$$
$$z = 3$$

Substitute x = -1 and y = 2 in equation (1) to find z.

The solution set is $\{(-1, 2, 3)\}$.

C

25. $\begin{cases} 3x - 7y + 5z = 52 & (1) \\ -2x + 6y - 3z = -41 & (2) \\ 4x - 3y + 2z = 8 & (3) \end{cases}$

$$\begin{array}{r} -4x + 12y - 6z = -82 \\ \underline{4x - 3y + 2z = 8} \\ 9y - 4z = -74 \quad (4) \end{array}$$

Multiply equation (2) by 2 and add to equation (3) to eliminate x.

$$\begin{array}{r} -6x + 18y - 9z = -123 \\ \underline{6x - 14y + 10z = 104} \\ 4y + z = -19 \quad (5) \end{array}$$

Multiply equation (2) by 3 and equation (1) by 2 to eliminate x.

$$\begin{array}{r} 9y - 4z = -74 \quad (4) \\ 4y + z = -19 \quad (5) \end{array}$$

$$\begin{array}{r} 16y + 4z = -76 \\ \underline{9y - 4z = -74} \\ 25y = -150 \\ y = -6 \end{array}$$

Multiply equation (5) by 4 and add to equation (4) to eliminate z.

$$\begin{array}{r} 9(-6) - 4z = -74 \\ -54 - 4z = -74 \\ -4z = -20 \\ z = 5 \end{array}$$

Replace y with -6 in equation (4) and solve for z.

$$\begin{array}{r} 3x - 7(-6) + 5(5) = 52 \\ 3x + 42 + 25 = 52 \\ 3x + 67 = 52 \\ 3x = -15 \\ x = -5 \end{array}$$

Now replace y with -6 and z with 5 in equation (1) to solve for x.

The solution set is $\{(-5, -6, 5)\}$.

29. $\begin{cases} 10x + 15y + 10z = 1 & (1) \\ 15x - 20y + 15z = 27 & (2) \\ 2x + 5y - 3z = -5 & (3) \end{cases}$

$$\begin{array}{r} 8x + 20y - 12z = -20 \\ \underline{15x - 20y + 15z = 27} \\ 23x + 3z = 7 \quad (4) \end{array}$$

Multiply equation (3) by 4 and add to equation (2) to eliminate y.

$$\begin{array}{r} -6x - 15y + 9z = 15 \\ \underline{10x + 15y + 10z = 1} \\ 4x + 19z = 16 \quad (5) \end{array}$$

Multiply equation (3) by -3 and add to equation (1) to eliminate y.

$$\begin{array}{r} -437x - 57z = -133 \\ \underline{12x + 57z = 48} \\ -425x = -85 \\ x = \frac{1}{5} \end{array}$$

Multiply equation (4) by -19 and multiply equation (5) by 3. Add to eliminate z.

$$4\left(\frac{1}{5}\right) + 19z = 16$$

Replace x with $\frac{1}{5}$ in equation (5) to solve for z.

$$\frac{4}{5} + 19z = 16$$
$$4 + 95z = 80$$
$$95z = 76$$
$$z = \frac{4}{5}$$

Clear the fraction by multiplying by 5.

$$10\left(\frac{1}{5}\right) + 15y + 10\left(\frac{4}{5}\right) = 1$$
$$2 + 15y + 8 = 1$$
$$15y = -9$$
$$y = -\frac{3}{5}$$

Replace x with $\frac{1}{5}$, and z with $\frac{4}{5}$ in equation (1).

The solution set is $\left\{\left(\frac{1}{5},\ -\frac{3}{5},\ \frac{4}{5}\right)\right\}$.

D

33. The Civic Theater in Tucson, Arizona, sold 900 tickets to its production of "The Rainmaker." The tickets were priced at $8, $6.50, and $5. The total income from the sale of tickets was $5400. If twice as many $6.50 tickets were sold than $8 tickets, how many of each price tickets were sold?

Simpler word form:

$$\left(\begin{array}{c}\text{Income from}\\ \text{\$8 tickets}\end{array}\right) + \left(\begin{array}{c}\text{Income from}\\ \text{\$6.50 tickets}\end{array}\right) + \left(\begin{array}{c}\text{Income from}\\ \text{\$5 tickets}\end{array}\right) = \$5400$$

$$\left(\begin{array}{c}\text{Number of}\\ \text{\$8 tickets}\\ \text{sold}\end{array}\right) + \left(\begin{array}{c}\text{Number of}\\ \text{\$6.50 tickets}\\ \text{sold}\end{array}\right) + \left(\begin{array}{c}\text{Number of}\\ \text{\$5 tickets}\\ \text{sold}\end{array}\right) = \begin{array}{c}900\\ \text{tickets}\end{array}$$

$$2\left(\begin{array}{c}\text{Number of}\\ \text{\$8 tickets}\\ \text{sold}\end{array}\right) = \begin{array}{c}\text{Number of}\\ \text{\$6.50 tickets sold}\end{array}$$

Select variables:
Let x represent the number of $8 tickets sold, y represent the number of $6.50 tickets sold, and z represent the number of $5 tickets sold.

Translate to algebra:

$$8x + 6.5y + 5z = 5400 \qquad (1)$$
$$x + y + z = 900 \qquad (2)$$
$$2x - y = 0 \text{ or } 2x = y \qquad (3)$$

$\begin{aligned}-5x - 5y - 5z &= -4500\\ \underline{8x + 6.5y + 5z} &= \underline{5400}\\ 3x + 1.5y &= 900 \quad (4)\end{aligned}$	Multiply equation (2) by -5 and add to equation (1).
$\begin{aligned}-6x + 3y &= 0\\ \underline{6x + 3y} &= \underline{1800}\\ 6y &= 1800\\ y &= 300\end{aligned}$	Multiply equation (3) by -3 and equation (4) by 2.
$\begin{aligned}3x + 1.5(300) &= 900\\ 3x + 450 &= 900\\ 3x &= 450\\ x &= 150\end{aligned}$	Replace y with 300 in equation (4).
$\begin{aligned}150 + 300 + z &= 900\\ z &= 450\end{aligned}$	Replace x with 150 and y with 300 in equation (2) and solve for z.

There were 150 \$8 tickets sold, 300 \$6.50 tickets sold, and 450 \$5 tickets sold.

37. The currents at various points in the circuit below are given by the following equations:

$$I_1 + I_3 = I_2$$
$$4I_1 + 3I_2 = 20$$
$$3I_2 + 2I_3 = 16$$

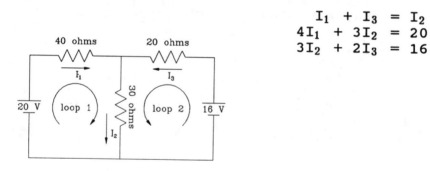

Solve for the values of the currents I_1, I_2, and I_3 (the units of current is amps).

$\begin{aligned}I_1 - I_2 + I_3 &= 0 \quad (1)\\ 4I_1 + 3I_2 &= 20 \quad (2)\\ 3I_2 + 2I_3 &= 16 \quad (3)\end{aligned}$	Rewrite the first equation given with the variables on the left.
$\begin{aligned}-3I_2 - 2I_3 &= -16\\ \underline{4I_1 + 3I_2} &\underline{= 20}\\ 4I_1 - 2I_3 &= 4 \quad (4)\end{aligned}$	Multiply equation (3) by -1 and add to equation (2) to eliminate I_2.

$$3I_1 - 3I_2 + 3I_3 = 0$$
$$\underline{4I_1 + 3I_2 \qquad\quad = 20}$$
$$7I_1 \qquad\quad + 3I_3 = 20 \qquad (5)$$

Multiply equation (1) by 3 and add to equation (2) to eliminate I_2 again.

$$12I_1 - 6I_3 = 12$$
$$\underline{14I_1 + 6I_3 = 40}$$
$$26I_1 = 52$$
$$I_1 = 2$$

Multiply equation (4) by 3 and equation (5) by 2 to eliminate I_3.

$$4(2) - 2I_3 = 4$$
$$8 - 2I_3 = 4$$
$$-2I_3 = -4$$
$$I_3 = 2$$

Replace I_1 with 2 in equation (4) and solve for I_3.

$$3I_2 + 2(2) = 16$$
$$3I_2 + 4 = 16$$
$$3I_2 = 12$$
$$I_2 = 4$$

Replace I_3 with 2 in equation (3) and solve for I_2.

The values of the currents are: $I_1 = 2$ amps, $I_2 = 4$ amps, and $I_3 = 2$ amps.

STATE YOUR UNDERSTANDING

41. Could three planes have as their common solution (intersection) an entire plane? Explain.

When the equations representing the three planes form a dependent system, they may intersect in an entire plane having all points in common. (See illustration (d) of Figure 7.4 in the text.)

CHALLENGE EXERCISES

45. Even though the following system is not a three-by-three system, solve the system using linear combiantions and notice the denominators for both the x- and y-values!

$$\begin{cases} ax + by = c & (1) \\ dx + ey = f & (2) \end{cases}$$

$$aex + bey = ce$$
$$\underline{-bdx - bey = -bf}$$
$$(ae-bd)x = ce - bf$$
$$x = \frac{ce - bf}{ae - bd}$$

Multiply equation (1) by e and equation (2) by -b. Add to eliminate y.

$$-adx - bdy = -cd$$
$$\underline{adx + aey = af}$$
$$(ae-bd)y = af - cd$$
$$y = \frac{af - cd}{ae - bd}$$

Multiply equation (1) by -d and equation (2) by a. Add to eliminate x.

The solution set is $\left\{\left(\dfrac{ce - bf}{ae - bd}, \dfrac{af - cd}{ae - bd}\right)\right\}$. The fractions forming the ordered pair have a common denominator, ae − bd.

MAINTAIN YOUR SKILLS

Write in standard form:

49.   3x − 18 = y + 7
      3x − y − 18 = 7   Subtract y from both sides.
          3x − y = 25   Add 18 to both sides.

Draw the graph of the line described:

53. The line with slope − $\dfrac{3}{5}$ and y-intercept (0, 6).

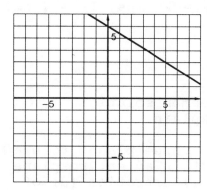

First, plot the given point (0, 6) on the y-axis. Then go down 3, right 5, and plot the point (5, 3).

EXERCISES 7.4   EVALUATING DETERMINANTS

A

Evaluate the following determinants:

1. $\begin{vmatrix} 1 & -1 \\ -1 & 1 \end{vmatrix}$

    a = 1, b = −1, c = −1, d = 1   Identify a, b, c, and d.

    $\begin{vmatrix} 1 & -1 \\ -1 & 1 \end{vmatrix} = 1(1) - (-1)(-1)$ Substitute in the formula ad − bc, and simplify.

       = 1 − 1
       = 0

300

5. $\begin{vmatrix} 1 & -1 \\ 2 & 3 \end{vmatrix}$

   $a = 1, \; b = -1, \; c = 2, \; d = 3$ Identify a, b, c, d.

   $\begin{vmatrix} 1 & -1 \\ 2 & 3 \end{vmatrix} = 1(3) - (-1)(2)$ Substitute in the formula ad - bc.

   $\phantom{\begin{vmatrix} 1 & -1 \\ 2 & 3 \end{vmatrix}} = 3 - (-2)$ Simplify.

   $\phantom{\begin{vmatrix} 1 & -1 \\ 2 & 3 \end{vmatrix}} = 5$

9. $\begin{vmatrix} 1 & 0 \\ 3 & -5 \end{vmatrix}$

   $a = 1, \; b = 0, \; c = 3, \; d = -5$ Identify a, b, c, d.

   $\begin{vmatrix} 1 & 0 \\ 3 & -5 \end{vmatrix} = (1)(-5) - (0)(3)$ Substitute in the formula, ad - bc.

   $\phantom{\begin{vmatrix} 1 & 0 \\ 3 & -5 \end{vmatrix}} = -5 - 0$

   $\phantom{\begin{vmatrix} 1 & 0 \\ 3 & -5 \end{vmatrix}} = -5$

13. $\begin{vmatrix} 1.1 & -3.5 \\ 2.6 & -0.4 \end{vmatrix}$

   $a = 1.1, \; b = -3.5, \; c = 2.6, \; d = -0.4$

   $\begin{vmatrix} 1.1 & -3.5 \\ 2.6 & -0.4 \end{vmatrix} = (1.1)(-0.4) - (-3.5)(2.6)$ ad - bc

   $\phantom{\begin{vmatrix} 1.1 & -3.5 \\ 2.6 & -0.4 \end{vmatrix}} = -0.44 - (-9.1)$

   $\phantom{\begin{vmatrix} 1.1 & -3.5 \\ 2.6 & -0.4 \end{vmatrix}} = 8.66$

Solve for a:

17. $\begin{vmatrix} a & 1 \\ 1 & 1 \end{vmatrix} = 2$

   $\begin{vmatrix} a & 1 \\ 1 & 1 \end{vmatrix} = a \cdot 1 - 1 \cdot 1 = a - 1$ Expand the 2 × 2 determinant and set it equal to 2.

   $a - 1 = 2$

   $a = 3$

   The solution set is {3}.

B

Evaluate the following determinants:

21. $\begin{vmatrix} 4 & -7 \\ 3 & -6 \end{vmatrix}$

a = 4, b = -7, c = 3, d = -6

$\begin{vmatrix} 4 & -7 \\ 3 & -6 \end{vmatrix}$ = 4(-6) - (-7)(3)

$= -24 - (-21)$

$= -3$

25. $\begin{vmatrix} -13 & 11 \\ 5 & -4 \end{vmatrix}$

a = -13, b = 11, c = 5, d = -4

$\begin{vmatrix} -13 & 11 \\ 5 & -4 \end{vmatrix}$ = (-13)(-4) - (11)(5)

$= 52 - 55$

$= -3$

29. $\begin{vmatrix} 3a & 2b \\ -4a & 5b \end{vmatrix}$ = (3a)(5b) - (2b)(-4a)

$= 15ab - (-8ab)$

$= 23ab$

33. $\begin{vmatrix} -3 & -1 & 2 \\ 2 & 2 & -1 \\ -1 & 1 & 1 \end{vmatrix}$ Expand about the first column. The entries are -3, 2, and -1. The signs associated with the entries are +1, -1, and +1, respectively.

$= (-3)(+1)\begin{vmatrix} 2 & -1 \\ 1 & 1 \end{vmatrix} + 2(-1)\begin{vmatrix} -1 & 2 \\ 1 & 1 \end{vmatrix} + (-1)(+1)\begin{vmatrix} -1 & 2 \\ 2 & -1 \end{vmatrix}$

$= -3[2(1)-(-1)(1)] - 2[(-1)(1)-(2)(1) - 1[(-1)(-1-)(2)(2)]$

$= -3(2 + 1) - 2(-1 - 2) - 1(1 - 4)$

$= -3(3) - 2(-3) - 1(-3)$

$= -9 + 6 + 3$

$= 0$

Solve for a:

37. $\begin{vmatrix} -9 & a \\ -6 & 7 \end{vmatrix} = 66$

$\begin{vmatrix} -9 & a \\ -6 & 7 \end{vmatrix} = (-9)(7) - (a)(-6)$ Expand the 2 × 2 determinant and set it equal to 66.

$= -63 + 6a$

$6a - 63 = 66$
$\qquad 6a = 129$
$\qquad\ \ a = \dfrac{129}{6}$
$\qquad\ \ a = \dfrac{43}{2}$

The solution set is $\left\{\dfrac{43}{2}\right\}$.

C

Evaluate the following determinants:

41. $\begin{vmatrix} -3 & -3 & 1 \\ 4 & 1 & 4 \\ 2 & 2 & -3 \end{vmatrix}$ Expand about the first row. The entries are -3, -3, and 1. The signs associated with the entries are +1, -1, and +1.

$= (-3(+1) \begin{vmatrix} 1 & 4 \\ 2 & -3 \end{vmatrix} + (-3)(-1) \begin{vmatrix} 4 & 4 \\ 2 & -3 \end{vmatrix} + (1)(+1) \begin{vmatrix} 4 & 1 \\ 2 & 2 \end{vmatrix}$

$= -3[(1)(-3)-(4)(2)] + 3[(4)(-3)-(4)(2)] + 1[(4)(2)-(1)(2)]$
$= -3(-3 - 8) + 3(-12 - 8) + (8 - 2)$
$= -3(-11) + 3(-20) + 6$
$= 33 - 60 + 6$
$= -21$

45. $\begin{vmatrix} 0 & 2 & -5 \\ 1 & 3 & 6 \\ 0 & 4 & 1 \end{vmatrix}$ Expand about the first column.

$= 0(+1) \begin{vmatrix} 3 & 6 \\ 4 & 1 \end{vmatrix} + 1(-1) \begin{vmatrix} 2 & -5 \\ 4 & 1 \end{vmatrix} + 0(+1) \begin{vmatrix} 2 & -5 \\ 3 & 6 \end{vmatrix}$
$= 0 - 1[(2)(1) - (-5)(4)] + 0$
$= -(2 + 20)$
$= -22$

49. $\begin{vmatrix} 12 & 5 & 0 \\ -6 & -4 & 0 \\ 2 & 3 & -4 \end{vmatrix}$ Expand about the third column.

$= 0 + 0 + (-4)(+1)\begin{vmatrix} 12 & 5 \\ -6 & -4 \end{vmatrix}$

$= 0 - 4[(-48) - (-30)]$

$= -4(-18)$

$= 72$

53. $\begin{vmatrix} 0 & 1 & 3 & 0 & 1 \\ 0 & 2 & 1 & 3 & 0 \\ 1 & 0 & 0 & -2 & 1 \\ 3 & 6 & -1 & 0 & 0 \\ 0 & 2 & 1 & 1 & 1 \end{vmatrix}$ Expand about the fifth column to take advantage of the zeros.

$= 1(+1)\begin{vmatrix} 0 & 2 & 1 & 3 \\ 1 & 0 & 0 & -2 \\ 3 & 6 & -1 & 0 \\ 0 & 2 & 1 & 1 \end{vmatrix} + 0 + 1(+1)\begin{vmatrix} 0 & 1 & 3 & 0 \\ 0 & 2 & 1 & 3 \\ 3 & 6 & -1 & 0 \\ 0 & 2 & 1 & 1 \end{vmatrix}$

$+ 0 + 1(+1)\begin{vmatrix} 0 & 1 & 3 & 0 \\ 0 & 2 & 1 & 3 \\ 1 & 0 & 0 & -2 \\ 3 & 6 & -1 & 0 \end{vmatrix}$

Now expand each 4 × 4 determinant about column one.

$= 1\left[0 + 1(-1)\begin{vmatrix} 2 & 1 & 3 \\ 6 & -1 & 0 \\ 2 & 1 & 1 \end{vmatrix} + 3(+1)\begin{vmatrix} 2 & 1 & 3 \\ 0 & 0 & -2 \\ 2 & 1 & 1 \end{vmatrix} + 0 \right]$

$+ 0 + 1\left[0 + 0 + 3(+1)\begin{vmatrix} 1 & 3 & 0 \\ 2 & 1 & 3 \\ 2 & 1 & 1 \end{vmatrix} + 0 \right]$

$+ 0 + 1\left[0 + 0 + 1(+1)\begin{vmatrix} 1 & 3 & 0 \\ 2 & 1 & 3 \\ 6 & -1 & 0 \end{vmatrix} + 3(-1)\begin{vmatrix} 1 & 3 & 0 \\ 2 & 1 & 3 \\ 0 & 0 & -2 \end{vmatrix} \right]$

Now evaluate the 3 × 3 determinants.

$= 0 -1[(-2+0+18) - (-6+0+6)] + 3[(0-4+0) - (0-4+0)] + 3[(1+18+0) - (0+3+6)] + [(0+54+0) - (0-3+0)] - 3[(-2+0+0) - (0+0-12)]$

$= -1(16) + 3(0) + 3(10) + (57) - 3(10)$

$= -16 + 30 + 57 - 30$

$= 41$

304

Solve for a:

57.
$$\begin{vmatrix} -1 & -2 & 1 \\ a & 0 & 0 \\ -1 & 1 & 1 \end{vmatrix} = 3$$

$$\begin{vmatrix} -1 & -2 & 1 \\ a & 0 & 0 \\ -1 & 1 & 1 \end{vmatrix} = (-1)(a)\begin{vmatrix} -2 & 1 \\ 1 & 1 \end{vmatrix} + 0\begin{vmatrix} -1 & 1 \\ -1 & 1 \end{vmatrix} - 0\begin{vmatrix} -1 & -2 \\ -1 & 1 \end{vmatrix}$$

Expand about the second row.

$$= -a(-2 - 1) + 0 - 0$$
$$= -a(-3)$$
$$= 3a$$
$$3a = 3$$
$$a = 1$$

Set 3a equal to 3 and solve for a.

The solution set is {1}.

D

61. Find the area of a triangle with vertices at (3, 4), (6, 0), and (2, 1).

$$A = \frac{1}{2}\begin{vmatrix} x_1 & y_1 & 1 \\ x_2 & y_2 & 1 \\ x_3 & y_3 & 1 \end{vmatrix}$$

Formula for the area of a triangle with vertices at (x_1, y_1), (x_2, y_2) and (x_3, y_3).

$$A = \frac{1}{2}\begin{vmatrix} 3 & 4 & 1 \\ 6 & 0 & 1 \\ 2 & 1 & 1 \end{vmatrix}$$

Replace (x_1, y_1) with (3, 4), (x_2, y_2) with (6, 0) and (x_3, y_3) with (2, 1).

$$= \frac{1}{2}\left[4(-1)\begin{vmatrix} 6 & 1 \\ 2 & 1 \end{vmatrix} + 0 + 1(-1)\begin{vmatrix} 3 & 1 \\ 6 & 1 \end{vmatrix}\right]$$

Expand about the second column.

$$= \frac{1}{2}[-4(6 - 2) - 1(3 - 6)]$$

$$= \frac{1}{2}[-4(4) - 1(-3)]$$

$$= \frac{1}{2}(-16 + 3)$$

$$= \frac{1}{2}(-13)$$

$$= -\frac{13}{2}$$

The area of the triangle is $\frac{13}{2}$ square units.

The formula is for the absolute value of A.

65. Evaluate the determinant of the following matrix using minors:

$$\begin{vmatrix} 3 & -4 & 3 \\ 1 & 7 & 1 \\ 2 & 18 & 2 \end{vmatrix}$$ Expand about the first column.

$$= 3(+1)\begin{vmatrix} 7 & 1 \\ 18 & 2 \end{vmatrix} + 1(-1)\begin{vmatrix} -4 & 3 \\ 18 & 2 \end{vmatrix} + 2(+1)\begin{vmatrix} -4 & 3 \\ 7 & 1 \end{vmatrix}$$

$= 3[14 - 18] - 1[-8 - (54)] + 2[-4 - (21)]$
$= 3(-4) - 1(-62) + 2(-25)$
$= -12 + 62 - 50$
$= 0$

69. What value of b will make the determinant equal to 10?

$$\begin{vmatrix} b & -1 \\ 1 & b \end{vmatrix} = b^2 - (-1)$$ Evaluate the determinant, and set it equal to 10.

$b^2 + 1 = 10$ Solve for b.
$b^2 = 9$
$b = \pm 3$

MAINTAIN YOUR SKILLS

Solve:

73. $\dfrac{1}{x - 3} - \dfrac{1}{2} = \dfrac{1}{x}$

$2x - 1(x)(x - 3) = 2(x - 3)$ Multiply both sides by the common denominator:
$\quad 2x - x^2 + 3x = 2x - 6$ $2x(x - 3)$
$\quad\quad -x^2 + 5x = 2x - 6$
$\quad\quad\quad\quad\quad 0 = x^2 - 3x - 6$
$a = 1, \ b = -3, \ c = -6$

$x = \dfrac{3 \pm \sqrt{9 - 4(1)(-6)}}{2}$ Use the quadratic formula to solve for x.

$\quad = \dfrac{3 \pm \sqrt{33}}{2}$

The solution set is $\left\{\dfrac{3 \pm \sqrt{33}}{2}\right\}$.

77.
$$x + 1 \leq \frac{4}{x + 4}$$

$$\frac{(x + 1)(x + 4)}{x + 4} - \frac{4}{x + 4} \leq 0$$

$$\frac{x^2 + 5x + 4 - 4}{x + 4} \leq 0$$

$$\frac{x(x + 5)}{x + 4} \leq 0 \qquad \text{Standard form.}$$

The critical numbers are 0, -4, -5.

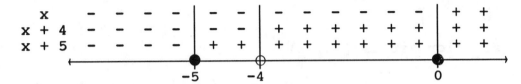

The solution set is $(-\infty, -5] \cup (-4, 0]$.

EXERCISES 7.5   SOLVING SYSTEMS USING CRAMER'S RULE

A

Solving using Cramer's Rule:

1. $\begin{cases} 2x + 5y = 15 \\ 3x - 4y = 11 \end{cases}$

$D = \begin{vmatrix} 2 & 5 \\ 3 & -4 \end{vmatrix}$, $D_x = \begin{vmatrix} 15 & 5 \\ 11 & -4 \end{vmatrix}$, $D_y = \begin{vmatrix} 2 & 15 \\ 3 & 11 \end{vmatrix}$

Write the determinants D, D_x, and D_y. For D_x replace the first column of D with the constant terms. For D_y replace the second column of D with the constant terms.

$D = 2(-4) - 3(5) = -8 - 15 = -23$
$D_x = 15(-4) - 11(5) = -60 - 55 = -115$
$D_y = 2(11) - 3(15) = 22 - 45 = -23$

$x = \dfrac{D_x}{D} = \dfrac{-115}{-23} = 5$

Use Cramer's rule to solve for x and y.

$y = \dfrac{D_y}{D} = \dfrac{-23}{-23} = 1$

The solution set is $\{(5, 1)\}$.

307

5. $\begin{cases} 3x + 4y = 10 \\ 2x - 3y = 1 \end{cases}$

$D = \begin{vmatrix} 3 & 4 \\ 2 & -3 \end{vmatrix}$, $D_x = \begin{vmatrix} 10 & 4 \\ 1 & -3 \end{vmatrix}$, $D_y = \begin{vmatrix} 3 & 10 \\ 2 & 1 \end{vmatrix}$

Write the determinants D, D_x, D_y. Expand D, D_x, D_y.

$D = 3(-3) - 2(4) = -9 - 8 = -17$

$D_x = 10(-3) - 1(4) = -30 - 4 = -34$

$D_y = 3(1) - 2(10) = 3 - 20 = -17$

$x = \dfrac{D_x}{D} = \dfrac{-34}{-17} = 2$

Solve for x and y.

$y = \dfrac{D_y}{D} = \dfrac{-17}{-17} = 1$

The solution set is $\{(2, 1)\}$.

9. $\begin{cases} x - 3y = 4 \\ 3x - 9y = 5 \end{cases}$

$D = \begin{vmatrix} 1 & -3 \\ 3 & -9 \end{vmatrix}$, $D_x = \begin{vmatrix} 4 & -3 \\ 5 & -9 \end{vmatrix}$, $D_y = \begin{vmatrix} 1 & 4 \\ 3 & 5 \end{vmatrix}$

$D = 1(-9) - 3(-3) = -9 + 9 = 0$

$D = 0$; therefore, there is no unique solution.

The system has no unique solution.

B

13. $\begin{cases} x + y + z = 1 \\ x - y + z = 2 \\ x - y - z = 3 \end{cases}$

Find the value of D, D_x, D_y, and D_z.

$D = \begin{vmatrix} 1 & 1 & 1 \\ 1 & -1 & 1 \\ 1 & -1 & -1 \end{vmatrix} = 1(2) - 1(0) + 1(2) = 4$

Expand about the first column. The value of each minor is in parentheses.

$D_x = \begin{vmatrix} 1 & 1 & 1 \\ 2 & -1 & 1 \\ 3 & -1 & -1 \end{vmatrix} = 1(2) - 2(0) + 3(2) = 8$

$D_y = \begin{vmatrix} 1 & 2 & 1 \\ 1 & 2 & 1 \\ 1 & 3 & -1 \end{vmatrix} = 1(-5) - 1(-4) + 1(-1) = -2$

$$D_z = \begin{vmatrix} 1 & 1 & 1 \\ 1 & -1 & 2 \\ 1 & -1 & 3 \end{vmatrix} = 1(-1) - 1(4) + 1(3) = -2$$

$$x = \frac{D_x}{D} = \frac{8}{4} = 2$$

Use Cramer's rule to find x, y, and z.

$$y = \frac{D_y}{D} = \frac{-2}{4} = -\frac{1}{2}$$

$$z = \frac{D_z}{D} = \frac{-2}{4} = -\frac{1}{2}$$

The solution set is $\left\{ \left(2, -\frac{1}{2}, -\frac{1}{2}\right) \right\}$.

C

17. $\begin{cases} 2x - 3y + z = -5 \\ 2x + 5y - 4z = -4 \\ 3x + 2y - 2z = -5 \end{cases}$

Find the value of D, D_x, D_y, and D_z.

$$D = \begin{vmatrix} 2 & -3 & 1 \\ 2 & 5 & -4 \\ 3 & 2 & -2 \end{vmatrix} = 2(-2) - 2(4) + 3(7) = 9$$

$$D_x = \begin{vmatrix} -5 & -3 & 1 \\ -4 & 5 & -4 \\ -5 & 2 & -2 \end{vmatrix} = -5(-2) + 4(4) - 5(7) = -9$$

$$D_y = \begin{vmatrix} 2 & -5 & 1 \\ 2 & -4 & -4 \\ 3 & -5 & -2 \end{vmatrix} = 2(-12) - 2(15) + 3(24) = 18$$

$$D_z = \begin{vmatrix} 2 & -3 & -5 \\ 2 & 5 & -4 \\ 3 & 2 & -5 \end{vmatrix} = 2(-17) - 2(25) + 3(37) = 27$$

$$x = \frac{D_x}{D} = \frac{-9}{9} = -1$$

$$y = \frac{D_y}{D} = \frac{18}{9} = 2$$

$$z = \frac{D_z}{D} = \frac{27}{9} = 3$$

The solution set is $\{(-1, 2, 3)\}$.

21. $\begin{cases} 2x + 3y - 4z = 0 \\ x + 6y - 2z = 3 \\ 3x + 3y + 6z = 8 \end{cases}$ Find D, D_x, D_y and D_z.

$$D = \begin{vmatrix} 2 & 3 & -4 \\ 1 & 6 & -2 \\ 3 & 3 & 6 \end{vmatrix} = 2(42) - 1(30) + 3(18) = 108$$

Expand about the first column.

$$D_x = \begin{vmatrix} 0 & 3 & -4 \\ 3 & 6 & -2 \\ 8 & 3 & 6 \end{vmatrix} = 0 - 3(30) + 8(18) = 54$$

$$D_y = \begin{vmatrix} 2 & 0 & -4 \\ 1 & 3 & -2 \\ 3 & 8 & 6 \end{vmatrix} = 2(34) - 1(32) + 3(12) = 72$$

$$D_z = \begin{vmatrix} 2 & 3 & 0 \\ 1 & 6 & 3 \\ 3 & 3 & 8 \end{vmatrix} = 2(39) - 1(24) + 3(9) = 81$$

Find x, y, and z using Cramer's rule.

$$x = \frac{D_x}{D} = \frac{54}{108} = \frac{1}{2}$$

$$y = \frac{D_y}{D} = \frac{72}{108} = \frac{2}{3}$$

$$z = \frac{D_z}{D} = \frac{81}{108} = \frac{3}{4}$$

The solution set is $\left\{ \left(\frac{1}{2}, \frac{2}{3}, \frac{3}{4} \right) \right\}$.

25. $\begin{cases} x + y = 1 \\ y + z = 9 \\ x + z = -6 \end{cases}$ Find the value of D, D_x. D_y, and D_z.

$$D = \begin{vmatrix} 1 & 1 & 0 \\ 0 & 1 & 1 \\ 1 & 0 & 1 \end{vmatrix} = 1(1) + 0 + 1(1) = 2$$

Expand about the first column.

$$D_x = \begin{vmatrix} 1 & 1 & 0 \\ 9 & 1 & 1 \\ -6 & 0 & 1 \end{vmatrix} = 1(1) - 9(1) - 6(1) = -14$$

$$D_y = \begin{vmatrix} 1 & 1 & 0 \\ 0 & 9 & 1 \\ 1 & -6 & 1 \end{vmatrix} = 1(15) + 0 + 1(1) = 16$$

$$D_z = \begin{vmatrix} 1 & 1 & 0 \\ 0 & 1 & 9 \\ 1 & 0 & -6 \end{vmatrix} = 1(-6) + 0 + 1(8) = 2$$

Find x, y, and z using
Cramer's rule.

$$x = \frac{D_x}{D} = \frac{-14}{2} = -7$$

$$y = \frac{D_y}{D} = \frac{16}{2} = 8$$

$$z = \frac{D_z}{D} = \frac{2}{2} = 1$$

The solution set is $\{(-7, 8, 1)\}$.

29. $\begin{cases} 5x - 7y \quad\quad = -11 \\ \quad\quad 5y + 6z = -15 \\ 6x \quad\quad - 5z = 37 \end{cases}$

Expand the determinants
about the first column.

$$D = \begin{vmatrix} 5 & -7 & 0 \\ 0 & 5 & 6 \\ 6 & 0 & -5 \end{vmatrix} = 5(-25) + 0 + 6(-42) = -377$$

$$D_x = \begin{vmatrix} -11 & -7 & 0 \\ -15 & 5 & 6 \\ 37 & 0 & -5 \end{vmatrix} = 11(-25) + 15(35) + 37(-42) = -754$$

$$D_y = \begin{vmatrix} 5 & -11 & 0 \\ 0 & -15 & 6 \\ 6 & 37 & -5 \end{vmatrix} = 5(-147) + 0 + 6(-66) = -1131$$

$$D_z = \begin{vmatrix} 5 & -7 & -11 \\ 0 & 5 & -16 \\ 6 & 0 & 37 \end{vmatrix} = 5(185) + 0 + 6(160) = 1885$$

$$x = \frac{D_x}{D} = \frac{-754}{-377} = 2$$

$$y = \frac{D_y}{D} = \frac{-1131}{-377} = 3$$

$$z = \frac{D_z}{D} = \frac{1885}{-377} = -5$$

The solution set is $\{(2, 3, -5)\}$.

33. $\begin{cases} x + y + z = 2 \\ 2y - z - w = 2 \\ x - w = 1 \\ 2x + 3z + 2w = 0 \end{cases}$

$D = \begin{vmatrix} 1 & 1 & 1 & 0 \\ 0 & 2 & -1 & -1 \\ 1 & 0 & 0 & -1 \\ 2 & 0 & 3 & 2 \end{vmatrix}$ Expand about the third row to take advantage of the zeros.

$= 1\begin{vmatrix} 1 & 1 & 0 \\ 2 & -1 & -1 \\ 0 & 3 & 2 \end{vmatrix} + 0 + 0 + 1\begin{vmatrix} 1 & 1 & 1 \\ 0 & 2 & -1 \\ 2 & 0 & 3 \end{vmatrix}$

Expand about the first column.

$= 1[1(1) - 2(2) + 0] + 0 + 1[1(6) + 0 + 2(-3)]$
$= 1(-3) + 1(0)$
$= -3$

$D_x = \begin{vmatrix} 2 & 1 & 1 & 0 \\ 2 & 2 & -1 & -1 \\ 1 & 0 & 0 & -1 \\ 0 & 0 & 3 & 2 \end{vmatrix}$ Expand about the second column.

$= -1\begin{vmatrix} 2 & -1 & -1 \\ 1 & 0 & -1 \\ 0 & 3 & 2 \end{vmatrix} + 2\begin{vmatrix} 2 & 1 & 0 \\ 1 & 0 & -1 \\ 0 & 3 & 2 \end{vmatrix} + 0 + 0$

Expand about the first column.

$= -1[2(3) - 1(1) + 0] + 2[2(3) - 1(2) + 0]$
$= -1(5) + 2(4)$
$= 3$

$D_y = \begin{vmatrix} 1 & 2 & 1 & 0 \\ 0 & 2 & -1 & -1 \\ 1 & 1 & 0 & -1 \\ 2 & 0 & 3 & 2 \end{vmatrix}$ Expand about the first column.

$= 1\begin{vmatrix} 2 & -1 & -1 \\ 1 & 0 & -1 \\ 0 & 3 & 2 \end{vmatrix} + 0 + 1\begin{vmatrix} 2 & 1 & 0 \\ 2 & -1 & -1 \\ 0 & 3 & 2 \end{vmatrix} - 2\begin{vmatrix} 2 & 1 & 0 \\ 2 & -1 & -1 \\ 1 & 0 & -1 \end{vmatrix}$

$= 1[2(3)-1(1)+0] + 1[2(1)-2(2)+0] - 2[2(1)-2(-1)+1(-1)]$
$= 1(5) + 1(-2) - 2(3)$
$= -3$

$$D_z = \begin{vmatrix} 1 & 1 & 2 & 0 \\ 0 & 2 & 2 & -1 \\ 1 & 0 & 1 & -1 \\ 2 & 0 & 0 & 2 \end{vmatrix}$$

Expand about the second column.

$$= -1\begin{vmatrix} 0 & 2 & -1 \\ 1 & 1 & -1 \\ 2 & 0 & 2 \end{vmatrix} + 2\begin{vmatrix} 1 & 2 & 0 \\ 1 & 1 & -1 \\ 2 & 0 & 2 \end{vmatrix} + 0 + 0$$

Expand each 3 × 3 determinant about the first row.

$$= -1[0 - 2(4) - 1(-2)] + 2[1(2) - 2(4) + 0]$$
$$= 6 + (-12)$$
$$= -6$$

$$D_w = \begin{vmatrix} 1 & 1 & 1 & 2 \\ 0 & 2 & -1 & 2 \\ 1 & 0 & 0 & 1 \\ 2 & 0 & 3 & 0 \end{vmatrix}$$

Expand about the second column.

$$= -1\begin{vmatrix} 0 & -1 & 2 \\ 1 & 0 & 1 \\ 2 & 3 & 0 \end{vmatrix} + 2\begin{vmatrix} 1 & 1 & 2 \\ 1 & 0 & 1 \\ 2 & 3 & 0 \end{vmatrix} + 0 + 0$$

$$= -1[0 - 1(-6) + 2(-1)] + 2[1(-3) - 1(-6) + 2(1)]$$
$$= -1(4) + 2(5)$$
$$= 6$$

$$x = \frac{D_x}{D} = \frac{-3}{3} = -1$$

$$y = \frac{D_y}{D} = \frac{-3}{-3} = 1$$

$$z = \frac{D_z}{D} = \frac{-6}{-3} = 2$$

$$w = \frac{D_w}{D} = \frac{6}{-3} = -2$$

The solution set is {(-1, 1, 2, -2)}.

Solve using Cramer's rule. Assume that a and b are nonzero constants, a ≠ b:

37. $\begin{cases} ax + by = a \\ bx + ay = b \end{cases}$ Find D, D_x, and D_y.

$$D = \begin{vmatrix} a & b \\ b & a \end{vmatrix} = a^2 - b^2$$

$$D_x = \begin{vmatrix} a & b \\ b & a \end{vmatrix} = a^2 - b^2$$

$$D_y = \begin{vmatrix} a & a \\ b & b \end{vmatrix} = 0$$

$$x = \frac{D_x}{D} = \frac{a^2 - b^2}{a^2 - b^2} = 1$$

$$y = \frac{D_y}{D} = \frac{0}{a^2 - b^2} = 0$$

The solution set is {(1, 0)}.

D

41. The sum of two numbers is 1459. The second number is 85 more than the first. What are the two numbers?

Simpler word form:
A first number + a second number = 1459.
The second number = the first number + 85.

Select variables:
Let x represent the first number and y represent the second number.

Translate to algebra:
$\begin{cases} x + y = 1459 & (1) \\ y = x + 85 & (2) \end{cases}$

Solve:

$$x + y = 1459 \qquad (1)$$
$$-x + y = 85 \qquad (2)$$

Rewrite the second equation.

$$D = \begin{vmatrix} 1 & 1 \\ -1 & 1 \end{vmatrix} = 1 - (-1) = 2$$

Write the determinants D, D_x, and D_y

$$D_x = \begin{vmatrix} 1459 & 1 \\ 85 & 1 \end{vmatrix} = 1459 - 85 = 1374$$

$$D_y = \begin{vmatrix} 1 & 1459 \\ -1 & 85 \end{vmatrix} = 85 - (-1459) = 15444$$

$$x = \frac{D_x}{D} = \frac{1374}{2} = 687$$

$$y = \frac{D_y}{D} = \frac{1544}{2} = 772$$

Use Cramer's rule to solve for x and y.

The solution set is {(687, 772)}.

Answer:
The numbers are 687 and 772.

45. A stock broker buys two stocks, one at $37.50 a share and the other at $14.75 a share. If the total purchase consisted of 650 shares at a cost of $14,137.50, how many shares were purchased at each price?

Simpler word form:

$$\begin{pmatrix} \text{Number of} \\ \$37.50 \\ \text{shares} \end{pmatrix} + \begin{pmatrix} \text{Number of} \\ \$14.75 \\ \text{shares} \end{pmatrix} = 650 \text{ shares}$$

$$\begin{pmatrix} \text{Cost of} \\ \$37.50 \\ \text{shares} \end{pmatrix} + \begin{pmatrix} \text{Cost of} \\ \$14.75 \\ \text{shares} \end{pmatrix} = \$14,137.50$$

Select variables:
Let x represent the number of shares purchased at $37.50 a share, and y represent the number of shares at $14.75 a share.

Translate to algebra:
$$x + y = 650 \qquad (1)$$
$$37.50x + 14.75y = 14137.50 \qquad (2)$$

$$x + y = 650 \qquad (1)$$
$$3750x + 1475y = 1413750 \qquad (2)$$

Multiply equation (2) by 100 to eliminate the decimals.

$$D = \begin{vmatrix} 1 & 1 \\ 3750 & 1475 \end{vmatrix} = -2275$$

Write the determinants D, D_x, and D_y.

$$D_x = \begin{vmatrix} 650 & 1 \\ 1413750 & 1475 \end{vmatrix} = -455000$$

$$D_y = \begin{vmatrix} 1 & 650 \\ 3750 & 1413750 \end{vmatrix} = -1023750$$

$$x = \frac{D_x}{D} = \frac{-455000}{-2275} = 200$$

Use Cramer's rule to solve for x and y.

$$y = \frac{D_y}{D} = \frac{-1023750}{-2275} = 450$$

The solution set is {(200, 450)}.

Answer:
At $37.50 a share, 200 shares were purchased, and 450 shares were purchased at $14.75 a share.

49. The following current loop diagram leads to the following equations:

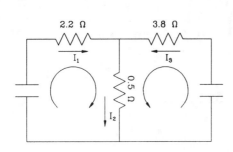

$$I_1 + I_3 = I_2 \qquad (1)$$
$$2.2I_1 + 0.5I_2 = 12.6 \qquad (2)$$
$$3.8I_3 + 0.5I_2 = 10.5 \qquad (3)$$

Find, to the nearest tenth, I_1, I_2, and I_3 using Cramer's rule.

$$I_1 - I_2 + I_3 = 0$$
$$2.2I_1 + 0.5I_2 = 12.6$$
$$0.5I_2 + 3.8I_3 = 10.5$$

Rewrite equation (1).

Evaluate the determinants involved.

316

$$D = \begin{vmatrix} 1 & -1 & 1 \\ 2.2 & 0.5 & 0 \\ 0 & 0.5 & 3.8 \end{vmatrix} = 1(1.9) - 2.2(-4.3) + 0 = 11.36$$

Expand about the first column.

$$D_{I_1} = \begin{vmatrix} 0 & -1 & 1 \\ 12.6 & 0.5 & 0 \\ 10.5 & 0.5 & 3.8 \end{vmatrix} = 0 - 12.6(-4.3) + 10.5(-0.5) = 48.93$$

$$D_{I_2} = \begin{vmatrix} 1 & 0 & 1 \\ 2.2 & 12.6 & 0 \\ 0 & 10.5 & 3.8 \end{vmatrix} = 1(47.88) - 2.2(-10.5) + 0 = 70.98$$

$$D_{I_3} = \begin{vmatrix} 1 & -1 & 0 \\ 2.2 & 0.5 & 12.6 \\ 0 & 0.5 & 10.6 \end{vmatrix} = 1(-1.05) - 2.2(-10.5) + 0 = 22.05$$

$$I_1 = \frac{D_{I_1}}{D} = \frac{48.93}{11.36} \approx 4.3$$

$$I_2 = \frac{D_{I_2}}{D} = \frac{70.98}{11.36} \approx 6.2$$

$$I_3 = \frac{D_{I_3}}{D} = \frac{22.05}{11.36} \approx 1.9$$

To the nearest tenth, $I_1 \approx 4.3$ amps, $I_2 \approx 6.2$ amps, and $I_3 = \approx 1.9$ amps.

CHALLENGE EXERCISES

53. In a 2 × 2 system, if the value of the determinant D is zero, the system is either dependent or independent and inconsistent. Using Cramer's rule and the determinant D, what value for "a" would make one of these two occur?

$$\begin{cases} ax + 3y = 5 \\ 2x - y = -7 \end{cases}$$

$$D = \begin{vmatrix} a & 3 \\ 2 & -1 \end{vmatrix} = -a - 6$$

Find D and set it equal to zero.

$$-a - 6 = 0$$
$$-6 = a$$

When a = -6, the value of the determinant D is zero and the system is either dependent or independent and inconsistent.

57. In a 3 × 3 system, it is mentioned in this section that if the value of the determinant D is zero, we say there is no unique solution. What value of "a" would make this happen.

$$\begin{cases} ax + 3y - z = 4 \\ 2y - 3z = 5 \\ ax - 3z = 1 \end{cases}$$

Find D and set it equal to zero.

$$\begin{vmatrix} a & 3 & -1 \\ 0 & 2 & 3 \\ a & 0 & -3 \end{vmatrix} = a(-6) + 0 + a(11) = 5a$$

$$5a = 0$$
$$a = 0$$

When a = 0, the determinant D is zero and there is no unique solution.

MAINTAIN YOUR SKILLS

Solve:

61. $|x - 7| + x < 3$
$|x - 7| < 3 - x$
$ -3 + x < x - 7 < 3 - x$
$ -3 + x < x - 7 \text{ and } x - 7 < 3 - x$
$ -3 < -7 \text{ and } 2x < 10 -3 < -7 \text{ is a}$
$ \text{and } x < 5 \text{contradiction.}$

The solution set is the empty set since there is no number to satisfy the contradiction.

65. $y < -\dfrac{2}{3}x + 4$

$ y = -\dfrac{2}{3}x + 4$

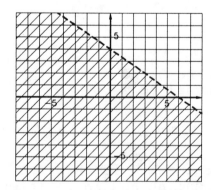

First draw the graph of $y = -\dfrac{2}{3}x + 4$ using a broken line since the inequality symbol is <.

Second, test the point (0, 0).

$0 < -\dfrac{2}{3}(0) + 4$

$0 < 4$ True.

The graph is the open half plane that contains the origin.

318

A

Solve each of the following systems using matrices:

1. $\begin{cases} x + y = -1 & (1) \\ 2x + 5y = -14 & (2) \end{cases}$

$\begin{bmatrix} 1 & 1 & | & -1 \\ 2 & 5 & | & -14 \end{bmatrix}$ First form the augmented matrix.

$\begin{bmatrix} -2 & -2 & | & 2 \\ 2 & 5 & | & -14 \end{bmatrix}$ Multiply the first row by -2 so that the numbers in the first column are opposites.

$\begin{bmatrix} -2 & -2 & | & 2 \\ 0 & 3 & | & -12 \end{bmatrix}$ Add the first row to the second row.

$\begin{bmatrix} 1 & 1 & | & -1 \\ 0 & 1 & | & -4 \end{bmatrix}$ Multiply the first row by $-\frac{1}{2}$. Multiply the second row by $\frac{1}{3}$.

$\begin{aligned} x + y &= -1 \quad (1) \\ y &= -4 \\ x + (-4) &= -1 \\ x &= 3 \end{aligned}$ Write the corresonding system. Solve for x using -4 to replace y in the first equation.

The solution set is $\{(3, -4)\}$.

5. $\begin{cases} 2x + 7y = 3 \\ x + y = 4 \end{cases}$

$\begin{bmatrix} 2 & 7 & | & 3 \\ 1 & 1 & | & 4 \end{bmatrix}$ Form the augmented matrix.

$\begin{bmatrix} 2 & 7 & | & 3 \\ -2 & -2 & | & -8 \end{bmatrix}$ Multiply the second row by -2 and add it to the first row.

$\begin{bmatrix} 2 & 7 & | & 3 \\ 0 & 5 & | & -5 \end{bmatrix}$

$\begin{bmatrix} 1 & \frac{7}{2} & | & \frac{3}{2} \\ 0 & 1 & | & -1 \end{bmatrix}$ Multiply the first row by $\frac{1}{2}$ and the second row by $\frac{1}{5}$.

$$\begin{cases} x + \dfrac{7}{2}y = \dfrac{3}{2} & (1) \\ \quad\quad\ y = -1 & (2) \end{cases}$$

Write the system of equations that corresponds to the matrix.

$$x + \frac{7}{2}(-1) = \frac{3}{2}$$

Replace -1 for y in equation (1).

$$\begin{aligned} 2x - 7 &= 3 \\ 2x &= 10 \\ x &= 5 \end{aligned}$$

Clear fractions.

The solution set is $\{(5, -1)\}$.

9. $$\begin{cases} 2x + 6y = 13 \\ \ x + 3y = \ \ 5 \end{cases}$$

$$\begin{bmatrix} 2 & 6 & | & 13 \\ 1 & 3 & | & 5 \end{bmatrix}$$

Form the augmented matrix.

$$\begin{bmatrix} 2 & 6 & | & 13 \\ -2 & -6 & | & -10 \end{bmatrix}$$

Multiply the second row by -2.

$$\begin{bmatrix} 2 & 6 & | & 13 \\ 0 & 0 & | & 3 \end{bmatrix}$$

Add the second row to the first.

$$2x + 6y = 13$$

Write the corresponding system.

$$0 = 3$$

Contradiction.

The solution set is the empty set, $\emptyset$.

B

13. $$\begin{cases} 4x - \ y = \ \ \ 3 \\ 8x + 3y = -29 \end{cases}$$

$$\begin{bmatrix} 4 & -1 & | & 3 \\ 8 & 3 & | & -29 \end{bmatrix}$$

Augmented matrix.

$$\begin{bmatrix} -8 & 2 & | & -6 \\ 8 & 3 & | & -29 \end{bmatrix}$$

Multiply the first row by -2.

$$\begin{bmatrix} -8 & 2 & | & -6 \\ 0 & 5 & | & -35 \end{bmatrix}$$

Add to row two.

$$\begin{bmatrix} 1 & -\dfrac{1}{4} & | & \dfrac{3}{4} \\ 0 & 1 & | & -7 \end{bmatrix}$$

Multiply the first row by $-\dfrac{1}{8}$ and the second row by $\dfrac{1}{5}$.

$$\begin{cases} x - \frac{1}{4}y = \frac{3}{4} & (1) \\ \quad\quad y = -7 & (2) \end{cases}$$

Corresponding system of equations.

$$x - \frac{1}{4}(-7) = \frac{3}{4}$$

Replace y in equation (1) with -7 and solve for x.

$$4x + 7 = 3$$
$$4x = -4$$
$$x = -1$$

Clear the fractions by multiplying both sides by 4. Simplify.

$$(-1) - \frac{1}{4}y = \frac{3}{4}$$
$$-4 - y = 3$$
$$-y = 7$$
$$y = -7$$

Replace -1 for x and solve for y in equation (1).

The solution set is {(-1, -7)}.

$$17. \quad \begin{cases} \frac{1}{3}x - \frac{2}{3}y = 4 \\ \frac{2}{3}x + \frac{1}{3}y = 3 \end{cases}$$

$$\begin{bmatrix} \frac{1}{3} & -\frac{2}{3} & \Big| & 4 \\ \frac{2}{3} & \frac{1}{3} & \Big| & 3 \end{bmatrix}$$

Augmented matrix.

$$\begin{bmatrix} -\frac{2}{3} & \frac{4}{3} & \Big| & -8 \\ \frac{2}{3} & \frac{1}{3} & \Big| & 3 \end{bmatrix}$$

Multiply the first row by -2.

$$\begin{bmatrix} -\frac{2}{3} & \frac{4}{3} & \Big| & -8 \\ 0 & \frac{5}{3} & \Big| & -5 \end{bmatrix}$$

Add to row two.

$$\begin{bmatrix} 1 & -2 & \Big| & 12 \\ 0 & 1 & \Big| & -3 \end{bmatrix}$$

Multiply the first row by $-\frac{3}{2}$, and the second row by $\frac{3}{5}$.

$$x - 2y = 12$$
$$y = -3$$
$$x - 2(-3) = 12$$
$$x + 6 = 12$$
$$x = 6$$

Corresponding system.

Replace y with -3 and solve for x.

The solution set is {(6, -3)}.

C

21. $\begin{cases} x + y + z = 3 \\ 2x + 3y + z = 6 \\ 2x - y - z = 0 \end{cases}$

$$\begin{bmatrix} 1 & 1 & 1 & | & 3 \\ 2 & 3 & 1 & | & 6 \\ 2 & -1 & -1 & | & 0 \end{bmatrix}$$

Form the augmented matrix.

$$\begin{bmatrix} 1 & 1 & 1 & | & 3 \\ 2 & 3 & 1 & | & 6 \\ 0 & -3 & -3 & | & -6 \end{bmatrix}$$

Multiply row one by -2 and add to row three.

$$\begin{bmatrix} 1 & 1 & 1 & | & 3 \\ 0 & 1 & -1 & | & 0 \\ 0 & -3 & -3 & | & -6 \end{bmatrix}$$

Multiply row one by -2 and add to row two.

$$\begin{bmatrix} 1 & 1 & 1 & | & 3 \\ 0 & 1 & -1 & | & 0 \\ 0 & 0 & -6 & | & -6 \end{bmatrix}$$

Multiply row two by 3 and add to row three.

$$\begin{bmatrix} 1 & 1 & 1 & | & 3 \\ 0 & 1 & -1 & | & 0 \\ 0 & 0 & 1 & | & 1 \end{bmatrix}$$

Divide row three by -6.

$$\begin{array}{rcl} x + y + z &=& 3 \\ y - z &=& 0 \\ z &=& 1 \end{array}$$

Write the corresponding system of equations.

$$\begin{array}{rcl} y - 1 &=& 0 \\ y &=& 1 \end{array}$$

Use the second equation to solve for y.

$$\begin{array}{rcl} x + 1 + 1 &=& 3 \\ x &=& 1 \end{array}$$

Use the first equation to solve for x.

The solution set is $\{(1, 1, 1)\}$.

25. $\begin{cases} x + 2y \quad\;\;\; = 6 \\ x \quad\;\; + 3z = -7 \\ \quad\; y + 3z = 2 \end{cases}$

$$\begin{bmatrix} 1 & 2 & 0 & | & 6 \\ 1 & 0 & 3 & | & -7 \\ 0 & 1 & 3 & | & 2 \end{bmatrix}$$

Augmented matrix.

$$\begin{bmatrix} 1 & 2 & 0 & | & 6 \\ 0 & -2 & 3 & | & -13 \\ 0 & 1 & 3 & | & 2 \end{bmatrix}$$

Multiply row one by -1 and add to row two.

$$\begin{bmatrix} 1 & 2 & 0 & | & 6 \\ 0 & 1 & 3 & | & 2 \\ 0 & -2 & 3 & | & -13 \end{bmatrix}$$

Exchange rows two an three.

$$\begin{bmatrix} 1 & 2 & 0 & 6 \\ 0 & 1 & 3 & 2 \\ 0 & 0 & 9 & -9 \end{bmatrix}$$

Multiply row two by 2 and add to row three.

$$\begin{bmatrix} 1 & 2 & 0 & 6 \\ 0 & 1 & 3 & 2 \\ & & 1 & -1 \end{bmatrix}$$

Divide row three by 9.

$$\begin{cases} x + 2y & = 6 \\ y + 3z & = 2 \\ z & = -1 \end{cases}$$

Write the corresponding system of equations.

$$\begin{aligned} y + 3(-1) &= 2 \\ y - 3 &= 2 \\ y &= 5 \end{aligned}$$

Use the second equation to solve for y.

$$\begin{aligned} x + 2(5) &= 6 \\ x + 10 &= 6 \\ x &= -4 \end{aligned}$$

Use the first equation to solve for x.

The solution set is $\{(-4, 5, -1)\}$.

29. $\begin{cases} 2x - 5y - 7z = 1 \\ 3x - 2y + z = -10 \\ 4x + y + 5z = -17 \end{cases}$

$$\begin{bmatrix} 2 & -5 & -7 & 1 \\ 3 & -2 & 1 & -10 \\ 4 & 1 & 5 & -17 \end{bmatrix}$$

Augmented matrix.

$$\begin{bmatrix} 2 & -5 & -7 & 1 \\ 3 & -2 & 1 & -10 \\ 0 & 11 & 19 & -19 \end{bmatrix}$$

Multiply row one by -2, and add it to row three.

$$\begin{bmatrix} 2 & -5 & -7 & 1 \\ 0 & -11 & -23 & -23 \\ 0 & 11 & 19 & -19 \end{bmatrix}$$

Multiply row one by 3 and row two by -2. Add.

$$\begin{bmatrix} 2 & -5 & -7 & 1 \\ 0 & -11 & -23 & 23 \\ 0 & 0 & -4 & 4 \end{bmatrix}$$

Add rows two and three.

$$\begin{bmatrix} 1 & -\dfrac{5}{2} & -\dfrac{7}{2} & \dfrac{1}{2} \\ 0 & 1 & \dfrac{23}{11} & -\dfrac{23}{11} \\ 0 & 0 & 1 & -1 \end{bmatrix}$$

Divide row one by 2, row two by -11, and row three by -4.

$$\begin{cases} x - \dfrac{5}{2}y - \dfrac{7}{2}z = \dfrac{1}{2} \\ y + \dfrac{23}{11}z = -\dfrac{23}{11} \\ z = -1 \end{cases}$$

Corresponding system.

$$y + \frac{23}{11}(-1) = -\frac{23}{11}$$
$$y - \frac{23}{11} = -\frac{23}{11}$$
$$y = 0$$

Use the second equation to solve for y.

$$x - \frac{5}{2}(0) - \frac{7}{2}(-1) = \frac{1}{2}$$
$$x + \frac{7}{2} = \frac{1}{2}$$
$$x = -\frac{6}{2}$$
$$x = -3$$

Use the first equation to solve for x.

The solution set is {(-3, 0, -1)}.

CHALLENGE EXERCISES

33. Solve using the augmented matrix:

$$\begin{cases} ax + y = 3 \\ x - 2y = -4 \end{cases}$$

$$\begin{bmatrix} a & 1 & 3 \\ 1 & -2 & -4 \end{bmatrix}$$

Augmented matrix.

$$\begin{bmatrix} 1 & -2 & -4 \\ a & 1 & 3 \end{bmatrix}$$

Exchange the rows.

$$\begin{bmatrix} 1 & -2 & -4 \\ 0 & 2a+1 & 4a+3 \end{bmatrix}$$

Multiply row one by -1 and add to row two.

$$\begin{bmatrix} 1 & -2 & -4 \\ 0 & 1 & \frac{4a+3}{2a+1} \end{bmatrix}$$

Divide row two by 2a + 1.

$$x - 2y = -4$$
$$y = \frac{4a+3}{2a+1}$$

Corresponding system.

$$x - 2\left(\frac{4a+3}{2a+1}\right) = -4$$

Use the first equation to solve for x.

$$x(2a + 1) - 2(4a + 3) = -4(2a + 1)$$
$$2ax + x - 8a - 6 = -8a - 4$$
$$x(2a + 1) = 2$$
$$x = \frac{2}{2a+1}$$

The solution set is $\left\{\left(\frac{2}{2a+1}, \frac{4a+3}{2a+1}\right)\right\}$.

37. Write the coordinates of the x- and y-intercepts of the graph of

$y = \frac{1}{5}x - 9$

$0 = \frac{1}{5}x - 9$ Let y = 0 to find the x-intercept.

$0 = x - 45$ Clear fractions
$45 = x$

$y = \frac{1}{5}(0) - 9$ Let x = 0 to find the y-intercept.

$y = -9$

The x-intercept is (45, 0) and the y-intercept is (0, -9).

Solve:

41. $(x + 3)^2 + 2(x + 23) = (3x + 5)^2$
       $x^2 + 6x + 9 + 2x + 46 = 9x^2 + 30x + 25$ Multiply.
            $x^2 + 8x + 55 = 9x^2 + 30x + 25$ Simplify.
                     $0 = 8x^2 + 22x - 30$
                     $0 = 2(4x^2 + 11x - 15)$
                     $0 = 2(4x - 15)(x + 1)$ Factor completely.
      $4x + 15 = 0$ or $x - 1 = 0$ Zero-product property.
          $x = -\frac{15}{4}$ or $x = 1$

The solution set is $\left\{ -\frac{15}{4}, 1 \right\}$.

CHAPTER 8

SECOND-DEGREE EQUATIONS IN TWO VARIABLES

EXERCISES 8.1 PARABOLAS

A

Find the vertex, determine whether the vertex is a maximum or minimum, find the x- and y-intercepts, determine the equation of the axis of symmetry, and graph.

1. $f(x) = x^2 - 2$
    $a = 1, b = 0, c = -2$

$h = -\dfrac{0}{2 \cdot 1} = 0$

The x-coordinate of the vertex (h) is found by letting $h = -\dfrac{b}{2a}$.

$f(0) = -2$

Determine the y-coordinate of the vertex by substituting 0 for x in the equation.

The vertex is at (0, -2).

The vertex is a minimum point. The curve opens upward.

$a > 0$.

$\begin{aligned} 0 &= x^2 - 2 \\ 2 &= x^2 \\ \pm\sqrt{2} &= x \end{aligned}$

Substitute 0 for f(x) and solve for x to find the x-intercepts.

The x-intercepts are $(-\sqrt{2}, 0)$ and $(\sqrt{2}, 0)$.

$f(0) = -2$

Substitute 0 for x to find the y-intercept.

The y-intercept is (0, -2).

Since the vertex is at (0, -2), $x = 0$ is the equation of the axis of symmetry.

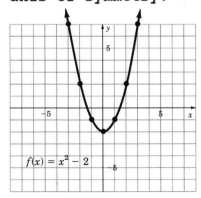

$f(x) = x^2 - 2$

Plot the graph. Three points are known:

$(0, -2)$, $(-\sqrt{2}, 0)$, and $(\sqrt{2}, 0)$. Find two other points such as $(-3, 7)$ and $(3, 7)$.

327

5. $y = (x + 1)^2 - 2$
    $a = 1$, $h = -1$, and $k = -2$

    Identify a, h, and k in $f(x) = a(x - h)^2 + k$.

    The vertex is $(-1, -2)$

    Identify the vertex (h, k).

    The vertex is a minimum point. The curve opens upward.

    $a > 0$.

    $0 = (x + 1)^2 - 2$
    $0 = x^2 + 2x + 1 - 2$
    $0 = x^2 + 2x - 1$

    Substitute 0 for y and solve for x to find the x-intercepts.

    $x = \dfrac{-2 \pm \sqrt{4 - 4(1)(-1)}}{2}$

    Quadratic formula.

    $\quad = \dfrac{-2 \pm 2\sqrt{2}}{2}$

    $\quad = -1 + \sqrt{2}$

    The x-intercepts are $(-1 - \sqrt{2}, 0)$ and $-1 + \sqrt{2}, 0)$.

    $y = (0 + 1)^2 - 2$
    $\quad = 1 - 2$
    $\quad = -1$

    Substitute 0 for x to find the y-intercept.

    The y-intercept is $(0, -1)$.

    Since the vertex is at $(-1, -2)$, $x = -1$ is the equation of the axis of symmetry.

    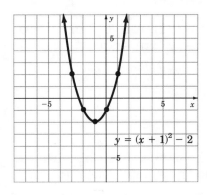

    Plot the graph. Three points are known:

    $(-1, -2)$, $(-1 -\sqrt{2}, 0)$, and $(-1 +\sqrt{2}, 0)$. Find two more such as $(2, 7)$ and $(-3, 2)$.

9. $y = -x^2 + 3$
   $a = -1$, $b = 0$, $c = 3$

   $h = -\dfrac{0}{2(-1)} = 0$

   The x-coordinate of the vertex (h) is found by letting $h = -\dfrac{b}{2a}$.

   $y = -0^2 + 3 = 3$

   Substitute 0 for x to find the y-coordinate of the vertex.

   The vertex is at (0, 3).

   The vertex is a maximum point. The curve opens downward.

   $a < 0$.

   $0 = -x^2 + 3$
   $x^2 = 3$
   $x = \pm\sqrt{3}$

   Substitute 0 for y and solve for x to find the x-intercepts.

   The x-intercepts are $(-\sqrt{3},\ 0)$ and $\sqrt{3},\ 0)$.

   $y = -0^2 + 3$
   $y = 3$

   Substitute 0 for x and solve to find the y-intercept.

   The y-intercept is (0, 3).

   Since the vertex is at (0, 3), $x = 0$ is the equation of the axis of symmetry.

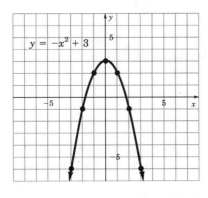

   Plot the graph. Three points are known: (0, 3), $(-\sqrt{3},\ 0)$, and $(\sqrt{3},\ 0)$. Find two more, such as (-2, -1) and (2, -1).

329

Graph the relation:

13. $x = y^2 - 6$

The parabola opens right. $a = 1.$

The vertex is at (-6, 0) $-\dfrac{b}{2a} = -\dfrac{0}{2} = 0$

$x = 0^2 - 6 = -6$ $x = 0^2 - 6 = -6$
The x-intercept is (-6, 0).   x-intercept, $y = 0$
Notice that this point is also the vertex.

$\begin{aligned} 0 &= y^2 - 6 \\ 6 &= y^2 \end{aligned}$ y-intercept, $x = 0$

$\pm\sqrt{6} = y$

The y-intercepts are $(0, \pm\sqrt{6})$ or $(0, \pm 2.4)$.

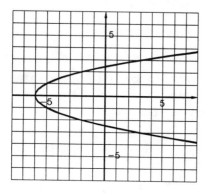

x	y
-2	±2
3	±3

Plot the points.

B

Find the vertex, determine whether the vertex is a maximum or minimum point, find the x- and y-intercepts, determine the equation of the axis of symmetry, and graph:

17. $y = -\dfrac{1}{3}x^2 + 2x - 5$

$-\dfrac{b}{2a} = \dfrac{-2}{-\frac{2}{3}} = \dfrac{2}{\frac{2}{3}} = 3$

The x-coordinate of the vertex is found by using $-\dfrac{b}{2a}.$

$\begin{aligned} y &= -\tfrac{1}{3}(3)^2 + 2(3) - 5 \\ &= -\tfrac{1}{3}(9) + 6 - 5 \\ &= -3 + 6 - 5 \\ &= -2 \end{aligned}$

Substitute 3 for x in the equation to determine the y-coordinate of the vertex.

The vertex is at (3, -2).

330

The vertex is a maximum point.

Since a < 0, the graph has a maximum point.

$0 = -\frac{1}{3}x^2 + 2x - 5$
$0 = -x^2 + 6x - 15$
$0 = x^2 - 6x + 15$

$x = \dfrac{6 \pm \sqrt{36 - 4(1)(15)}}{2}$

$x = \dfrac{6 \pm \sqrt{-24}}{2}$

There are no x-intercepts.

Substitute 0 for y, and solve the equation for x to find the x-intercepts.

No x-intercepts since $\sqrt{-24}$ is not a real number.

$y = -\frac{1}{3}(0)^2 + 2(0) - 5$
$y = -5$

The y-intercept is (0, -5).

Substitute 0 for x to find the y-intercept.

Since the vertex is at (3, -2), x = 3 is the equation of the axis of symmetry.

The vertical line that passes through the vertex is the axis of symmetry.

By symmetry, determine that (6, -5) is a point on the graph.

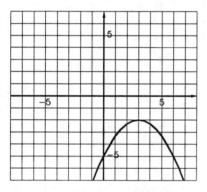

21. $y = x^2 - 4x + 7$

$-\dfrac{b}{2a} = -\dfrac{-4}{2} = 2$

Evaluate $-\dfrac{b}{2a}$ to determine the x-coordinate of the vertex.

$y = (2)^2 - 4(2) + 7$
$\quad = 4 - 8 + 7$
$\quad = 3$

Replace x with 2 to find the y-coordinate of the vertex.

The vertex is at (2, 3).

The vertex is a minimum point.

Since a > 0, the graph has a minimum point.

331

$0 = x^2 - 4x + 7$

$x = \dfrac{4 \pm \sqrt{16 - 4(1)(7)}}{2}$

$ = \dfrac{4 \pm \sqrt{-12}}{2}$

To find the x-intercepts, substitute 0 for y. Use the quadratic formula to solve for x.

$\sqrt{-12}$ is not a real number.

There are no x-intercepts.

$y = (0)^2 - 4(0) + 7$
$ = 7$

To find the y-intercept, substitute 0 for x.

The y-intercept is (0, 7).

Since the vertex is at (2, 3), x = 2 is the equation of the line of symmetry.

The vertical line that passes through the vertex is the axis of symmetry.

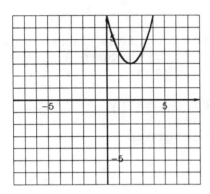

x	y
1	4
2	3
3	4
0	7
4	7

C

25. $y = x^2 - 2x - 5$

$-\dfrac{b}{2a} = -\dfrac{-2}{2} = 1$

x-coordinate of the vertex.

$y = (1)^2 - 2(1) - 5$
$ = 1 - 2 - 5$
$ = -6$

y-coordinate of the vertex.

The vertex is at (1, -6).

The vertex is a minimum point.

a > 0, so the graph has a minimum point.

$$0 = x^2 - 2x - 5 \qquad \text{x-intercepts, } y = 0$$

$$x = \frac{2 \pm \sqrt{4 - 4(1)(-5)}}{2}$$

$$= \frac{2 \pm \sqrt{24}}{2}$$

$$= \frac{2 \pm 2\sqrt{6}}{2}$$

$$= 1 \pm \sqrt{6}$$

$$1 + \sqrt{6} \approx 3.4 \qquad 1 - \sqrt{6} \approx -1.4$$

The x-intercepts are $(1 \pm \sqrt{6}, 0)$.

$$Y = 0^2 - 2(0) - 5 \qquad \text{y-intercept, } x = 0$$
$$= -5$$

The y-intercept is $(0, -5)$.

The equation of the line of symmetry is $x = 1$, which is the equation of the vertical line passing through the vertex.

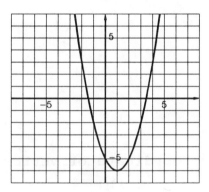

x	y
1	-2
3	-2
4	3

29. $y = -3x^2 - 6x - 3$

$$-\frac{b}{2a} = -\frac{-6}{-6} = -1$$

$$y = -3(-1)^2 - 6(-1) - 3$$
$$= -3(1) + 6 - 3$$
$$= -3 + 6 - 3$$
$$= 0$$

The vertex is at $(-1, 0)$.

The vertex is a maximum point since $a < 0$.

$0 = -3x^2 - 6x - 3$ x-intercepts, y = 0

$$x = \frac{6 \pm \sqrt{36 - 4(-3)(-3)}}{2(-3)}$$

$$= \frac{6 \pm \sqrt{0}}{-6}$$

$$= -1$$ A double root.

The x-intercept is (-1, 0).

$y = -3(0)^2 - 6(0) - 3$ y-intercept, x = 0
$= -3$

The y-intercept is (0, -3)

The equation of the axis of
symmetry is x = -1, the line that
passes through the vertex vertically.

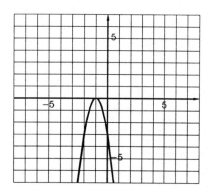

x	y
-2	-3
0	-3
-1	0

33. $y = \frac{1}{4}x^2 + \frac{1}{2}x - \frac{11}{4}$

$$-\frac{b}{2a} = -\frac{\frac{1}{2}}{\frac{1}{2}} = -1$$ x-coordinate of vertex.

$y = \frac{1}{4}(-1)^2 + \frac{1}{2}(-1) - \frac{11}{4}$ y-coordinate of the vertex.

$$= \frac{1}{4} - \frac{1}{2} - \frac{11}{4}$$

$$= -\frac{12}{4}$$

$$= -3$$

The vertex is at (-1, -3).

The vertex is a minimum point since a > 0.

$$0 = \frac{1}{4}x^2 + \frac{1}{2}x - \frac{11}{4}$$

x-intercepts, y = 0.

$$0 = x^2 + 2x - 11$$

Multiply by 4 to clear the fractions.

$$x = \frac{-2 \pm \sqrt{4 - 4(1)(-11)}}{2}$$

$$= \frac{-2 \pm \sqrt{48}}{2}$$

$$= \frac{-2 \pm 4\sqrt{3}}{2}$$

$$= -1 \pm 2\sqrt{3}$$

$-1 + 2\sqrt{3} \approx 2.5$ and $-1 - 2\sqrt{3} \approx -4.5$

The x-intercepts are $(-1 \pm 2\sqrt{3}, 0)$.

$$y = \frac{1}{4}(0)^2 + \frac{1}{2}(0) - \frac{11}{4}$$

y-intercept, x = 0

$$= -\frac{11}{4}$$

The y-intercept is $\left(0, -\frac{11}{4}\right)$.

The equation of the axis of symmetry is x = -1, the equation of the vertical line that passes through the vertex.

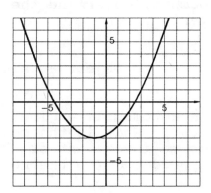

x	y
-5	1
3	1
-3	-2
1	-2

D

37. If the demand curve of a manufacturer of T-shirts is $y = -\frac{1}{2}x^2 + 4x + 1500$, find the week of highest demand and the number of T-shirts needed to meet the demand.

$-\dfrac{b}{2a} = -\dfrac{4}{-1} = 4$

To find the maximum point (of highest demand) determine the vertex.

$y = -\frac{1}{2}(4)^2 + 4(4) + 1500$
$= -8 + 16 + 1500$
$= 1508$

Vertex is at (4, 1508).

At the vertex, y has its maximum value, 1508.

So during the 4th week, 1508 T-shirts are needed to meet demands.

41. The height of an object thrown upward from a height of 6 ft at a velocity of 96 ft/sec at any time (t) is given by $h = 96t - 16t^2 + 6$. What is the maximum height reached by the object? In how many seconds will the object reach its maximum height?

$h = \dfrac{-b}{2a} = -\dfrac{96}{-32} = 3$
$k = 96(3) - 16(3)^2 + 6 = 150$

To find the maximum height and the second in which it occurs, determine the vertex.

The vertex is at (3, 150).

The maximum height reached is 150 ft. The object attains this height in 3 seconds.

45. The cost (in dollars) of producing N automobile engine blocks is given by $C = 1.7N^2 + 135,000$. What is the cost of producing 15,000 engine blocks? For an investment of $550,000,000 how many blocks can be produced (to the nearest whole block)?

$C = 1.7(15000)^2 + 135,000$
$= 382,500,000 + 135,000$
$= 38,263,500$

To find the cost of producing 1500 blocks, substitute 1500 for N. Solve for C, the cost.

$550,000,000 = 1.7N^2 + 135,000$
$549,865,000 = 1.7N^2$
$323,450,000 = N^2$
$17,985 \approx N$

To find how many blocks, N, can be produced with $550,000,000 replace C 550,000,000, and solve for N.

The cost of producing 15000 blocks is $38,263,500. Approximately 17,985 blocks can be produced for an investment of $550,000,000.

49. If $y = ax^2 + bx + c$ is the general equation of a parabola that opens up or down, what can be said about the parabola if (a) $b^2 - 4ac > 0$? (b) If $b^2 - 4ac = 0$? (c) If $b^2 - 4ac < 0$?

   a)   If $b^2 - 4ac > 0$, there are two x-intercepts.

   b)   If $b^2 - 4ac = 0$, there is one x-intercept.

   c)   If $b^2 - 4ac < 0$, there are no x-intercepts.

CHALLENGE EXERCISES

53. Given $x = ay^2 + by + c$, find the equation of a parabola that opens left or right that passes through (-2, -1), (4, 5) and (4, -3).

	Form a system of equations. Generate three equations in a, b, and c, by replacing the x and y with their respective values from each ordered pair given.
$-2 = a(-1)^2 + b(-1) + c$ $-2 = a - b + c$ (1)	From the ordered pair (-2, -1).
$4 = a(5)^2 + b(5) + c$ $4 = 25a + 5b + c$ (2)	From the ordered pair (4, 5).
$4 = a(-3)^2 + b(-3) + c$ $4 = 9a - 3b + c$ (3)	From the ordered pair (4, -3).
$a - b + c = -2$ (1) $25a + 5b + c = 4$ (2) $9a - 3b + c = 4$ (3)	System of equations. Solve the system by linear combinations.
$-a + b - c = 2$ $\underline{25a + 5b + c = 4}$ $24a + 6b = 6$ (4)	Multiply equation (1) by -1. Add to equation (2) to eliminate c.
$-a + b - c = 2$ $\underline{9a - 3b + c = 4}$ $8a - 2b = 6$ (5)	Multiply equation (1) by -1 and add to equation (3) to again eliminate c.
$24a + 6b = 6$ (4) $8a - 2b = 6$ (5)	

$$24a - 6b = 18$$
$$\underline{24a + 6b = 6}$$
$$48a = 24$$
$$a = \frac{1}{2}$$

Multiply equation (5) by 3 and add to equation (4) to elimniate a.

$$24\left(\frac{1}{2}\right) + 6b = 6$$
$$6b = -6$$
$$b = -1$$

Replace a in equation (4) by $\frac{1}{2}$. Solve for b.

$$\frac{1}{2} + 1 + c = -2$$
$$c = -\frac{7}{2}$$

Replace -1 for b and $\frac{1}{2}$ and a in equation (1) and solve for c.

$$x = ay^2 + by + c$$
$$x = \frac{1}{2}y^2 - y - \frac{7}{2}$$

Original equation. Replace a with $\frac{1}{2}$, b with -1, and c with $-\frac{7}{2}$.

The equation $x = \frac{1}{2}y^2 - y - \frac{7}{2}$ passes through the points (-2, -1), (4, 5), and (4, -3).

MAINTAIN YOUR SKILLS

Given $x^2 + y^2 = 400$. Find the real value(s) of y for the given value of x:

57. x = 12

$$(12)^2 + y^2 = 400$$
$$y^2 = 400 - 144$$
$$y^2 = 256$$
$$y = \pm 16$$

Replace x with 12.

Given $2x^2 - y^2 = 8$. Find the real value(s) of y for the given value of x:

61. x = 3

$$2(3)^2 - y^2 = 8$$
$$2(9) - y^2 = 8$$
$$18 - 8 = y^2$$
$$10 = y^2$$
$$\pm\sqrt{10} = y$$

Replace x with 3.

A

Write the coordinates of the center and the length of the radius
and graph the following:

1. $x^2 + y^2 = 25$
     Center $(0, 0)$

     Radius 5

This is the standard form
of an equation whose graph
is a circle with center at
the origin.
The radius is the square
root of the constant term.

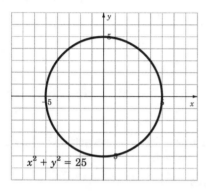

5. $2x^2 + 2y^2 = 8$
     $x^2 + y^2 = 4$

     Center $(0, 0)$
     Radius 2

Divide both sides by the
common factor, 2.
From the standard form.
The radius is the square
root of the constant term.

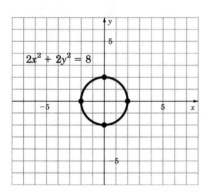

Write the equation in standard form of the circle with the given
coordinates as center and the given value of r as radius:

9. (2, 4), r = 7
    $(x - h)^2 + (y - k)^2 = r^2$ Standard form of a circle
                                         with center at (h, k) and
                                         radius, r.
    $(x - 2)^2 + (y - 4)^2 = 49$ Substitute.

13. (1, 1), r = 6
    $(x - 1)^2 + (y - 1)^2 = 36$ Substitute 1 for h, 1 for
                                         k, and 6 for r is the
                                         standard form of a circle
                                         with center at (h, k) and
                                         radius, r.

B

Write the coordinates of the center and the length of the radius
and graph each of the following:

17. $(x + 5)^2 + (y - 3)^2 = 4$ From the standard form:
      $[x - (-5)]^2 + (y - 3)^2 = 2^2$ h = −5 and k = 3.

                                         Radius = $\sqrt{4}$.
      Center (−5, 3)
      Radius 2

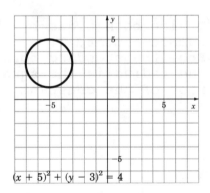

$(x + 5)^2 + (y - 3)^2 = 4$

21. $(x - 1)^2 + (y - 2)^2 = \dfrac{9}{4}$

Center (1, 2)   From the standard form.

Radius $\dfrac{3}{2}$ $\sqrt{\dfrac{9}{4}} = \dfrac{3}{2}$

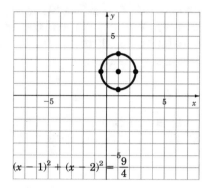

$(x - 1)^2 + (x - 2)^2 = \dfrac{9}{4}$

Write the equation in standard form of the circle with the given
coordinates as center and the given value of r as radius:

25. (-3, -8), r = 16
     $(x - h)^2 + (y - k)^2 = r^2$ Standard form of the
                                            equation of a circle with
                                            center at (h, k) and
                                            radius r.
     $[x - (-3)]^2 + [y - (-8)]^2 = (16)^2$ Substitute -3 for h, -8
                                             for k and 16 for r.
         $(x + 3)^2 + (y + 8)^2 = 256$

C

Write the coordinates of the center and the length of the radius
of the following:

29. $4x^2 + 4y^2 = 9$

                                            Divide by 4 on both sides.
                                            Standard form of a circle
     $x^2 + y^2 = \dfrac{9}{4}$ with center at (0, 0).
     Center (0, 0)
     Radius $\dfrac{3}{2}$

341

Write the equation in standard form of the circle with the given coordinates as the center and the given value of r as the radius:

33.   (1, 1) $r = \dfrac{7}{2}$

$$(x - h)^2 + (y - k)^2 = r^2$$ Standard form with center at (h, k) and radius r.

$$(x - 1)^2 + (y - 1)^2 = \dfrac{49}{4}$$ Substitute 1 for h, 1 for k, and $\dfrac{7}{2}$ for r.

The following is an equation whose graph is a circle. Write it in standard form, then write the coordinates of the center and the length of the radius:

37. $x^2 + y^2 - 10x + 4y + 17 = 0$

$$x^2 - 10x \qquad + y^2 + 4y \qquad = -17$$
$$x^2 - 10x + 25 + y^2 + 4y + 4 = -17 + 25 + 4$$ Complete the square in both x and y.
$$(x - 5)^2 + (y + 2)^2 = 12$$

Center (5, -2)

Radius $2\sqrt{3}$

h = 5, k = -2, and $r = 2\sqrt{3}$ (or $\sqrt{12}$).

Given the center and the radius of a circle, write an equation in the form $ax^2 + by^2 + cx + dy + e = 0$:

41.   (-2, -7), r = 1.2

$$(x - h)^2 + (y - k)^2 = r^2$$ Standard form.
$$[x - (-2)]^2 + [y - (-7)]^2 = (1.2)^2$$ Substitute -2 for h, -7 for k, and 1.2 for r.

$$(x + 2)^2 + (y + 7)^2 = 1.44$$
$$x^2 + 4x + 4 + y^2 + 14y + 49 = 1.44$$ Multiply.
$$x^2 + y^2 + 4x + 14y + 51.56 = 0$$ Subtract 1.44 from both sides.

D

45. If $(x - h)^2 + (y - k)^2 = 0$ is the equation of the degenerate circle (i.e., the point (h, k)), what is the graph of $x^2 + y^2 - 4x + 8y + 20 = 0$?

$x^2 + y^2 - 4x + 8y + 20 = 0$
$x^2 - 4x + y^2 + 8y + 20 = 0$ Regroup.
To copmlete the square for $x^2 - 4x$ the constant 4 is needed, while to complete the square for $y^2 + 8x$ the constant 16 is needed.

$x^2 - 4x + 4 + y^2 + 8y + 16 = 0$ Replace 20 with $4 + 16$.
        $(x - 2)^2 + (y + 4)^2 = 0$ Factor.
or $(x - 2)^2 + [y - (-4)] = 0$ The "center" of the degenerate circle is $(2, -4)$.

The graph of $x^2 + y^2 - 4x + 8y + 20 = 0$ is the point $(2, -4)$.

49. A tangent line is a line that intersects a circle at exactly one point. The tangent line is also perpendicular to the radius at the point where the tangent intersects the circle. Given the circle $x^2 + y^2 = 25$, find the equation of the tangent line at the point $(3, -4)$.

$x^2 + y^2 = 25$
The center of this circle is $(0, 0)$ with radius 5.

$$m_1 = \frac{y_2 - y_1}{x_2 - x_1} = \frac{-4 - 0}{3 - 0}$$
     Find the slope, m, of a line from the center $(0, 0)$ to the point $(3, -4)$.

$$= -\frac{4}{3}$$

$$m_2 = \frac{3}{4}$$
     The slope, m_2, of the tangent line is the negative reciprocal of m_1. Use the point-slope

$$y - y_1 = m(x - x_1)$$
$$y - (-4) = \frac{3}{4}(x - 3)$$
     formula with $m = \frac{3}{4}$ and $(x_1, y_1) = (3, -4)$.

$4y + 16 = 3x - 9$
    $25 = 3x - 4y$
     Clear the fraction by multiplying both sides by 4.

The equation of the tangent line at the point $(3, -4)$ is $3x - 4y = 25$.

53. If you graph $x^2 + y^2 = 25$ and $x^2 + y^2 = 9$ on the same graph, what is the area of the region between the circles?

$x^2 + y^2 = 25$ has a radius 5.

$A = \pi r^2$
$A = 25\pi$ Replace r with 5.

$x^2 + y^2 = 9$ has a radius 3.

$A = \pi r^2$
$A = 9\pi$ Replace r with 3.

Area between the circles = $25\pi - 9\pi$ The area between the
                        = 16π circles is the difference of the two areas.

The area between the circles is 16π.

57. Find the equation of the line tangent to the circle $(x - 3)^2 + (y + 1)^2 = 25$ at the point $(0, 3)$.

$(x - 3)^2 + (y + 1)^2 = 25$

The center is at $(3, -1)$ and the radius is 5.

$$m_1 = \frac{y_2 - y_1}{x_2 - x_1}$$

First find the slope, m_1, of the line segment (radius) from the center of the circle $(3, -1)$ to the point on the circle $(0, 3)$.

$$= \frac{-1 - 3}{3 - 0} = \frac{-4}{3}$$

$$= -\frac{4}{3}$$

The line tangent to the circle is perpendicular to the radius whose slope is $-\frac{4}{3}$.

$$m_2 = \frac{3}{4}$$

Slope, m_2, of the tangent line is the negative reciprocal of $-\frac{4}{3}$.

$$y - y_1 = m(x - x_1)$$

Use the point-slope formula to find the equation of the tangent line.

$$y - 3 = \frac{3}{4}(x - 0)$$

$4y - 12 = 3x$
$3x - 4y = -12$

The equation of the line tangent to the given circle at the point $(0, 3)$ is $3x - 4y = 12$.

Solve:

61. $4\sqrt{x - 1} + 2 = 1 - 4x$

$\quad\quad\quad 4\sqrt{x - 1} = -1 - 4x$ Isolate the radical.
$\quad\quad 16(x - 1) = 1 + 8x + 16x^2$ Square both sides.
$\quad\quad 16x - 16 = 16x^2 + 8x + 1$
$\quad\quad\quad\quad\quad 0 = 16x^2 - 8x + 17$
$a = 16,\ b = -8,\ c = 17$ Use quadratic formula to solve for x.

$$x = \frac{8 \pm \sqrt{8^2 - 4(16)(17)}}{2 \cdot 16}$$

$$= \frac{8 \pm \sqrt{-1024}}{32}$$ No real-number solutions.

The solution set is the empty set, $\emptyset$.

65. $\quad 2y^2 - 47y + 66 = 0$
$\quad (2y - 3)(y - 22) = 0$ Factor.

$2y - 3 = 0$ or $y - 22 = 0$
$\quad y = \frac{3}{2}$ or $\quad\quad y = 22$

The solution set is $\left\{\frac{3}{2},\ 22\right\}$.

EXERCISES 8.3 ELLIPSES AND HYPERBOLAS

A

Draw the graphs of the following conic sections:

1. $x^2 - y^2 = 1$ Original equation.

$\dfrac{x^2}{1} - \dfrac{y^2}{1} = 1$ Write in standard form. This is the standard form of a hyperbola with intercepts at (1, 0) and (-1, 0).

Table of values.

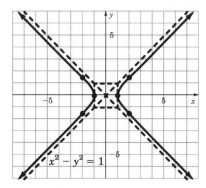

x	y
± 1	0
$\pm\sqrt{2}$	± 1
± 2	$\pm\sqrt{3}$

345

5. $\dfrac{x^2}{4} - \dfrac{y^2}{4} = 1$

Standard form, with intercepts at (2, 0) and (-2, 0).

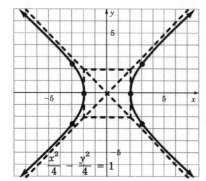

Table of values:

x	y
±2	0
±4	$\pm 2\sqrt{3}$
±6	$\pm 4\sqrt{2}$

Write the equation of this conic section in standard form:

9.

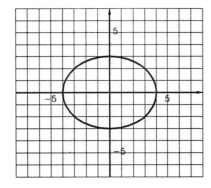

The x-intercepts are (4, 0) and (-4, 0), so a = 4.
The y-intercepts are (0, 3) and (0, -3), so b = 3.

$\dfrac{x^2}{16} + \dfrac{y^2}{9} = 1$

Substitute 4 for a and 3 for b in the standard form.

B

Draw the graphs of the following conic sections:

13. $\dfrac{x^2}{4} - \dfrac{y^2}{9} = 1$

a = 2 and b = 3.

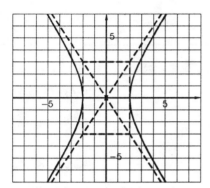

Since $a^2 = 4$, and $b^2 = 9$.

Sketch the rectangle with vertices at (2, 3), (2, -3), (-2, 3) and (-2, -3). Extend the diagonals.
Fit the graph to the extended diagonals by plotting the intercepts, and additional points from a table of values.

Table of values:

x	y
±2	0
±4	±5.20
±6	±8.49

17. $\dfrac{x^2}{4} + \dfrac{y^2}{16} = 1$

a = 2, b = 4.

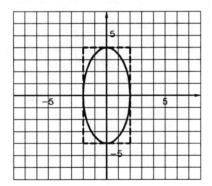

Since $a^2 = 4$, and $b^2 = 16$.

Draw the rectangle that is the boundary of the ellipse. The rectangle has vertices at (2, 4), (-2, 4), (2, -4) and (-2, -4).
The graph of the ellipse is inscribed within the rectangle.
Plot additional points from a table of values.

x	y
0	±4
±1	±3.46
±2	0

C

21. $\dfrac{x^2}{27} + \dfrac{y^2}{3} = 3$

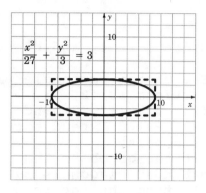

Standard form of an ellipse with intercepts at $(\pm 3\sqrt{3},\ 0)$ and $(0,\ \pm\sqrt{3})$.

Table of values:

x	y
$\pm 3\sqrt{3} \approx \pm 5.2$	0
0	$\pm\sqrt{3} \approx \pm 1.7$
± 1	$\pm\dfrac{4\sqrt{5}}{3} \approx \pm 3.0$
± 3	$\pm 2\sqrt{2} \approx \pm 2.8$

25. $5x^2 - 6y^2 = 60$

   $\dfrac{x^2}{12} - \dfrac{y^2}{10} = 1$

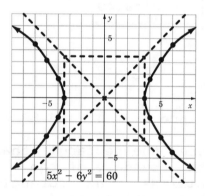

Original equation.

Write in standard form. Divide both sides by 60. This is the standard form of a hyperbola with intercepts at $(\pm 2\sqrt{3},\ 0)$.

Table of values:

x	y
$\pm 2\sqrt{3} \approx \pm 3.5$	0
± 4	$\approx \pm 1.8$
± 5	$\approx \pm 3.3$
± 6	$\approx \pm 4.5$

Write the equation of this conic section in standard form:

29.

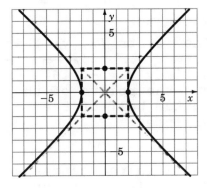

The x-intercepts are (2, 0) and (-2, 0).

The y-intercepts of the rectangle are (0, 2) and (0, -2).

$$\frac{x^2}{4} - \frac{y^2}{4} = 1$$

From the graph, write the coordinates of the x-intercepts of the hyperbola. The x-intercepts are ($\pm a$, 0). Then write the y-intercepts of the rectang.e The y-intercepts of the rectangle are at (0, $\pm b$). Substitute the values of a and b in the standard form $\frac{x^2}{a^2} - \frac{y^2}{b^2} = 1$.

D

33. If we solve $4x^2 + y^2 = 16$ (which is an ellipse) for y, we get $y = \pm\sqrt{16 - 4x^2}$ OR $y = \pm 2\sqrt{4 - x^2}$. What is the graph of $y = +\sqrt{16 - 4x^2}$? What is the graph of $y = -\sqrt{16 - 4x^2}$?

$y = +\sqrt{16 - 4x^2}$

x	y
± 2	0
0	4
± 1	3.46

The points (x, y) that satisfy $y = +\sqrt{16 - 4x^2}$ are the points of the top half of the ellipse.

349

$$y = -\sqrt{16 - 4x^2}$$

x	y
0	-4
±2	0
±1	-3.46

The points (x, y) that satisfy $y = -\sqrt{16 - 4x^2}$ are the points of the bottom half of the ellipse.

37. A circular plate has an elliptical hole in the center as in the diagram. Write the equation of the plate and the hole.

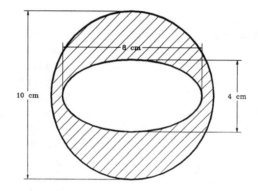

Equation of the plate: (circle)

Since the diameter indicated is 10 cm, the radius is 5 cm.

$x^2 + y^2 = 5^2$
$x^2 + y^2 = 25$

Substitute r = 5 in the standard form of a circle.

Equation of the hole: (ellipse)

Since the vertical opening is 8 cm, a = 4. The horizontal opening is 4 cm, so b = 2.

$\dfrac{x^2}{a^2} + \dfrac{y^2}{b^2} = 1$

Standard form of an ellipse.

$\dfrac{x^2}{4^2} + \dfrac{y^2}{2^2} = 1$

Substitute 4 for a and 2 for b.

$\dfrac{x^2}{16} + \dfrac{y^2}{4} = 1$

41. An ellipse is the set of all points such that the sum of the distances from two fixed points is a constant. Place two small nails in a board about 6 inches apart. Make a loop of string that will fit over the two nails with a little slack. Place a pencil inside the loop and move the pencil around the two nails keeping the loop taut. Why is the curve an ellipse?

The string is a constant length. The two nails are the fixed points. No matter what position the pencil is at, it will be kept at a constant distance from the two fixed points by the string, and so the curve formed is satisfied by the conditions in the definition of an ellipse.

CHALLENGE EXERCISES

45. Given $\dfrac{x^2}{4} = \dfrac{y^2}{9} = 1$, write the equations of the two asymptotes (the diagonals of the rectangle you use to graph the hyperbola). Are the diagonals perpendicular?

a = 2, b = 3

Since $a^2 = 4$, a = 2. And since $b^2 = 9$, b = 3.

Two points on the diagonal having a positive slope are (2, 3) and (-2, -3), in opposite corners of the rectangle.

$$\text{slope} = \frac{y_2 - y_1}{x_2 - x_1} = \frac{-3 - 3}{-2 - 2} = \frac{-6}{-4} = \frac{3}{2}$$

Use (2, 3) to replace (x_1, y_1) and (-2, -3) to replace (x_2, y_2).

$$y - 3 = \frac{3}{2}(x - 2)$$
$$y - 3 = \frac{3}{2}x - 3$$
$$y = \frac{3}{2}x$$

Use the point-slope formula to generate the equation of the diagonal. Substitute (2, 3) for (x_1, y_1) and $\frac{3}{2}$ for m.

Now in the same way, determine the equation of the second diagonal, using the two points: (-2, 3) and (2, -3).

351

$$\text{Slope} = \frac{y_2 - y_1}{x_2 - x_1} = \frac{-3 - 3}{2 - (-2)} = -\frac{6}{4} = -\frac{3}{2}$$

$$y - y_1 = m(x - x_1)$$
$$y - 3 = -\frac{3}{2}[x - (-2)]$$

Replace (x_1, y_1) with $(-2, 3)$ and m with $-\frac{3}{2}$.

$$y - 3 = -\frac{3}{2}x - 3$$
$$y = -\frac{3}{2}x$$

The equations of the two asymptotes are: $y = \frac{3}{2}x$ and $y = -\frac{3}{2}x$.

Since $\left(\frac{3}{2}\right)\left(-\frac{3}{2}\right) \neq -1$, these two lines are not perpendicular.

MAINTAIN YOUR SKILLS

Solve:

49. $\dfrac{b + 7}{2b^2 + 3b} = \dfrac{b + 1}{2b^2 - b - 6}$

Note the denominators:
$2b^2 + 3b = b(2b + 3)$ and
$2b^2 - b - 6 = (2b+3)(b-2)$.

$$(b + 7)(b - 2) = b(b + 1)$$
$$b^2 + 5b - 14 = b^2 + b$$
$$4b = 14$$
$$b = \frac{7}{2}$$

Multiply both fractions by the L.C.D.:
$b(2b + 3)(b - 2)$.

The solution set is $\left\{\dfrac{7}{2}\right\}$.

Solve for w:

53. $w + a(w + b) = w + 1$
$w + aw + ab = w + 1$
$aw + ab = 1$
$aw = 1 - ab$

Multiply.
Subtract w from both sides. Subtract ab from both sides.

$$w = \frac{1 - ab}{a}$$

Divide both sides by a.

352

A

Solve:

1. $\begin{cases} y = x^2 & (1) \\ y = 4x & (2) \end{cases}$

$$y = 4x$$
$$x^2 = 4x$$
$$x^2 - 4x = 0$$
$$x(x - 4) = 0$$
$$x = 0 \text{ or } x = 4$$

Equation (1) is solved for y. Substitute x^2 for y in equation (2).

$$y = 0^2 = 0$$
$$(0, 0)$$
$$y = 4^2 = 16$$
$$(4, 16)$$

Substitute both values of x in equation (1).
x = 0
Ordered pair.
x = 4
Ordered pair.

The solution set is {(0, 0), (4, 16)}.

5. $\begin{cases} x^2 + y^2 = 16 & (1) \\ x + y = 4 & (2) \end{cases}$

$$x + y = 4$$
$$x = -y + 4$$

Solve equation (2) for x.

$$x^2 + y^2 = 16$$
$$(-y + 4)^2 + y^2 = 16$$
$$y^2 - 8y + 16 + y^2 = 16$$
$$2y^2 - 8y = 0$$
$$2y(y - 4) = 0$$
$$y = 0 \text{ or } y = 4$$

Substitute $-y + 4$ for x in equation (1).

Substitute both values of y in equation (2).
y = 0

$$x + 0 = 4$$
$$x = 4$$

$$(0, 4)$$

Ordered pair.

The solution set is {(0, 4), (4, 0)}.

9. $\begin{cases} x^2 - y^2 = 1 & (1) \\ \quad\ y = 2 & (2) \end{cases}$

$x^2 - 2^2 = 1$
$\qquad x^2 = 5$

$\qquad\ x = \pm\sqrt{5}$

Equation (2) is solved for y, so substitute 2 for y in equation (1).

The solution set is $\{(-\sqrt{5},\ 2),\ (\sqrt{5},\ 2)\}$.

B

13. $\begin{cases} x^2 - y^2 = 4 & (1) \\ x\ +\ y = 1 & (2) \end{cases}$

$x + y = 1$
$\quad\ x = -y + 1$

Solve equation (2) for x.

$x^2 - y^2 = 4$
$(-y + 1)^2 - y^2 = 4$
$y^2 - 2y + 1 - y^2 = 4$
$\qquad\qquad -2y = 4$

$\qquad\qquad\ \ y = -\dfrac{3}{2}$

Substitute $-y + 1$ for x in equation (1).

$x + y = 1$

$x - \dfrac{3}{2} = 1$

$\quad\ x = \dfrac{5}{2}$

Substitute $-\dfrac{3}{2}$ for y in

equation (2).

The solution set is $\left\{ \left(\dfrac{5}{2},\ -\dfrac{3}{2} \right) \right\}$.

17. $\begin{cases} \quad\ y = x^2 - 4x + 2 & (1) \\ y - x = 2 & (2) \end{cases}$

$y - x = 2$
$y - 2 = x$

Solve equation (2) for x.

$y = (y - 2)^2 - 4(y - 2) + 2$

Substitute $y - 2$ for x in equation (1).

$y = y^2 - 4y + 4 - 4y + 8 + 2$
$0 = y^2 - 9y + 14$
$0 = (y - 7)(y - 2)$
$y = 7 \text{ or } y = 2$

Standard form.
Solve for y.

$y - 2 = x$
$7 - 2 = x$
$\quad\ 5 = x$

Substitute 7 for y in equation (2).

$$y - 2 = x$$
$$2 - 2 = x$$
$$0 = x$$

Substitute 2 for y in equation (2).

The solution set is $\{(5, 7), (0, 2)\}$.

C

Solve:

21. $\begin{cases} 3x^2 + 4y^2 = 12 & (1) \\ 3x + 2y = 0 & (2) \end{cases}$

$$2y = -3x$$
$$y = -\frac{3x}{2}$$

Solve equation (2) for y.

$$3x^2 + 4\left(-\frac{3}{2}x\right)^2 = 12$$

Substitute $y = -\frac{3}{2}x$ in equation (1).

$$3x^2 + 4\left(\frac{9}{4}x^2\right) = 12$$
$$3x^2 + 9x^2 = 12$$
$$12x^2 = 12$$
$$x^2 = 1$$
$$x = \pm 1$$

$$3(1) + 2y = 0$$
$$2y = -3$$
$$y = -\frac{3}{2}$$

Substitute $x = 1$ in equation (2).

$$3(-1) + 2y = 0$$
$$2y = 3$$
$$y = \frac{3}{2}$$

Substitute $x = -1$ in equation (2).

The solution set is $\left\{\left(1, -\frac{3}{2}\right)\right\}, \left\{\left(-1, \frac{3}{2}\right)\right\}$.

25. $\begin{cases} 4x^2 - y^2 = 15 & (1) \\ 2x - y + 3 = 0 & (2) \end{cases}$

$$2x + 3 = y$$

Solve equation (2) for y.

$$4x^2 - (2x + 3)^2 = 15$$
$$4x^2 - [4x^2 + 12x + 9] = 15$$
$$4x^2 - 4x^2 - 12x - 9 = 15$$
$$-12x = 24$$
$$x = -2$$

Substitute $y = 2x + 3$ in equation (1) and solve for x.

$$2(-2) - y + 3 = 0$$
$$-4 - y + 3 = 0$$
$$-1 = y$$

Substitute x = -2 in equation (2) and solve for y.

The solution set is {(-2, -1)}.

29. $\begin{cases} x^2 + y^2 = 9 & (1) \\ y - x = 3\sqrt{2} & (2) \end{cases}$

$$y = x + 3\sqrt{2}$$

Solve equation (2) for y.

$$x^2 + (x + 3\sqrt{2})^2 = 9$$

Substitute $y = x + 3\sqrt{2}$ in equation (1).

$$x^2 + x^2 + 6\sqrt{2}x + 18 = 9$$
$$2x^2 + 6\sqrt{2}x + 9 = 0$$

$$x = \frac{-6\sqrt{2} \pm \sqrt{72 - 4(2)(9)}}{4}$$

$$= \frac{-6\sqrt{2}}{4}$$

$$= -\frac{3\sqrt{2}}{2}$$

$$y - \left(-\frac{3\sqrt{2}}{2}\right) = 3\sqrt{2}$$

Substitute $x = -\frac{3\sqrt{2}}{2}$ in equation (2).

$$2y + 3\sqrt{2} = 6\sqrt{2}$$

Clear the fraction.

$$2y = 3\sqrt{2}$$

Multiply by 2.

$$y = \frac{3\sqrt{2}}{2}$$

The solution set is $\left\{\left(-\frac{3\sqrt{2}}{2}, \frac{3\sqrt{2}}{2}\right)\right\}$.

D

33. The cost of producing n units of steel at the Jimeo Corporation is given by c = 10n. The money received from the sale of n units of steel is given by $c = \frac{1}{10}n^2$. Find the number of units that Jimeo Corporation needs to produce to break even.

$10n = \frac{1}{10}n^2$

To break even, costs must equal sales. Set 10n (cost) equal to $\frac{1}{10}n^2$ (sales).

$100n = n^2$

Clear the fraction.

$n^2 - 100n = 0$
$n(n - 100) = 0$
$n = 0$ or $n - 100 = 0$
$\qquad\qquad n = 100$

Reject n = 0, since the number of units sold will be a positive number.

The Jimeo Corporation must sell 100 units to break even.

37. Is it possible to find the real numbers whose difference is 1, and whose product is 1? If so, find all the possibilities.

Let x represent one number, so x + 1 represents the other.

$x(x + 1) = 1$

The product of x and (x + 1) is 1.

$x^2 + x - 1 = 0$

$x = \dfrac{-1 \pm \sqrt{1 - 4(1)(-1)}}{1}$

$\quad = \dfrac{-1 \pm \sqrt{5}}{2}$

If $x = \dfrac{-1 + \sqrt{5}}{2}$

then $x + 1 = \dfrac{-1 + \sqrt{5}}{2} + 1$

$\qquad\qquad = \dfrac{-1 + \sqrt{5} + 2}{2} = \dfrac{1 + \sqrt{5}}{2}$

And if $x = \dfrac{-1 - \sqrt{5}}{2}$

then $x + 1 = \dfrac{-1 - \sqrt{5}}{2} + 1$

$= \dfrac{-1 - \sqrt{5} + 2}{2} = \dfrac{1 - \sqrt{5}}{2}$

So there are two pairs of numbers that satisfy the given conditions: $\dfrac{-1 + \sqrt{5}}{2}$ and $\dfrac{1 + \sqrt{5}}{2}$ is one pair and $\dfrac{-1 - \sqrt{5}}{2}$ and $\dfrac{1 - \sqrt{5}}{2}$ is the other.

STATE YOUR UNDERSTANDING

41. In a system containing two equations, one a hyperbola and the other a line, how many different possibilities are there for the number of points of intersection?

    The line and the hyperbola may intersect in one point, two points, or none at all.

CHALLENGE EXERCISES

45. Graph the following system and estimate the solution:
    $$\begin{cases} x^2 + y^2 + 4x - 6y = 7 \\ 2x + y = 5 \end{cases}$$

    $x^2 + 4x + y^2 - 6y = 7$

    $(x^2 + 4x + 4) + (y^2 - 6y + 9) = 7 + 13$ Complete the square in both x and y.

    $(x + 2)^2 + (y - 3)^2 = 20$ Standard equation of a circle, with center at (-2, 3) and radius $\sqrt{20}$.

358

$$2x + y = 5$$
$$y = -2x + 5$$

Slope-intercept form.

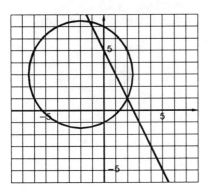

From the graph the solutions can be estimated at (-1, 7) and (2, 1).

MAINTAIN YOUR SKILLS

Evaluate:

49. $\begin{vmatrix} 11 & -3 & 4 \\ 0 & -2 & 1 \\ 6 & 1 & 3 \end{vmatrix}$

Use the first column and expand by minors.

$$= 11 \begin{vmatrix} -2 & 1 \\ 1 & 3 \end{vmatrix} - 0 + 6 \begin{vmatrix} -3 & 4 \\ -2 & 1 \end{vmatrix}$$
$$= 11[-6 - 1] + 6[-3 + 8]$$
$$= 11(-7) + 6(5)$$
$$= -77 + 30$$
$$= -47$$

Solve by linear combinations:

53. $\begin{cases} 3x + 8y = 9 & (1) \\ 2x + 3y = 5 & (2) \end{cases}$

$$\begin{array}{r} -6x - 16y = -18 \\ 6x + 9y = 15 \\ \hline -7y = -3 \end{array}$$

Multiply equation (1) by -2 and equation (2) by 3.

$$y = \frac{3}{7}$$

$$3x + 8\left(\frac{3}{7}\right) = 9$$

Substitute $y = \frac{3}{7}$ in equation (1).
Clear the fraction.

$$21x + 24 = 63$$
$$21x = 39$$
$$x = \frac{39}{21}$$
$$x = \frac{13}{7}$$

The solution set is $\left\{\left(\frac{13}{7}, \frac{3}{7}\right)\right\}$.

EXERCISES 8.5   SOLVING SYSTEMS OF EQUATIONS INVOLVING QUADRATIC EQUATIONS II

A

Solve:

1. $\begin{cases} x^2 + y^2 = 12 & (1) \\ x^2 - y^2 = 4 & (2) \end{cases}$

$$\begin{aligned} x^2 + y^2 &= 12 \\ \underline{x^2 - y^2} &= \underline{4} \\ 2x^2 &= 16 \\ x^2 &= 8 \\ x &= \pm\sqrt{8} = \pm 2\sqrt{2} \end{aligned}$$

Add equations (1) and (2).

$$\begin{aligned} (\pm 2\sqrt{2})^2 + y^2 &= 12 \\ 8 + y^2 &= 12 \\ y^2 &= 4 \\ y &= \pm 2 \end{aligned}$$

Substitute $x = \pm 2\sqrt{2}$ in equation (1).

The solution set is $\{(2\sqrt{2},\ 2),\ (2\sqrt{2},\ -2),\ (-2\sqrt{2},\ 2),\ (-2\sqrt{2},\ -2)\}$.

5. $\begin{cases} x^2 + y^2 = 4 & (1) \\ 4x^2 + y^2 = 4 & (2) \end{cases}$

$$\begin{aligned} -x^2 - y^2 &= -4 \\ \underline{4x^2 + y^2} &= \underline{4} \\ 3x^2 &= 0 \\ x &= 0 \end{aligned}$$

Multiply equation (1) by -1, and form a combination by adding the result to equation (2).

$$\begin{aligned} 0^2 + y^2 &= 4 \\ y^2 &= 4 \\ y &= \pm 2 \end{aligned}$$

Substitute $x = 0$ in equation (1).

The solution set is $\{(0,\ 2),\ (0,\ -2)\}$.

B

9. $\begin{cases} 2x^2 + y^2 = 6 & (1) \\ 4x^2 + 3y^2 = 16 & (2) \end{cases}$

$\begin{aligned} -4x^2 - 2y^2 &= -12 \\ \underline{4x^2 + 3y^2} &= \underline{16} \\ y^2 &= 4 \\ y &= \pm 2 \end{aligned}$

Form a combination by multiplying equation (1) by -2 and adding to equation (2). The terms containing x^2 will disappear.

$\begin{aligned} 2x^2 + (\pm 2)^2 &= 6 \\ 2x^2 + 4 &= 6 \\ 2x^2 &= 2 \\ x^2 &= 1 \\ x &= \pm 1 \end{aligned}$

Substitute $y = \pm 2$ in equation (1).

The solution set is $\{(1, 2), (1, -2), (-1, 2), (-1, -2)\}$.

13. $\begin{cases} x^2 + y^2 = 9 & (1) \\ y = x^2 + 3 & (2) \end{cases}$

$x^2 + (x^2 + 3)^2 = 9$

Substitute $y = x^2 + 3$ in equation (1).

$\begin{aligned} x^2 + x^4 + 6x^2 + 9 &= 9 \\ x^4 + 7x^2 &= 0 \\ x^2(x^2 + 7) &= 0 \end{aligned}$

$x^2 = 0$ or $x^2 + 7 = 0$
$x = 0$ or $x^2 = -7$

Contradiction.

$\begin{aligned} y &= 0^2 + 3 \\ y &= 3 \end{aligned}$

Substitute $x = 0$ in equation (2).

The solution set is $\{(0, 3)\}$.

C

17. $\begin{cases} 3x^2 + 4y^2 = 52 & (1) \\ 4x^2 - 5y^2 = 59 & (2) \end{cases}$

Solve by linear combinations. Multiply equation (1) by -4 and equation (2) by 3.

$\begin{aligned} -12x^2 - 16y^2 &= -208 \\ \underline{12x^2 - 15y^2} &= \underline{177} \\ -31y^2 &= -31 \\ y^2 &= 1 \\ y &= \pm 1 \end{aligned}$

$$3x^2 + 4(1) = 52$$
$$3x^2 = 48$$
$$x^2 = 16$$
$$x = \pm 4$$

Substitute $y = \pm 1$ in equation (1).

The solution set is $\{(\pm 4, \pm 1)\}$ or $\{(4, 1, (4, -1), (-4, 1), (-4, -1)\}$.

21.
$$\begin{cases} x^2 + y^2 = 2 & (1) \\ y = x^2 & (2) \end{cases}$$

$$x^2 + x^2 = 2$$
$$2x^2 = 2$$
$$x^2 = 1$$
$$x = \pm 1$$

Substitute $y = x^2$ in equation (1).

$$y = (\pm 1)^2 = 1$$

Substitute $x = \pm 1$ in equation (2).

The solution set is $\{(\pm 1, 1)\}$ or $\{(1, 1), (-1, 1)\}$.

D

25. In Example 4, if the garden was represented by the equation $4x^2 + y^2 = 400$, where would the pine trees be placed?

EXAMPLE 4

A landscape architect wants to plant four pine trees 13 feet from the center of an ellipse but on the perimeter of the ellipse. If the elliptical garden is represented by the equation $4x^2 + 10y^2 = 1000$, where on the graph will he indicate the position of the trees? (*Hint*: If the trees are 13 feet from the center of the ellipse, they are on the circle whose equation is $x^2 + y^2 = 169$.)

$$\begin{cases} 4x^2 + y^2 = 400 & (1) \\ x^2 + y^2 = 169 & (2) \end{cases}$$

Solve the system of equations.

$$4x^2 + y^2 = 400$$
$$\underline{- x^2 - y^2 = -169}$$
$$3x^2 \qquad = 231$$
$$x^2 = 77$$

Multiply equation (2) by -1, and add to equation (1).

$$x = \pm\sqrt{77} \text{ or approximately } \pm 8.8.$$

$$4(77) + y^2 = 400$$
$$y^2 = 92$$

Substitute $x^2 = 77$ in equation (1).

$$y = \pm\sqrt{92} = \pm 2\sqrt{23} \text{ or approximately } \pm 9.6.$$

The trees should be placed at the points whose coordinates are those given here:

SOLUTIONS	
Exact	Approximate
$(\sqrt{77},\ 2\sqrt{23})$	$(8.8,\ 9.6)$
$(\sqrt{77},\ -2\sqrt{23})$	$(8.8,\ -9.6)$
$(-\sqrt{77},\ 2\sqrt{23})$	$(-8.8,\ 9.6)$
$(-\sqrt{77},\ -2\sqrt{23})$	$(-8.8\ -9.6)$

29. An orbiting body follows the path $4x^2 + y^2 = 16$ (where all the units are in 100,000 miles). A comet follows the path $y = x^2 - 4$. Will the two paths intersect? Where?

$$\begin{cases} 4x^2 + y^2 = 16 & (1) \\ \qquad y = x^2 - 4 & (2) \end{cases}$$
Solve the system of equations.

$$4x^2 + (x^2 - 4)^2 = 16$$
$$4x^2 = x^4 - 8x^2 + 16 = 16$$
$$x^4 - 4x^2 = 0$$
$$x^2(x^2 - 4) = 0$$
$$x^2(x + 2)(x - 2) = 0$$
$$x = 0 \text{ or } x = -2 \text{ or } x = 2$$

Substitute $y = x^2 - 4$ in equation (1).

$$x = 0$$
$$y = 0 - 4$$
$$y = -4$$

Substitute each replacement for x in equation (2).

$$x = 2$$
$$y = 2^2 - 4 = 0$$

$$x = -2$$
$$y = (-2)^2 - 4 = 0$$

Yes, the two paths will cross at $(0, -4)$, $(2, 0)$, and $(-2, 0)$.

33. Two circular gears (one smaller than the other) are designed to touch at exactly one point. If their equations are $x^2 + y^2 = 25$ and $x^2 + y^2 - 14x + 45 = 0$, will they fit the requirement? Where?

$$\begin{cases} \qquad\quad x^2 + y^2 = 25 & (1) \\ x^2 + y^2 - 14x + 45 = 0 & (2) \end{cases}$$
Solve the system of equations.

$$\begin{array}{r} -x^2 - y^2 \qquad\qquad\quad = -25 \\ \underline{x^2 + y^2 - 14x + 45 = \quad 0} \\ -14x - 45 = -25 \\ -14x = -70 \\ x = 5 \end{array}$$

Multiply equation (1) by -1, and add to equation (1).

$$25 + y^2 = 25$$
$$y^2 = 0$$
$$y = 0$$

Substitute x = 5 in equation (1).

Since the two equations have one intersection at (5, 0), the two gears will fit the requirement.

CHALLENGE EXERCISES

37. Graph the "logo" in Exercise 32 to verify your points of intersection.

$$\begin{cases} x^2 + y^2 = 25 & (1) \\ x^2 + 4y^2 = 100 & (2) \end{cases}$$

System given in Exercise 32.

$$x^2 + y^2 = 25 \qquad (1)$$

Standard form of a circle with center at (0, 0) and radius 5.

$$x^2 + 4y^2 = 100 \qquad (2)$$
$$\frac{x^2}{100} + \frac{y^2}{25} = 1$$

Divide both sides by 100. Standard form of an ellipse where a = 10 and b = 5.

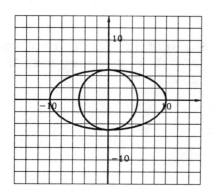

From the graph, the two intersections are (0, 5) and (0 -5).

364

Combine. Reduce if possible:

41. $\dfrac{5}{x + 3} - \dfrac{x}{x - 5} + \dfrac{x^2 - 2}{x^2 - 2x - 15}$

$= \dfrac{5}{x + 3} - \dfrac{x}{x - 5} + \dfrac{x^2 - 2}{(x + 3)(x - 5)}$ Factor.

$= \dfrac{5}{x + 3} \cdot \dfrac{x - 5}{x - 5} - \dfrac{x}{x - 5} \cdot \dfrac{x + 3}{x + 3} + \dfrac{x^2 - 2}{(x + 3)(x - 5)}$

Build each fraction to
have a common denominator:
$(x + 3)(x - 5)$.

$= \dfrac{5(x - 5) - x(x + 3) + x^2 - 2}{(x + 3)(x - 5)}$

$= \dfrac{5x - 25 - x^2 - 3x + x^2 - 2}{(x + 3)(x - 5)}$

$= \dfrac{2x - 27}{(x + 3)(x - 5)}$

Simplify:

45. $\dfrac{\dfrac{1}{x - 1} - \dfrac{1}{x + 2}}{\dfrac{1}{x + 2} + \dfrac{1}{x - 1}}$

$= \left(\dfrac{\dfrac{1}{x - 1} - \dfrac{1}{x + 2}}{\dfrac{1}{x + 2} + \dfrac{1}{x - 1}}\right) \left(\dfrac{(x + 2)(x - 1)}{(x + 2)(x - 1)}\right)$

Multiply by the
common denominator
$(x + 2)(x - 1)$.

$= \dfrac{\left(\dfrac{1}{x - 1}\right)(x + 2)(x - 1) - \left(\dfrac{1}{x + 2}\right)(x + 2)(x - 1)}{\left(\dfrac{1}{x + 2}\right)(x + 2)(x - 1) + \left(\dfrac{1}{x - 1}\right)(x + 2)(x - 1)}$

$= \dfrac{(x + 2) - (x - 1)}{(x - 1) + (x + 2)}$

$= \dfrac{x + 2 - x + 1}{x - 1 + x + 2}$

$= \dfrac{3}{2x + 1}$

CHAPTER 9

FUNCTIONS

EXERCISES 9.1 RELATIONS AND FUNCTIONS

A

State the range and domain of each relation:

1. {(1, 1), (2, 4), (3, 9), (4, 16)}

    Domain: {1, 2, 3, 4}   The set of first numbers
                                    (x-values) of the ordered
                                    pairs.

    Range: {1, 4, 9, 16}   The set of second numbers
                                    (y-values) of the ordered
                                    pairs.

5. {-4, 8), (4, 8), (5, 8), (0, 8), (1, 8)}

    Domain: {-4, 0, 1, 4, 5}   List in order the x-values
                                        of the ordered pairs.

    Range: {8}   The only value used as a
                                      second number is 8.

Write each of the following relations as a set of ordered pairs, and state its range:

9. $\{(x, y) \mid y = |5x - 6| + 4, \ x = -3, -1, 1, 3\}$

    If $x = -3$, $y = |5 \cdot -3 - 6| + 4$ Find a value of y for each
                    $= 21 + 4$ value of x.
                    $= 25$

      $x = -1$, $y = |5 \cdot -1 - 6| + 4$
                 $= 11 + 4$
                 $= 15$

      $x = 1$, $y = |5 \cdot 1 - 6| + 4$
              $= 1 + 4$
              $= 5$

      $x = 3$, $y = |5 \cdot 3 - 6| + 4$
              $= 9 + 4$
              $= 13$

    The set is {(-3, 25), (-1, 15), (1, 5), (3, 13)}.

    Range: {5, 13, 15, 25}   Set of y-values.

From the graphs, state the range and domain of each relation.
Identify those that are functions:

13.

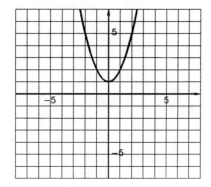

Domain: {x|x ∈ ℝ}

Range: {y|y ≥ 1}

There is a point
corresponding to every
point on the x-axis.
The lowets value of the
graph is at y = 1.

The graph is a function by the vertical line test.

17.

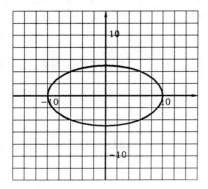

Domain: {x|-10 ≤ x ≤ 10}

Range: {y|-5 ≤ y ≤ 5}

There are points
corresponding to x-values
between -10 and 10.
The highest y-value is 5,
and the lowest is -5.

The graph does not represent a function because it fails the
vertical line test.

B

State the range and domain of the relation defined by each of the
following:

21. $y = \sqrt{100 - x^2}$

    Domain: $\{x \mid -10 \leq x \leq 10\}$ Restrict x-values since
                                              the principal square root
                                            of a negative number is
                                            not real.

    Range: $\{y \mid 0 \leq y \leq 10\}$ The range is nonnegative
                                                and less than or equal to
                                            10.

    The relation is a function since for each value of x, there
    is exactly one y-value, in this case, a principal square
    root.

25. $y^2 = 3x^2 + 1$

    Domain: $\{x \mid x \in \mathbb{R}\}$ There are no restrictions
                                              on x.

    Range: $\{y \mid \leq -1 \text{ or } y \geq 1\}$ When the x-value is zero,
                                              the y-values are ±1. For
                                            each other value of x, the
                                            y-values are either
                                            greater than 1 or less
                                          than -1.

    This relation is not a function since for every value of x,
    there are two y-values.

29. $y = |x| - 7$

    Domain: $\{x \mid x \in \mathbb{R}\}$ No restrictions on x.
    Range: $\{y \mid y \geq -7\}$ The least y-value is -7.

    This relation is a function since for each x-value, there is
    exactly one y-value.

33. $y = x^3$

    Domain: $\{x \mid x \in \mathbb{R}\}$ No restrictions on x.
    Range: $\{y \mid y \in \mathbb{R}\}$ No restrictions on y.

    This relation is a function since for each x-value, there is
    exactly one y-value.

37. $y = \dfrac{1}{x}$

Domain: $\{x \mid x \ne 0\}$

Range: $\{y \mid y \ne 0\}$

If x = 0, the denominator is zero, but division by zero is not defined. Since the numerator cannot equal zero, the fraction cannot equal zero.

This relation is a function since for each x-value, there is exactly one y-value.

c

State the range and domain of the relation defined by each of the following and identify those that are functions.

41. $y = \dfrac{x - 2}{x^2 - 9}$

Domain: $\{x \mid x \ne \pm 3\}$

Range: $\{y \mid y \in \mathbb{R}\}$

Restrict x-values in the denominator, since division by zero is not defined. When x = ±3, the denominator is zero. No restrictions on y.

This relation is a function. For each x-value there is only one y-value.

45. $y = \dfrac{x - 7}{x^2 - x - 6} = \dfrac{x - 7}{(x - 3)(x + 2)}$

At x = 3 or x = -2, the denominator is zero.

Domain: $\{x \mid x \ne -2, 3\}$

Range: $\{y \mid y \in \mathbb{R}\}$

Restrict the x-values since division by zero is undefined.

This relation is a function. Each x-value has only one corresponding y-value.

Draw the graph of each of the following. Use the graph to
determine the range and domain. Use the vertical line test to
identify functions:

49.   y = |x - 5|

x	y
-1	6
0	5
1	4
2	3
3	2
4	1
5	0
6	1
7	2
8	3
9	4

First, find some ordered
pairs in the relation.

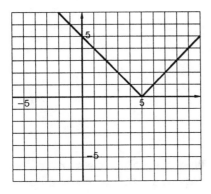

Plot the points and draw
the graph.

By the vertical line test
this relation is a
function.

Domain: $\{x \mid x \in \mathbb{R}\}$
Range: $\{y \mid y \geq 0\}$

From the graph.

A function.

53.   y = |3x - 6| -1

Make a table of values.

x	y
-1	8
0	5
1	2
2	-1
3	2
4	5
5	8

371

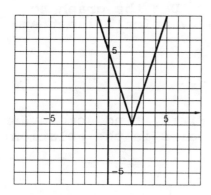

Plot the points, and draw
the graph.

The relation is a function
since its graph passes the
vertical line test.

Domain: $\{x \mid x \in \mathbb{R}\}$
Range: $\{y \mid y \geq -1\}$

A function.

From the graph.

57. $x^2 + y^2 = 16$

Standard form of a circle
with center at (0, 0) and
radius 4.

x	y
0	±4
±4	0
±2	±3.5
±1	3.9

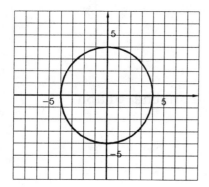

The graph does not
represent a function since
it does not pass the
vertical line test.

Domain: $\{x \mid -4 < x < 4\}$
Range: $\{y \mid -4 < y < 4\}$

Not a function.

372

61. $4x^2 + y^2 = 16$

$\dfrac{x^2}{4} + \dfrac{y^2}{16} = 1$

Divide by 16 to write the equation as standard form of an ellipse with a = 2, and b = 4.

x	y
±2	0
0	±4
±1	±3.5

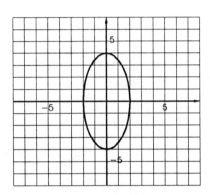

Domain: $\{x \mid -2 \leq x < 2\}$
Range: $\{y \mid -4 \leq y \leq 4\}$

From the graph.

D

65. During the growing season, the height (h) (in feet) of an elephant-ear bamboo is related to the number of good growing days (d) by the formula $h = \sqrt{\dfrac{1}{2}d^2 + 4}$. Find the height of the bamboo after 4, 6, 10, or 15 growing days. Express the answers as ordered pairs, rounding the height to the nearest tenth of a foot.

d = 4

Replace d by each value.

$h = \sqrt{\dfrac{1}{2}(4)^2 + 4}$

$= \sqrt{8 + 4}$

$= \sqrt{12}$

≈ 3.5

373

$$d = 6$$

$$h = \sqrt{\frac{1}{2}(6)^2 + 4}$$

$$= \sqrt{18 + 4}$$

$$= \sqrt{22}$$

$$\approx 4.7$$

$$d = 10$$

$$h = \sqrt{\frac{1}{2}(10)^2 + 4}$$

$$= \sqrt{50 + 4}$$

$$= \sqrt{54}$$

$$\approx 7.3$$

$$d = 15$$

$$h = \sqrt{\frac{1}{2}(15)^2 + 4}$$

$$= \sqrt{112.5 + 4}$$

$$= \sqrt{116.5}$$

$$\approx 10.8$$

The answers are $\{(4, 3.5), (6, 4.7), (10, 7.3), (15, 10.8)\}$.

69. The height of a ball thrown from the top of a 100 ft building is given by $h = -16t^2 + 40t + 100$. What is the height 3 seconds after the ball is thrown? What is the height 4 seconds ater the ball is thrown? Express answers as ordered pairs (t, h).

$$h = -16t^2 + 40t + 100$$
$$= -16(3)2 + 40(3) + 100 \qquad \text{Replace t with 3.}$$
$$= 76$$
$$h = -16(4)^2 + 40(4) + 100 \qquad \text{Replace t with 4.}$$
$$= 4$$

After 3 seconds the height of the ball is 76 ft.
After 4 seconds the height of the ball is 4 ft.
As ordered pairs (t, h), the answers can be expressed:
(3, 76), and (4, 4).

STATE YOUR UNDERSTANDING

73. What is the difference between a function and a relation? What is meant by the domain and range of a function?

Both functions and relations are sets of ordered pairs. The difference is that in sets representing functions no x-value has more than one y-value corresonding to it.

The set of x-values of the ordered pairs is the domain of the function, while the set of y-values of the ordered pairs is the range of the function.

CHALLENGE PROBLEMS

77. Find two ordered pairs for the equation $x^2 + y^2 = 25$, to prove that the graph of this circle is not a function.

x	y
0	5
0	-5

(0, 5)
(0, -5)

The x-value zero has two y-values, so the graph is not a function.

MAINTAIN YOUR SKILLS

81. Find the equation, in standard form, of the line through the points (6, -5) and (-3, -2).

$$m = \frac{y_2 - y_1}{x_2 - x_1} = \frac{-2 - (-5)}{-3 - 6}$$

Find the slope. Use $(6, -5) = (x_1, y_1)$ and $(-3, -2) = (x_2, y_2)$.

$$= -\frac{3}{9}$$

$$= -\frac{1}{3}$$

$$y - y_1 = m(x - x_1)$$

Replace m with $-\frac{1}{3}$, x with 6 and y with -5 in the point-slope formula.

$$y - (-5) = -\frac{1}{3}(x - 6)$$

$$y + 5 = -\frac{1}{3}(x - 6)$$

$$3y + 15 = -x + 6$$
$$x + 3y = -9$$

Clear the fraction.

Solve:

85.
$$y = x - 7 \quad (1)$$
$$2x + 3y - 4 = 0 \quad (2)$$

$$2x + 3(x - 7) - 4 = 0$$
$$2x + 3x - 21 - 4 = 0$$
$$5x = 25$$
$$x = 5$$

Replace y = x - 7 in equation (2) and solve for x.

$$y = 5 - 7$$
$$= -2$$

Replace x = 5 in equation (1).

The solution set is {(5, -2)}.

EXERCISES 9.2   FUNCTION NOTATION

A

If f(x) = {(0.5, 2), (1, 3), (2, 5), (3, 7), (4, 9), (5, 11) and
   g(x) = {(0.5, 0.5), (1, 2), (2, 5), (3, 8), (4, 11), (5, 14)},
find the following:

1. f(3) = 7   In the function f, x = 3
                                          is paired with y = 7.

5. f(5) = 11   In the function f, x = 5
                                          is paired with y = 11.

9. g(5) - f(4)
        14 - 9   In the function g, 5 is
                                          paired with 14, while in
                                          the function f, 4 is
                                          paired with 9.

        14 - 9 = 5

Find the indicated values of f(x):

13. f(x) = 16 - x; f(-6), f(6)

     f(-6) = 16 - (-6) = 22   Replace x by -6.
     f(6) = 16 - 6 = 10   Replace x by 6.

17. f(x) = x^2 + x; f(-2), f(7)

     f(-2) = $(-2)^2$ + (-2)   Replace x with -2.
           = 4 - 2
           = 2

     f(7) = $(7)^2$ + 7   Replace x with 7.
           = 49 + 7
           = 56

B

If f(x) = x^2 - 3x - 18 and g(x) = x - 6, evaluate the following:

21. f(6) = 6^2 - 3(6) - 18   Replace x with 6 in f(x).
          = 36 - 18 - 18
          = 0

25. f(g(9)) = f(9 - 6)   Replace x with 9 in g(x).
             = f(3)
             = 3^2 - 3(3) - 18   Replace x with 3 in f(x).
             = -18

29. $f(5) - g(4) = [5^2 - 3(5) - 18] - [4 - 6]$ Replace x with 5
        $= [25 - 15 - 18] - [-2]$ in f(x) and x with
        $= (-8) - (-2)$ 4 in g(x).
        $= -6$

If $f(x) = 7x = 13$, $g(x) = \sqrt{x + 9}$, and $h(x) = x^2 - 6x + 1$,
evaluate the following:

33. $g(16) - h(-2) = [\sqrt{16 + 9}] - [(-2)^2 - 6(-2) + 1]$
        $= [\sqrt{25}] - [4 + 12 + 1]$
        $= 5 - 17$
        $= -12$

37. $f(2) - g(27) = [7(2) + 13] - [\sqrt{27 + 9}]$
        $= [27] - [\sqrt{36}]$
        $= 27 - 6$
        $= 21$

C

Given $f(x) = 6 - x$, $g(x) = x^2 - 4x + 7$, and $h(x) = |14 - 3x|$,
evaluate the following:

41. $f(-27) + h(-23) = [6 - (-27)] + [|14 - 3(-23)|]$
        $= (33) + (83)$
        $= 116$

45. $f(a) = 6 - a$

49. $g(a) - f(a) = [a^2 - 4a + 7] - [6 - a]$
        $= a^2 - 4a + 7 - 6 + a$
        $= a^2 - 3a + 1$

53. $h(f(20)) = h(6 - 20)$ Replace x with 20 in f(x).
        $= h(-14)$
        $= |14 - 3(-14)|$ Replace x with -14 in
                    h(x).
        $= 56$

57. $h(a^2 - 1) = |14 - 3(a^2 - 1)|$ Replace x with a^2 - 1 in
                    h(x).
        $= |14 - 3a^2 + 3|$

        $= |17 - 3a^2|$

61.   What is the difference between y = 2x and f(x) = 2x?

      Since y and f(x) mean the same thing there is no difference.
      Both represent the y value of the function f.

CHALLENGE EXERCISES

65.   Given f(x) = 2x - 3, find f(x + h). Substitute what you
      found into the expression $\frac{f(x + h) - f(x)}{h}$ and simplify.

      f(x + h) = 2(x + h) - 3   Replace x with x + h in
               = 2x + 2h - 3   f(x).

      $\frac{f(x + h) - f(x)}{h}$ Replace f(x + h) with
                                       2x + 2h - 3 and simplify.

      $= \frac{[2x + 2y - 3] - [2x - 3]}{h}$ Replace f(x) with 2x - 3.

      $= \frac{2x + 2h - 3 - 2x + 3}{h}$

      $= \frac{2h}{h}$

      = 2

69.   Given the two functions f(x) = 2x - 3 and g(x) = 3x + 5, we
      say that (f + g)(x) = (2x - 3) + (3x + 5) which simplifies
      to (f + g)(x) = 5x + 2. Find f(2), g(2), and then find
      (f + g)(2). Does f(2) + g(2) = (f + g)(2)?

      f(2) = 2(2) - 3
           = 1

      g(2) = 3(2) + 5
           = 11

      f(2) + g(2) = 1 + 11 = 12

      (f + g)(2) = 5(2) + 2   Replace x with 2 in
                 = 12   5x + 2, which is
                                     (f + g)(x).

      Yes, f(2) + g(2) = (f + g)(2).

73. Given $g(x) = x^2 + 1$, find $g(1)$, $g(3)$, then find $g(1 + 3)$. Does $g(1) + g(3) = g(1 + 3)$?

$$g(1) = 1^2 + 1$$
$$= 2$$

$$g(3) = 3^2 + 1$$
$$= 10$$

$$g(1 + 3) = g(4) = 4^2 + 1$$
$$= 17$$

$$g(1) + g(3) = 2 + 10$$
$$= 12$$

No, $g(1) + g(3) \neq g(1 + 3)$.

MAINTAIN YOUR SKILLS

Given: $y = -x^2 + 12x - 32$.

77. Draw the graph.

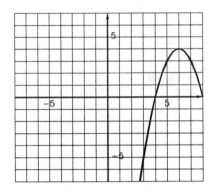

From exercises 74, 75, and 76, the vertex is at (6, 4), the axis of symmetry is x = 6, and the x-intercepts are (4, 0) and (8, 0).

Given: $y = 2x^2 + 8x - 3$.

81. Draw the graph.

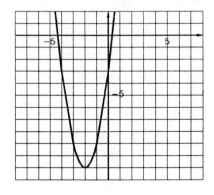

From exercises 78, 79, and 80, the vertex is at (-2, -11), the axis of symmetry is x = -2, and the x-intercepts are approximately (0.3, 0) and (-4.3, 0).

A

Clasify and graph each of the following functions:

1. $f(x) = 3x - 5$
     $y = 3x - 5$

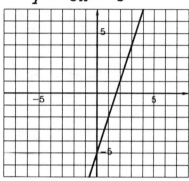

A linear function with m = 3 and y-intercept at -5.

x	y
-1	-8
0	-5
3	4

5. $f(x) = \frac{1}{2}x - 2$

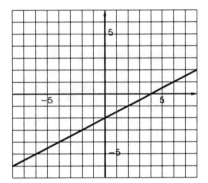

A linear function in slope-intercept form.

x	y
-2	-3
0	-2
2	-1
4	0
6	1

9. $f(x) = 2x^2 + 3$

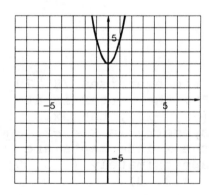

Quadratic function with a =2, b = 0, c = 3.

The graph of this function is a parabola opening upward with vertex at (0, 3).

B

13. $f(x) = x^2 + 5x + 6$

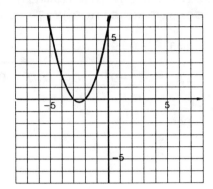

Quadratic function with
a = 1, b = 5, and c = 6.

The graph of this function
is a parabola opening
upward with vertex at
$\left(-\frac{5}{2},\ 0\right)$.

x	-3	-2	-4	-1
y	0	0	2	2

17. $f(x) = \sqrt{x^2 - 8x + 16}$

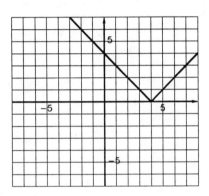

Square root function.

x	0	2	4	6	8
y	4	2	0	2	4

Observe that the given
equation is equivalent to

$y = \sqrt{(x - 4)^2}$.   Therefore
there is no restriction on
x, and $y \geq 0$.

C

21. $f(x) = \sqrt{x^2 + 4}$

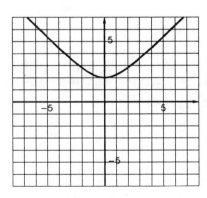

Square root function.

x	0	±2	±4
y	2	2.8	4.5

25. $f(x) = \dfrac{x}{x + 1}$

Rational function.

Notice that the domain is restricted since $x \neq -1$.

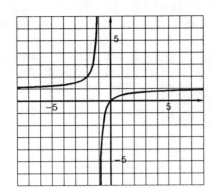

x	-5	-2	-1.1	-0.9	0	2	5
y	1.25	2	11	-9	0	0.7	0.8

29. $f(x) = |x - 4|$

Absolute value function.

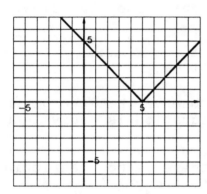

x	0	2	4	5	8
y	4	2	0	1	3

STATE YOUR UNDERSTANDING

33. Why, arithmetically, do the y-values get closest o 1 as the x-values get very large in Exercise 26?

$f(x) = \dfrac{x}{x - 1}$

As the x-values get very large, the ratio they form gets closer to the value one, since at very large values of x, the numerator and denominator differ only by 1.
For example, at x = 1000

$\dfrac{x}{x + 1} = \dfrac{1000}{1001} \approx 0.999$

and at x = 1000000

$\dfrac{x}{x + 1} = \dfrac{1000000}{1000001} \approx 0.999999$

For each of the following, make a table of values and graph.

37. $g(x) = -\sqrt{2 - x}$

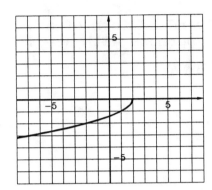

x	2	1	0	-1	-2
y	0	-1	$-\sqrt{2}$	$-\sqrt{3}$	-2

41. $g(x) = |x^2 - 4|$

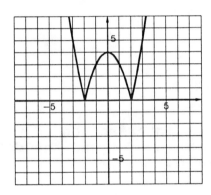

x	0	1	2	3	-1	-2	-3
y	4	3	0	5	3	0	5

45. Write the equation of the circle with center at the origin and a radius of 25.

$x^2 + y^2 = r^2$ Standard form.

$x^2 + y^2 = 25^2$ Replace r with 25.
$x^2 + y^2 = 625$

Identify the conic section whose equation is:

49. $9x^2 - 25y^2 = 144$

$$\frac{9x^2}{144} - \frac{25y^2}{144} = \frac{144}{144}$$ Divide by 144 and rewrite in standard form.

$$\frac{x^2}{16} - \frac{y^2}{\left(\frac{144}{25}\right)} = 1$$ Standard form of a hyperbola with a = 4 and b = $\frac{12}{5}$.

9.4   OPERATIONS WITH FUNCTIONS

A

Find $(f + g)(x)$, $(f - g)(x)$, $(f \cdot g)(x)$ and $\left(\frac{f}{g}\right)(x)$.   State the domain of each result:

1. $f(x) = x + 4;\ g(x) = x - 4$

$$(f + g)(x) = f(x) + g(x)$$
$$= (x + 4) + (x - 4)$$
$$= 2x$$

Domain of $f = \{x \mid x \in \mathbb{R}\}$
Domain of $g = \{x \mid x \in \mathbb{R}\}$
Domain of $f + g = \{x \mid x \in \mathbb{R}\}$ The intersection of the domain of f and the domain of g.

$$(f - g)(x) = f(x) - g(x)$$
$$= (x + 4) - (x - 4)$$
$$= x + 4 - x + 4$$
$$= 8$$

Domain of $f - g = \{x \mid x \in \mathbb{R}\}$

$$(f \cdot g)(x) = f(x) \cdot g(x)$$
$$= (x + 4)(x - 4)$$
$$= x^2 - 16$$

Domain of $f \cdot g = \{x \mid x \in \mathbb{R}\}$

$$\left(\frac{f}{g}\right)(x) = \frac{f(x)}{g(x)} = \frac{x + 4}{x - 4}$$

Domain of $\frac{f}{g} = \{x \mid x \in \mathbb{R}\ \text{and}\ x \neq 4\}$ Restrict x since division by zero is not defined.

384

5. $f(x) = 2x - 9$; $g(x) = x^2$

Domain of $f = \{x | x \in \mathbb{R}\}$

Domain of $g = \{x | x \in \mathbb{R}\}$

$$(f + g)(x) = f(x) + g(x)$$
$$= (2x - 9) + (x^2)$$
$$= x^2 + 2x - 9$$

Domain of $f + g = \{x | x \in \mathbb{R}\}$

$$(f - g)(x) = f(x) - g(x)$$
$$= (2x - 9) - (x^2)$$
$$= -x^2 + 2x - 9$$

Domain of $f - g = \{x | x \in \mathbb{R}\}$

$$(f \cdot g)(x) = f(x) \cdot g(x)$$
$$= (2x - 9)(x^2)$$
$$= 2x^3 - 9x^2$$

Domain of $f \cdot g = \{x | x \in \mathbb{R}\}$

$$\left(\frac{f}{g}\right)(x) = \frac{f(x)}{g(x)} = \frac{2x - 9}{x^2}$$

Domain of $\frac{f}{g} = \{x | x \in \mathbb{R} \text{ and } x \neq 0\}$ Division by zero is undefined.

Let $f(x) = 3x = 4$ and $g(x) = x^2 - 5x + 4$.
Find the following:

13. $(f \cdot g)(1) = f(1) \cdot g(1)$
$$= [3(1) + 4][1^2 - 5(1) + 4]$$
$$= (7)(0)$$
$$= 0$$

17. $(g/f) = \dfrac{g(1)}{f(1)} = \dfrac{1^2 - 5(1) + 4}{3(1) + 4} = \dfrac{0}{7} = 0$

B

Find the functions f and g such that the given operation on f and g will yield the given function:

21. $h(x) = 4x^2 - 5x + 2$; $f + g$

  If $f(x) = x^2 - 2x +$ and $g(x) = 3x^2 - 3x + 1$
  then $(f + g)(x) = (x^2 - 2x + 1) + (3x^2 - 3x + 1)$
  $= 4x^2 - 5x + 2$

25. $h(x) = x^2 + 8$; $f + g$

  If $f(x) = x^2 + x + 8$ and $g(x) = -x$
  then $(f + g)(x) = (x^2 + x + 8) + (-x)$
  $= x^2 + 8$

29. $h(x) = 4x - 3$; f/g

  If $f(x) = 4x^2 - 3x$ and $g(x) = x$, $x \neq 0$
  then $\dfrac{f}{g} = \dfrac{4x^2 - 3x}{x} = \dfrac{x(4x - 3)}{x}$
  $= 4x - 3$

Evaluate each of the following given that $f(x) = 2x - 3$ and $g(x) = 6x^2 - 13x + 6$:

33. $f(g(-1.5)) = f[6(-1.5)^2 - 13(-1.5) + 6]$ Replace x with
                                                          -1.5 in g(x)

  $= f(39)$
  $= 2(39) - 3$ Replace x with -45
  $= 75$ in f(x).

37. $f(g(0)) = f[6(0)^2 - 13(0) + 6]$ Replace x with 0 in g(x).
  $= f(6)$
  $= 2(6) - 3$ Reaplce x with 6 in f(x).
  $= 9$

C

Find (fog)(x) and (gof)(x) for each of the following:

41. $f(x) = 2x + 4$; $g(x) = x^2$

  $(fog)(x) = f[g(x)]$ Evaluate f at g(x).
  $= f[x^2]$ Replace g(x) with x^2.
  $= 2(x^2) + 4$ Evaluate f at x^2.
  $= 2x^2 + 4$

386

$$(gof)(x) = g[f(x)]$$
$$= g[2x + 4]$$
$$= (2x + 4)^2$$
$$= 4x^2 + 16x + 16$$

Evaluate g at f(x).
Replace 2x + 4 in g(x).

45. f(x) = x = 2; g(x) = x - 2

$$(fog)(x) = f[g(x)]$$
$$= f[x - 2]$$
$$= (x - 2) + 2$$
$$= x$$

Evaluate g(x).
Evaluate f at (x - 2).

$$(gof)(x) = g[f(x)]$$
$$= g[x + 2]$$
$$= (x + 2) - 2$$
$$= x$$

Evaluate f(x).
Evaluate g(x) at x + 2.

Let f(x) = 3x and g(x) = x - 1. Evaluate each of the following:

49. $(fog)(-1) = f[g(-1)]$
$$= f[-1 - 1]$$
$$= f(-2)$$
$$= 3(-2)$$
$$= -6$$

Evaluate g(-1).
Evaluate f at -2.

53. $(fof)(1) = f[f(1)]$
$$= f[3 \cdot 1]$$
$$= f(3)$$
$$= 3 \cdot 3$$
$$= 9$$

Evaluate f(1).
Evaluate f(3).

57. $(gof)(2) = g[f(2)]$
$$= g[3 \cdot 2]$$
$$= g(6)$$
$$= 6 - 1$$
$$= 5$$

Evaluate f(2).

Evaluate g(6).

Find (a) f(x + h), (b) f(x + h) - f(x), (c) $\dfrac{f(x + h) - f(x)}{h}$.
Item (c) is important in the study of calculus:

61. $f(x) = x^2$

(a) $f(x + h) = (x + h)^2 = x^2 + 2hx + h^2$

(b) $f(x + h) - f(x) = (x^2 + 2hx + h^2) - (x^2)$
$$= 2hx + h^2$$

(c) $\dfrac{f(x + h) - f(x)}{h} = \dfrac{2hx + h^2}{h}$
$$= \dfrac{h(2x + h)}{h}$$
$$= 2x + h$$

65. $f(x) = 2x^2 - 4x$

   (a) $f(x + h) = 2(x + h)^2 - 4(x + h)$
                    $= 2(x^2 + 2hx + h^2) - 4x - 4y$
                    $= 2x^2 + 4hx + 2h^2 - 4x - 4h$

   (b) $f(x + h) - f(x) = (2x^2 + 4hx + 2h^2 - 4x - 4h) - (2x^2 - 4x)$
                            $= 2x^2 + 4hx + 2h^2 - 4x - 4y - 2x^2 + 4x$
                            $= 4yx + 2h^2 - 4h$

   (c) $\dfrac{f(x + h) - f(x)}{h} = \dfrac{4hx + 2h^2 - 4h}{h}$

                                    $= \dfrac{h(4x + 2h - 4)}{h}$

                                    $= 4x + 2h - 4$

STATE YOUR UNDERSTANDING

69. If $f(x)$ and $g(x)$ are two functions, why is it true that
     $(f + g)(a) = f(a) + g(a)$ for any "a" in the domain of both
     but that $f/g(a) = \dfrac{f(a)}{g(a)}$ may not be true?

     $(f/g)(a)$ will not equal $\dfrac{f(a)}{g(a)}$ when $g(a) = 0$, since division
     by zero is not defined.

CHALLENGE EXERCISES

Given $g(x) = x - 3$ and $f(x) = \dfrac{1}{x}$, find:

73. $f(g(0))$

        $g(0) = 0 - 3 = -3$ First evaluate $g(0)$.
     $f(g(0)) = f(-3)$ Then evaluate f at $g(0)$.
             $= \dfrac{1}{-3}$ or $-\dfrac{1}{3}$

Given $g(x) = 2x - 3$ and $f(x) = \sqrt{x + 1}$, find:

77. $g(f(5))$

        $f(5) = \sqrt{5 + 1}$ Evaluate $f(5)$.
             $= \sqrt{6}$
     $g(f(5)) = g(\sqrt{6})$ Evaluate g at $f(5)$.
             $= 2(\sqrt{6}) - 3$
             $= 2\sqrt{6} - 3$

Given $g(x) = x + 2$ and $f(x) = \dfrac{x}{x-1}$, find:

81. $g(f(3))$

$$f(3) = \frac{3}{3-1}$$ Evaluate $f(3)$.

$$= \frac{3}{2}$$

$$g(f(3)) = g\left(\frac{3}{2}\right)$$ Evaluate g at $f(3)$.

$$= \frac{3}{2} + 2$$

$$= \frac{7}{2}$$

85. $g(f(-1))$

$$f(-1) = \frac{-1}{-1-1}$$ Evaluate $f(-1)$.

$$= \frac{-1}{-2}$$

$$= \frac{1}{2}$$

$$g(f(-1)) = g\left(\frac{1}{2}\right)$$ Evaluate g at $f(-1)$.

$$= \frac{1}{2} + 2$$

$$= \frac{5}{2}$$

MAINTAIN YOUR SKILLS

Solve:

89. $\dfrac{3y - 1}{4} - 4 = \dfrac{4y - 5}{5} + \dfrac{7y + 5}{10}$

$$5(3y - 1) - 80 = 4(4y - 5) + 2(7y + 5)$$ Multiply both
$$15y - 5 - 80 = 16y - 20 + 14y + 10$$ sides by 20
$$15y - 85 = 30y - 10$$ clearing all
$$-15y = 75$$ fractions.
$$y = -5$$

93. $x - 4 = \sqrt{7}$
$$x = -4 + \sqrt{7}$$ Subtract 4 from both
                         sides.

A

Find the inverse of the given function:

1. $\{(3, 2), (4, 3), (5, 4), (6, 5)\}$

    $f^{-1} = \{(2, 3), (3, 4), (4, 5), (5, 6)\}$ Interchange the
    x and y values.

5. $\{(9,3), (4,2), (1,1), (0,0), (-1,-1), (-4,-2), (-9,-3)\}$

    $f^{-1} = \{(3,9), (2,4), (1,1), (0,0), (-1,-1), (-2,-4),(-3,-9)\}$

Write the equation of the inverse of the following functions:

9. $f(x) = -x + 3$
    $$y = -x + 3$$ This equation defines f.

    $$x = -y + 3$$ Interchange x and y.
    $$x - 3 = -y$$
    $$y = -x + 3$$

    $$f^{-1} = -x + 3$$

13. $f(x) = 5x + 4$
    $$y = 5x + 4$$ The function f.

    $$x = 5y + 4$$ Interchange x and y.

    $$x - 4 = 5y$$
    $$y = \frac{1}{5}x - \frac{4}{5}$$

    $$f^{-1} = \frac{1}{5}x - \frac{4}{5}$$ The inverse of f.

B

17. $f(x) = \frac{2}{3}x - 5$

    $$y = \frac{2}{3}x - 5$$

    $$x = \frac{2}{3}y - 5$$ Interchange x and y.
    $$3x = 2y - 15$$
    $$3x + 15 = 2y$$
    $$f^{-1} = \frac{3}{2}x + \frac{15}{2}$$ The inverse of f.

21. $f(x) = x^2$
  $y = x^2$
  $x = y^2$ Interchange x and y.

  $y = \pm\sqrt{x}, \ x \geq 0$ Solve for y. Restrict the domain.

  $f^- = \pm\sqrt{x}, \ x \geq 0$

Determine whether or not each function graphed is a one-to-one function:

25.

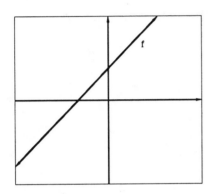

The graph is the graph of a one-to-one function since no horizontal line will intersect it more than one time.

29.

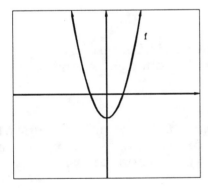

A horizontal line will intersect this function twice.
This is not the graph of a one-to-one function.

C

Each equation defines a function f. Write the equation of f^{-1}, sketch the graph of f and f^{-1} on the same set of axes, and state whether or not f^{-1} is a function:

33. $f(x) = 4x - 3$
           $y = 4x - 3$ Definition of f.
           $x = 4y - 3$ Interchange x and y.
    $x + 3 = 4y$
           $y = \frac{1}{4}x + \frac{3}{4}$

    $f^{-1} = \frac{1}{4}x + \frac{3}{4}$ The inverse of f.

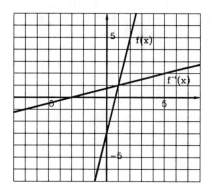

        Yes, f^{-1} is a function.

37. $f(x) = x^2 + 2$
           $y = x^2 + 2$ Definition of f.
           $x = y^2 + 2$ Interchange x and y.
    $x - 2 = y^2$
           $y = \pm\sqrt{x - 2}$

    $f^{-1}(x) = \pm\sqrt{x - 2}$ No, f^{-1} is not a function.
                                             For each $x \neq 2$, there are
                                             two values of y.

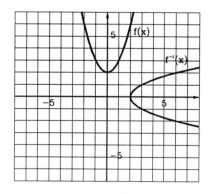

Find the composite function $(f \circ f^{-1})$:

41. $f(x) = x - 5$
       $y = x - 5$ Write the inverse of f.
       $x = y - 5$
       $x + 5 = y$
       $f^{-1}(x) = x + 5$
       $(f \circ f^{-1})(x) = f[f^{-1}(x)]$
                        $= f(x + 5)$
                        $= (x + 5) - 5$
                        $= x$

STATE YOUR UNDERSTANDING

45.   What is the inverse of a function, and when will the inverse
      be a function?

      The inverse of the function f is the set of ordered pairs
      formed by interchanging the x- and y-values of the ordered
      pairs of f.

      The inverse of f will be a function when the f is a
      one-to-one function; that is when for each y in the range of
      f there is only one x in its domain.

CHALLENGE EXERCISES

49.   In Exercise 13, you were asked to find the equation for f^{-1}.
      Using the result, find $f(f^{-1}(1))$ and $f^{-1}(f(1))$.

      $f(x) = 5x + 4$
      $f^{-1}(x) = \frac{1}{5}x - \frac{4}{5}$

      $f(f^{-1}(1)) = f\left[\frac{1}{5}(1) - \frac{4}{5}\right]$ Evaluate $f^{-1}(1)$.

                $= f\left(-\frac{3}{5}\right)$

                $= 5\left(-\frac{3}{5}\right) + 4$ Evaluate $f\left(-\frac{3}{5}\right)$.
                $= 1$

      $f^{-1}(f(1)) = f^{-1}[5 \cdot 1 + 4]$ Evaluate $f(1)$.
                $= f^{-1}(9)$ Evaluate $f^{-1}(9))$.
                $= \frac{1}{5}(9) - \frac{4}{5}$
                $= \frac{9}{5} - \frac{4}{5}$
                $= 1$

393

Using your best artistic skills, draw the inverse of the
following curve:

53.

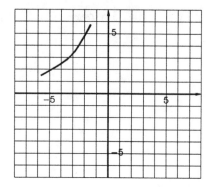

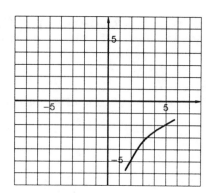

Since the points (5, -2)
and (2, -5) are on the
given graph, the points
(-2, 5) and (-5, 2) are on
the graph of its inverse.

57. Using the symmetry with respect to y = x, draw the inverse
of the curve you just drew in Exercise 56. Is it a
function? Is it one-to-one?

The inverse of the curve you drew for Exercise 56 is a
function, but it is not a one-to-one-function.

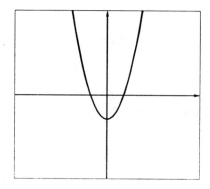

Solve:

61.
$$p^2 + \frac{1}{2}p = \frac{1}{8}$$

$$8p^2 + 4p = 1$$
Multiply by 8 to clear the fractions.

$$8p^2 + 4p - 1 = 0$$

$$p = \frac{-4 \pm \sqrt{16 - 4(8)(-1)}}{2 \cdot 8}$$
Use the quadratic formula to solve for p. a = 8, b = 4 and c = -1.

$$= \frac{-4 \pm \sqrt{48}}{16}$$

$$= \frac{-4 \pm 4\sqrt{3}}{16}$$

$$= \frac{4(-1 \pm \sqrt{3})}{4 \cdot 4}$$

$$= \frac{-1 \pm \sqrt{3}}{4}$$

The solution set is $\left\{\dfrac{-1 \pm \sqrt{3}}{4}\right\}$.

Solve for x:

65. $dx^2 - fx + 1 = 0$

$$x = \frac{f \pm \sqrt{f^2 - 4(d)(1)}}{2 \cdot d}$$
a = d, b = -f, and c = 1. Use the quadratic formula.

$$= \frac{f \pm \sqrt{f^2 - 4d}}{2d}$$

CHAPTER 10

EXPONENTIAL AND LOGARITHMIC FUNCTIONS

EXERCISES 10.1 EXPONENTIAL FUNCTIONS

A

Complete the following table of values:

1. $f(x) = 3^x$

x	-3	-2	-1	0	1	2	3
f(x)	$\frac{1}{27}$	$\frac{1}{9}$	$\frac{1}{3}$	1	3	9	27

$f(-3) = 3^{-3} = \dfrac{1}{3^3} = \dfrac{1}{27}$

$f(-2) = 3^{-2} = \dfrac{1}{3^2} = \dfrac{1}{9}$

$f(-1) = 3^{-1} = \dfrac{1}{3^1} = \dfrac{1}{3}$

$f(0) \quad = 3^0 = 1$

$f(1) \quad = 3^1 = 3$

$f(2) \quad = 3^2 = 9$

$f(3) \quad = 3^3 = 27$

5. $f(x) = \left(\dfrac{2}{3}\right)^x$

x	-3	-2	-1	0	1	2	3
f(x)	$\dfrac{27}{8}$	$\dfrac{9}{4}$	$\dfrac{3}{2}$	1	$\dfrac{2}{3}$	$\dfrac{4}{9}$	$\dfrac{8}{27}$

$f(-3) = \left(\dfrac{2}{3}\right)^{-3} = \left(\dfrac{3}{2}\right)^{3} = \dfrac{27}{8}$

$f(-2) = \left(\dfrac{2}{3}\right)^{-2} = \left(\dfrac{3}{2}\right)^{2} = \dfrac{9}{4}$

$f(-1) = \left(\dfrac{2}{3}\right)^{-1} = \left(\dfrac{3}{2}\right)^{1} = \dfrac{3}{2}$

$f(0) = \left(\dfrac{2}{3}\right)^{0} = 1$

$f(1) = \left(\dfrac{2}{3}\right)^{1} = \dfrac{2}{3}$

$f(2) = \left(\dfrac{2}{3}\right)^{2} = \dfrac{4}{9}$

$f(3) = \left(\dfrac{2}{3}\right)^{3} = \dfrac{8}{27}$

9. $f(x) = 10^x$

x	-3	-2	-1	0	1	2
f(x)	$\dfrac{1}{1000}$	$\dfrac{1}{100}$	$\dfrac{1}{10}$	1	10	100

Make a table of values.

$f(-3) = 10^{-3} = \dfrac{1}{10^3} = \dfrac{1}{1000}$

$f(-2) = 10^{-2} = \dfrac{1}{10^2} = \dfrac{1}{100}$

$f(-1) = 10^{-1} = \dfrac{1}{10^1} = \dfrac{1}{10}$

$f(0) = 10^0 = 1$
$f(1) = 10^1 = 10$
$f(2) = 10^2 = 100$

Solve:

13. $2^x = 32$ 32 can be written as a power of 2.

      $2^x = 2^5$ Substitute 2^5 for 32.

        $x = 5$ If $b^x = b^y$, then $x = y$.

B

Use the table of Exercise 1 to graph the function defined by the following equation. Write the domain and range for the function:

17. $f(x) = 3^x$

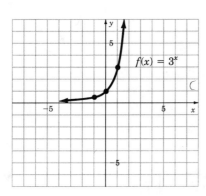

Sketch the graph by plotting the points from the table of values of Exercise 1.

Domain: $\{x \mid x \in \mathbb{R}\}$ or $\mathbb{R}$

Range: $\{y \mid y > 0\}$ or $\mathbb{R}^+$

21. $f(x) = \left(\dfrac{3}{4}\right)^{x}$

x	−3	−2	−1	0	1	2	3
f(x)	$\dfrac{64}{27}$	$\dfrac{16}{9}$	$\dfrac{4}{3}$	1	$\dfrac{3}{4}$	$\dfrac{9}{16}$	$\dfrac{27}{64}$

Make a table of values.

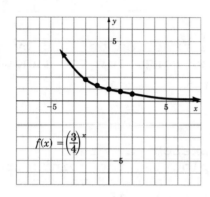

Sketch the graph by connecting the points with a smooth curve.

Domain: $\{x \mid x \in \mathbb{R}\}$

Range: $\{y \mid y > 0\}$

25. $\left(\dfrac{2}{3}\right)^{x} = \dfrac{16}{81}$

$\left[\dfrac{2}{3}\right]^{x} = \left[\dfrac{2}{3}\right]^{4}$

$x = 4$

$\dfrac{16}{81} = \left[\dfrac{2}{3}\right]^{4}$

Substitute $\left[\dfrac{2}{3}\right]^{4}$ for $\dfrac{16}{81}$.

If $b^{x} = b^{y}$, then $x = y$.

c

Graph the function defined by each of these equations. Write the domain and range of each:

29. $f(x) = \left(\dfrac{3}{2}\right)^x$

x	-3	-2	-1	0	1	2	3
f(x)	$\dfrac{8}{27}$	$\dfrac{4}{9}$	$\dfrac{2}{3}$	1	$\dfrac{3}{2}$	$\dfrac{9}{4}$	$\dfrac{27}{8}$

Make a table of values.

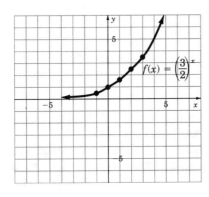

Sketch the graph.

Domain: $\{x \mid x \in \mathbb{R}\}$
Range: $\{y \mid y > 0\}$

33. $f(x) = (1.2)^x$

x	-2	-1	0	1	2	3	4
f(x)	0.69	0.83	1	1.2	1.44	1.73	2.07

Make a table of values

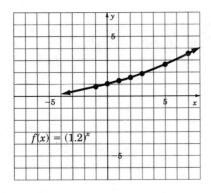

Sketch the graph.

Domain: $\{x \mid x \in \mathbb{R}\}$
Range: $\{y \mid y > 0\}$

401

Solve:

37. $2^{-x} = \left[\dfrac{1}{16}\right]$ $\dfrac{1}{16} = \dfrac{1}{2^4} = 2^{-4}$

   $2^{-x} = 2^{-4}$ Substitute 2^{-4} for $\dfrac{1}{16}$.

   $-x = -4$ If $b^x = b^y$, then $x = y$.
   $x = 4$

41. $\left[\dfrac{3}{4}\right]^{-x} = \dfrac{64}{27}$ $\dfrac{64}{27} = \left[\dfrac{4}{3}\right]^{3} = \left[\dfrac{3}{4}\right]^{-3}$

   $\left[\dfrac{3}{4}\right]^{-x} = \left[\dfrac{3}{4}\right]^{-3}$ Substitute $\left[\dfrac{3}{4}\right]^{-3}$ for $\dfrac{64}{27}$.

   $-x = -3$ If $b^x = b^y$, then $x = y$.
   $x = 3$

D

45. Based on the growth of the population for the past 20 years, the formula for the future world population is (approximately) $P = 4e^{0.2t} \times 10^9$, where t is the number of years and e ≈ 2.718. What is the population now (t = 0)? What will the population be (approximately) 10 years from now?

   $P = 4e^{0.2t} \times 10^9$ Formula.
   $P = 4e^{(0.2)0} \times 10^9$ Substitute t = 0.
   $\quad = 4e^0 \times 10^9$
   $\quad = 4 \times 10^9$ $e^0 = 1$.
   $\quad = 4,000,000,000$

   $P = 4e^{(0.2)(10)} \times 10^9$ Substitute t = 10.
   $\quad = 4e^2 \times 10^9$
   $\quad \approx 4(7.4) \times 10^9$ $e^2 \approx 7.4$
   $\quad \approx 29.6 \times 10^9$
   $\quad \approx 29,600,000,000$

   The population now is approximately 4,000,000,000. The population 10 years from now will be approximately 29,600,000,000.

49. The formula for interest compounded quarterly is
$A = \left[1 + \dfrac{r}{4}\right]^n$, where A is the total amount on deposit after n quarter-year periods if P dollars are invested at 4 percent per year.

Ms. Skinflint invests $1,000 at 8% interest compounded quarterly, which is to be given to her first great-grandchild upon its twenty-first birthday. If the great-granddaughter is 21 years of age 40 years after the money is invested (160 periods), how much does she receive? (Round to the nearest dollar.)

$A = \left[1 + \dfrac{r}{4}\right]^n$	Formula.
$= 1000 \left[1 + \dfrac{0.08}{4}\right]^{160}$	Substitute 1000 for P, 0.08 (8%) for r, and 160 for n.
$= 1000(1 + 0.02)^{160}$	
$= 1000(1.02)^{160}$	
≈ 23769.91	Use a calculator.
≈ 23770	Round to nearest dollar.

The granddaughter received approximately $23,770 from Ms. Skinflint.

53. The amount of bacteria in a culture is given by $N = N_0 e^{0.06t}$, where N_0 is the initial amount of bacteria, t is the time in hours, and N is the amount present after "t" hours. If the initial bacterial count is 3500, how many are there after 5 hours? Let $e \approx 2.718$. Round your answer to the nearest whole number.

$N = N_0 e^{0.06t}$	Original formula.
$N = 3500(2.718)^{(0.06)(5)}$	Substitute 3500 for N_0, 5 for t and 2.718 for e.
$N = 3500(2.718)^{0.3}$	
$N = 3500(1.3498)$	
$N = 4724$	

There are approximately 4,724 bacteria after 5 hours.

57. Over several years of testing, it has been found that the percentage (P) of a city's population that contributes to a charity fund-raiser is given by $P = 70(1 - e^{-0.04t})$ where t is the time in days that the campaign has run. What percentage of the population of the city will have contributed after 20 days? (nearest tenth) Let $e \approx 2.718$.

$$P = 70(1 - e^{-0.04t})$$ Original formula.
$$P = 70\left[1 - (2.718^{(-0.04)(20)})\right]$$ Substitute t = 20 and $e \approx 2.718$

$$= 70\left[1 - (2.718^{-0.8})\right]$$
$$= 70[1 - 0.4494]$$
$$= 70[0.5506]$$
$$= 38.5$$

Approximately 38.5% of the population will have contributed after 20 days.

STATE YOUR UNDERSTANDING

61. The graph of $f(x) = 3^x$ and the graph of $g(x) = -3^x$ are symmetrix to each other. What is the "line of symmetry?" Are they inverses?

The "line of symmetry" is the x-axis.

The line of symmetry for inverses is the line y = x, so the given graphs do not represent inverses.

CHALLENGE EXERCISES

65. $2^{x-2} = 4^x$

       $2^{x-2} = (2^2)^x$ Substitute 2^2 for 4.

       $2^{x-2} = 2^{2x}$

     $x - 2 = 2x$ If $b^x = b^y$, then x = y.
       $-x = 2$
         $x = -2$

69. $\left(\frac{1}{2}\right)^{x+1} = 4^x$ $4 = 2^2 = \left(\frac{1}{2}\right)^{-2}$

       $\left(\frac{1}{2}\right)^{x+1} = \left[\left(\frac{1}{2}\right)^{-2}\right]^x$ Replace 4 with $\left(\frac{1}{2}\right)^{-2}$.

       $\left(\frac{1}{2}\right)^{x+1} = \left(\frac{1}{2}\right)^{-2x}$

     $x + 1 = -2x$ If $b^x = b^y$, then x = y.
       $3x = -1$
        $x = -\frac{1}{3}$

73. Create two tables of values and graph both curves on the same graph:

(a) $f(x) = 3^x + 1$ and (b) $g(x) = 3^{x+1}$

$y = 3^x + 1$ $\qquad\qquad$ $y = 3^{x+1}$

x	-2	-1	0	1	2
y	$\dfrac{10}{9}$	$\dfrac{4}{3}$	2	4	10

$y = 3^{x+1}$

x	-2	-1	0	1	2
y	$\dfrac{1}{3}$	1	3	9	27

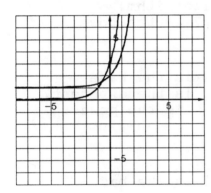

MAINTAIN YOUR SKILLS

Write the equation of the inverse of the following functions:

77. $f(x) = x + 4$
$\qquad y = x + 4$ $\qquad\qquad$ The equation that defines f.
$\qquad x = y + 4$ $\qquad\qquad$ Interchange x and y.
$x - 4 = y$ $\qquad\qquad$ Solve for y.

So $f^{-1}(x) = x - 4$.

81. $f(x) = \frac{1}{3}x - 4$

$y = \frac{1}{3}x - 4$

$x = \frac{1}{3}y - 4$ Interchange x and y.

$\frac{1}{3}y = x + 4$ Solve for y.

$y = 3x + 12$

So $f^{-1}(x) = 3x + 12$

EXERCISES 10.2   LOGARITHMIC FUNCTIONS

A

Write each logarithmic equation in exponential form:

1. $\log_3 27 = 3$
     $3^3 = 27$ 3 is the base.
                                3 is the exponent or
                                logarithm.
                                27 is the power.

5. $\log_x x^2 = 2$
     $x^2 = x^2$ x is the base.
                                2 is the exponent.
                                x^2 is the power.

9. $\log_{1/2} \frac{1}{8} = 3$
     $\left(\frac{1}{2}\right)^3 = \frac{1}{8}$ $\frac{1}{2}$ is the base.
                                3 is the exponent.
                                $\frac{1}{8}$ is the power.

Write each exponential equation in logarithmic form:

13. $2^6 = 64$
     $\log_2 64 = 6$ 2 is the base.
                                6 is the logarithm or
                                exponent.
                                64 is the power.

17. $8^{1/2} = \sqrt{8}$

   $\log_8 \sqrt{8} = \frac{1}{2}$

   8 is the base.

   $\frac{1}{2}$ is the logarithm.

   $\sqrt{8}$ is the power.

B

Find w:

21. $w = \log_6 36$

   $6^w = 36$

   Change from log form to exponential form.

   $6^w = 6^2$

   $w = 2$

   If $a^x = a^y$, then $x = y$.

25. $w = \log_{10} \frac{1}{10}$

   $10^w = \frac{1}{10}$

   Change from log form to exponential form.

   $10^w = 10^{-1}$

   $w = -1$

   If $a^x = a^y$, then $x = y$.

29. $w = \log_5 \frac{1}{625}$

   $5^w = \frac{1}{625}$

   Change from log form to exponential form.

   $5^w = 5^{-4}$

   $w = -4$

   If $a^x = a^y$, then $x = y$.

33. $w = \log_{1/4} 16$

   $\left(\frac{1}{4}\right)^w = 16$

   Change from log form to exponential form.

   $\left(\frac{1}{4}\right)^w = \left(\frac{1}{4}\right)^{-2}$

   $w = -2$

Complete the table of values:

37. $y = \log_{10} x$

x	$\frac{1}{100}$	$\frac{1}{10}$	1	100	1000
y					

Change $y = \log_{10} x$ to exponential form:

$x = 10^y$ and fill in the table of values.

$x = 10^y$

$\quad x = \frac{1}{100}$ $\qquad x = \frac{1}{10}$ $\qquad x = 1$

$\frac{1}{100} = 10^y$ $\qquad \frac{1}{10} = 10^y$ $\qquad 1 = 10^y$

$10^{-2} = 10^y$ $\qquad 10^{-1} = 10^y$ $\qquad 10^0 = 10^y$

$\quad y = -2$ $\qquad\qquad y = -1$ $\qquad\quad y = 0$

$\quad x = 100$ $\qquad\quad x = 1000$

$100 = 10^y$ $\qquad 1000 = 10^y$

$10^2 = 10^y$ $\qquad 10^3 = 10^y$

$\quad y = 2$ $\qquad\qquad y = 3$

x	$\frac{1}{100}$	$\frac{1}{10}$	1	100	1000
y	-2	-1	0	2	3

C

Find z:

41. $z = \log_4 8$

$\quad 4^z = 8$

Change to exponential form.

$(2^2)^z = 2^3$

Rewrite both sides using base 2.

$2^{2z} = 2^3$

$2z = 3$

$z = \frac{3}{2}$

If $a^x = a^y$, then $x = y$.

408

45. $z = \log_{32} 16$

$32^z = 16$ Change to exponential form.

$2^{5z} = 2^4$ Rewrite both sides using base 2.

$5z = 4$

$z = \dfrac{4}{5}$

49. $z = \log_{3/2}\left(\dfrac{8}{27}\right)$

$\left(\dfrac{3}{2}\right)^z = \dfrac{8}{27}$ Change to exponential form.

$\left(\dfrac{3}{2}\right)^z = \left(\dfrac{3}{2}\right)^{-3}$ Rewrite both sides using base $\dfrac{3}{2}$.

$z = -3$

Use the tables in Exercises 35 and 39 to sketch the following; state the domain and range:

53. $y = \log_4 x$

x	$\dfrac{1}{16}$	$\dfrac{1}{4}$	1	4	64
y	-2	-1	0	1	3

Table of values.

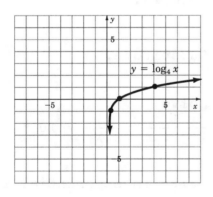

Sketch the graph by plotting the points and connecting them with a smooth curve.

Domain: $\{x \mid x > 0\}$ or $\mathbb{R}^+$
Range: $\{y \mid y \in \mathbb{R}\}$ or $\mathbb{R}$

57. $y = \log_{1/2} x$

x	4	2	1	$\frac{1}{2}$	$\frac{1}{4}$
y	-2	-1	0	1	2

Table of values.

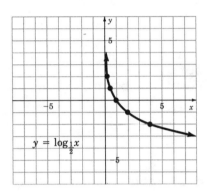

Sketch the graph.

Domain: $\{x \mid x > 0\}$ or $\mathbb{R}^+$
Range: $\{y \mid y \in \mathbb{R}\}$ or $\mathbb{R}$

61. Find the pH of a solution with an H^+ value of 1×10^5.

$$pH = \log_{10}\left(\frac{1}{H^+}\right)$$ Formula.

$$pH = \log_{10}\left(\frac{1}{1 \times 10^5}\right)$$ Substitute.

$$pH = \log_{10} 10^{-5}$$ Solve.
$$\frac{1}{1 \times 10^5} = \frac{1}{10^5} = 10^{-5}.$$

$$10^{pH} = 10^{-5}$$ Write in exponential form.

$$pH = -5$$

The pH of the solution is -5.

65. The age (t, in years) of an object is given by approximately $t = -18000 \log N_0$, where N_0 is the percentage (as a decimal) of the total carbon remaining that is Carbon 14. If 1% of the total carbon left in a bone is Carbon 14, what is the approximate age? What if the percentage is 0.1%, what is the approximate age?

$t = -1800 \log N_0$	Original formula.
$t = -1800 \log(1\%)$	Substitute 1% for N_0.
$\quad = -1800 \log (0.01)$	
$\quad = -1800(-2)$	Log 0.01 = -2.
$\quad = 36000$	
$t = -1800 \log N_0$	Original formula.
$t = -1800 \log (0.1\%)$	Substitute 0.1% for N_0.
$\quad = -1800 \log (0.001)$	
$\quad = -1800 (-3)$	Log 0.001 = -3.
$\quad = 54000$	

If N_0 = 1%, the approximate age is 36,000 years.
If N_0 = 0.1%, the approximate age is 54,000 years.

STATE YOUR UNDERSTANDING

69. What is the relationship between an exponential function and a logarithmic function?

These two functions are related as inverses. That is, the domain of the exponential function is the range of the logarithmic function and, likewise, the range of the exponential function is the domain of the logarithmic function.

CHALLENGE EXERCISES

73. $3 = \log_2 x$
$2^3 = x$ Rewrite in exponential form.

$\quad 8 = x$

The solution set is {8}.

77. $\log_5 5 = 2x - 4$
$\quad\quad 1 = 2x - 4$ Replace $\log_5 5$ with 1,
$\quad\quad 5 = 2x$ since $5^1 = 5$.
$\quad\quad \dfrac{5}{2} = x$

The solution set is $\left\{\dfrac{5}{2}\right\}$.

81. $\log_x 1 = 0$

$\qquad x^0 = 1$

Rewrite in exponential form. This is an identity as long as x conforms to the restrictions of any logarithmic base. So x must be greater than zero and not equal to one.

The solution set is $\{x \mid x \in \mathbb{R}, \ x > 0, \ x \neq 1\}$.

MAINTAIN YOUR SKILLS

83. $\quad (3a^2 b^3 c)(-15a^6 b^5 c^3)$

$\quad = -45a^8 b^8 c^4$

Multiply. Add exponents of like bases.

89. $(x + y)^2 (x + y)^4 = (x + y)^6$

Add exponents since the base is the same, x + y.

EXERCISES 10.3 PROPERTIES OF LOGARITHMS

A

Write the following as logarithms of a single number:

1. $\log 6_b + \log_b 8 = \log_b 48$

Property Five of logarithms.

5. $2\log_5 8 = \log_5 8^2 = \log_5 64$

Property Seven of logarithms.

9. $-2\log_4 3 = \log_4 3^{-2}$

$\qquad = \log_4 \left(\dfrac{1}{9}\right)$

Property Seven.

Write the following as a combination of logarithms of a single number:

13. $\log_b (132)^4 = 4\log_b 132$

Property Seven of logarithms.

Given $\log_{10} 5 = 0.69897$ and $\log_{10} 7 = 0.84510$, find the following:

17. $\log_{10} 25 = \log_{10} 5^2$ Write 25 as a power of 5.
$$= 2\log_{10} 5$$
Log of a base raised to a power.

$$= 2(0.69897)$$
$$= 1.39794$$

B

21. $3\log_b 4 + \log_b 3$

$$= \log_b 4^3 + \log_b 3$$
     Property Seven of logarithms.

$$= \log_b (4^3 \cdot 3)$$
     Property Five of logarithms.

$$= \log_b 192$$

25. $2\log_b 5 + \log_b 2 - \log_b 10$

$$= \log_b 5^2 + \log_b 2 - \log_b 10$$
     Property Seven of logarithms.

$$= \log_b (5^2 \cdot 2) - \log_b 10$$
     Property Five.

$$= \log_b \left(\frac{5^2 \cdot 2}{10}\right) = \log_b \left[\frac{50}{10}\right] = \log_b 5$$
     Property Six. Simplify.

Write the following as combinations of logarithms of single variables:

29. $\log_b \left(\frac{z}{x}\right) = \log_b z - \log_b x$
     Property Six.

33. $\log_5 \dfrac{x^2}{y^2} = \log_5 x^2 - \log_5 y^2$
     Property Six.

$$= 2\log_5 x - 2\log_5 y$$
     Property Seven.

Given $\log_{10} 2 = 0.30103$, $\log_{10} 3 = 0.47712$ and $\log_{10} 5 = 0.69897$, find the following:

37. $\log_{10} 40 = \log_{10} (2^3 \cdot 5^1)$
     Write 40 as the product of powers of 2 and 5.

$$= \log_{10} 2^3 + \log_{10} 5$$
Log of a product.
$$= 3\log_{10} 2 + \log_{10} 5$$
$$= 3(0.30103) + 0.69897$$
$$= 1.60206$$

C

Write the following as logarithms of single numbers:

41. $2\log_b 4 + 3\log_b 2 - 3\log_b 4$

    $= \log_b 4^2 + \log_b 2^3 - \log_b 4^3$ Property Seven.

    $= \log_b (4^2 \cdot 2^3) - \log_b 4^3$ Property Five.

    $= \log_b \left(\dfrac{4^2 \cdot 2^3}{4^3}\right) = \log_b \left(\dfrac{128}{64}\right) = \log_b 2$ Property Six.
Simplify.

45. $\dfrac{1}{3}\log_b 8 + \dfrac{1}{2}\log_b 49$

    $= \log_b 8^{1/3} + \log_b 49^{1/2}$ Property Seven.

    $= \log_b (8^{1/3} \cdot 49^{1/2})$ Property Five.

    $= \log_b (2 \cdot 7) = \log_b 14$ Simplify.

Write the following as combinations of logarithms of single variables:

49. $\log_c z\sqrt{xy} = \log_c z(xy)^{1/2}$

         $= \log_c zx^{1/2}y^{1/2}$

         $= \log_c z + \log_c x^{1/2} + \log_c y^{1/2}$

         $= \log_c z + \dfrac{1}{2}\log_c x + \dfrac{1}{2}\log_c y$

53. $\log_x \sqrt[3]{(a + b)^2} = \log_x (a + b)^{2/3}$

          $= \dfrac{2}{3}\log_x (a + b)$

Given $\log_{10} 2 = 0.30103$, $\log_{10} 3 = 0.47712$, $\log_{10} 5 = 0.69897$, and $\log_{10} 7 = 0.84510$, find the following:

57. $\log_{10} 84 = \log_{10} 2^2 \cdot 3 \cdot 7$ Factor 84 as the
product of powers of 2,
3 and 7.

    $= \log_{10} 2^2 + \log_{10} 3 + \log_{10} 7$ Log of a product.

    $= 2\log_{10} 2 + \log_{10} 3 + \log_{10} 7$

    $= 2(0.30103) + 0.47712 + 0.84510$

    $= 1.92428$

61. The magnitude on the Richter scale for a given earthquake is defined by $M(i) = \log_{10} \frac{i}{i_0}$, where i is the amplitude of the ground motion of the quake and i_0 is the amplitude of the ground motion of a "zero" quake. What is the magnitude on the Richter scale of an earthquake that is $10^{2.5}$ times stronger than the zero quake?

Formula:

$$M(i) = \log_{10} \frac{i}{i_0}$$

Substitute:

$$M(10^{2.5}) = \log_{10} \frac{10^{2.5} i_0}{i_0} \qquad\qquad i = 10^{2\ 5}$$

$$= \log_{10} 10^{2.5} \qquad\qquad \text{Reduce.}$$

$$= 2.5 \log_{10} 10$$

$$= 2.5 \qquad\qquad \log_{10} 10 = 1$$

Answer:
The magnitude of the earthquake is 2.5.

65. Find the pH of a solution of hydrochloric acid (a strong acid) if its hydrogen ion concentration (H^+) is 5×10^{-2}. (Use $\log_{10} 0.2 \approx -0.6990$, and round to the nearest tenth.) The formula for pH is $pH = \log_{10} \frac{1}{H^+}$.

Formula:

$$pH = \log_{10} \frac{1}{H^+}$$

Substitute:

$$pH = \log_{10} \frac{1}{5 \times 10^{-2}} \qquad\qquad H^+ \approx 5 \times 10^{-2}$$

Solve:

$$pH = \log_{10}\left(\frac{1}{5} \cdot \frac{1}{10^{-2}}\right)$$

$$= \log_{10}\frac{1}{5} + \log_{10}\frac{1}{10^{-2}}$$

$$= \log_{10}0.2 + \log_{10}10^2 \qquad \text{Substitute } \frac{1}{5} = -0.2 \text{ and}$$
$$\frac{1}{10^{-2}} = 10^2.$$

$$= \log_{10}0.2 + 2\log_{10}10$$

$$= -0.6990 + 2 \qquad \text{Substitute}$$
$$\log_{10}0.2 = -0.6990,$$
$$= 1.301 \qquad\qquad\qquad \text{and } \log_{10}10 = 1.$$
$$\approx 1.3 \text{ (to the nearest tenth)}$$

Answer:

The pH of the solution of hydrochloric acid is approximately 1.3.

69. Show that $\log_{10}(\log_2(\log_3 9)) = 0$

$$\log_{10}(\log_2(2)) = 0 \qquad \text{Replace } \log_3 9 \text{ with 2,}$$
$$\text{since } 3^2 = 9.$$
$$\log_{10}(1) = 0 \qquad\quad \text{Replace } \log_2 2 \text{ with 1,}$$
$$\text{since } 2^1 = 2.$$
$$0 = 0 \qquad\qquad\quad \text{Replace } \log_{10}1 \text{ with 0,}$$
$$\text{since } 10^0 = 1.$$

73. The age of a fossil, in years, is given by $t = -18,000\log_{10}R$ where R is the percentage of the remaining carbon in the sample that is Carbon 14. (Note: $\log_{10}1.5 \approx 0.17609$) What is the age of the fossil if 15% of the remaining carbon is Carbon 14. (Remember to convert 15% to a decimal in the formula.)

$t = -18,000\log_{10}R$ Formula.
  $= -18,000 \log_{10}0.15$ Replace R with 15% or 0.15.
                                $0.15 = 1.5 \times 10^{-1}$, so substitute 1.5×10^{-1}
  $= -18,000 \log_{10}[1.5 \times 10^{-1}]$ for 0.15.
  $= -18,000[\log_{10}1.5 + \log_{10}10^{-1}]$ Log of a product.
  $= -18,000[(0.17609 + (-1)]$ Substitute 0.17609 for $\log_{10}1.5$ and substitute -1 for $\log_{10}10^{-1}$ since $\log_b b^x = x.$
  $= -18,000(-0.82391)$
  $= 14830.38$

Rounded to the nearest year, the age of the fossil is 14,830 years.

416

77. Why is the following incorrect?

$$\frac{\log_b x^4}{\log_b y^3} = 4\log_b x - 3\log_b y$$

The <u>logarithm</u>, base b, <u>of the quotient</u> $\frac{x^4}{y^3}$ would equal the given expression:

$$\log_b \left(\frac{x^4}{y^3}\right) = 4\log_b x - 3\log_b y$$

There is no property relating the <u>quotient of logarithms</u> to any given expression.

CHALLENGE EXERCISES

If $\log_b 3 = 0.7925$, $\log_b 5 = 1.1610$, and $\log_b 7 = 1.4037$, where "b" is some unknown base, find:

81. $\log_b 45 = \log_b (3 \cdot 3 \cdot 5)$ Factor 45.
$\qquad\quad = \log_b 3 + \log_b 3 + \log_b 5$ Log of a product.
$\qquad\quad = 0.7925 + 0.7925 + 1.1610$
$\qquad\quad = 2.7460$

85. $\log_b \sqrt[3]{9} = \log_b (9)^{1/3}$
$\qquad\qquad = \log_b (3^2)^{1/3}$
$\qquad\qquad = \log_b 3^{2/3}$
$\qquad\qquad = \frac{2}{3}\log_b 3$
$\qquad\qquad = \frac{2}{3}(0.7925)$
$\qquad\qquad = 0.528\overline{3}$

MAINTAIN YOUR SKILLS

89. Find the coordinates of the center of the circle
$x^2 + 6x + y^2 - 1 = 0$

$x^2 + 6x + 9 + y^2 = 1 + 9$ Complete the square in x.
$\quad (x + 3)^2 + y^2 = 10$
Center: (-3, 0)   h = -3, k = 0

93. Solve

$$\begin{cases} x + 3y - 2z = 8 & (1) \\ x + y - z = 4 & (2) \\ 2x - y + z = -3 & (3) \end{cases}$$

$$\begin{array}{l} x + y - z = 4 \qquad (2) \\ \underline{2x - y + z = -3} \qquad (3) \\ 3x \qquad\quad = 1 \\ \qquad\quad x = \dfrac{1}{3} \end{array}$$

Eliminate the z term by adding (2) and (3).

$$\begin{array}{l} x + 3y - 2z = 8 \qquad (1) \\ \underline{4x - 2y + 2z = -6} \\ 5x + y \qquad\quad = 2 \qquad (4) \end{array}$$

Eliminate z by adding (1) to twice (3)

$$5\left(\dfrac{1}{3}\right) + y = 2$$

$$\dfrac{5}{3} + y = 2$$

$$y = \dfrac{1}{3}$$

Replace x in (4) with $\dfrac{1}{3}$ and solve for y.

$$2x - y + z = -3$$

$$2\left(\dfrac{1}{3}\right) - \dfrac{1}{3} + z = -3$$

$$\dfrac{1}{3} + z = -3$$

$$z = -\dfrac{10}{3}$$

Replace $\dfrac{1}{3}$ for x and $\dfrac{1}{3}$ for y in (3) and solve for z.

The solution set is $\left\{\left(\dfrac{1}{3}, \dfrac{1}{3}, -\dfrac{10}{3}\right)\right\}$.

EXERCISES 10.4   FINDING LOGARITHMS AND ANTILOGARITHMS

A

Find these logarithms without using tables or calculator:

1. $\log_{10} \dfrac{1}{10} = \log_{10} 10^{-1}$

    $= -1(\log_{10} 10)$

    $= -1$

$\dfrac{1}{10} = 10^{-1}$

Law Three of logarithms.

5. $\log_{10} \dfrac{1}{1000} = \log_{10} \dfrac{1}{10^3}$

$$= \log_{10} 10^{-3}$$
$$= -3\log_{10} 10$$
$$= -3$$

9. $\ln\sqrt{e}$

$x = \ln\sqrt{e}$	Set equal to x.
$e^x = \sqrt{e}$	Write in exponential form.
$e^x = e^{1/2}$	
$x = \dfrac{1}{2}$	If $b^x = b^y$, then $x = y$.

Find the antilogarithm without using a table or calculator:

13. $\log x = 4$ The base is 10.
         $10^4 = x$ Write in exponential form.
   $10,000 = x$

The antilogarithm is 10,000.

17. $\log x = 6$ The base is 10.
         $10^6 = x$ Write in exponential form.
  $1000000 = x$

The antilogarithm is 1,000,000.

21. $\ln x = 0$ The base is e.
         $e^0 = x$ Write in exponential form.
           $1 = x$

The antilogarithm is 1.

B

Find the logarithms:

25.   log 2.3

From the table:

log 2.3 ≈ 0.3617

By calculator:

ENTER	DISPLAY
2.3	2.3
log x	0.361727836

log 2.3 ≈ 0.3617.

29.   log 8.8

From table:
log 8.8 ≈ 0.9445

By calculator:

ENTER	DISPLAY
8.8	8.8
log x	0.9444826722

log 8.8 ≈ 0.9445

Using a calculator, find the logarithms:

33.   ℓn 16.2

ENTER	DISPLAY
16.2	16.2
ℓn x	2.785011242

ℓn 16.2 ≈ 2.7850

Find the antilogarithms:

37. log x = 2.2304

ENTER | DISPLAY

| 2.2304 | 2.2304

| 10ˣ | 169.9808513

The antilogarithm 2.2304 is approximately 169.9809.

41. log x = -4.2733

ENTER | DISPLAY

| -4.2733 | -4.2733

| 10ˣ | 5.32966608E - 05   Scientific notation.

≈ 0.000053296608
≈ 0.0001

The antilogarithm (-4.2733) is approximately 0.0001.

c

Find the logarithms:

45. log 230

ENTER | DISPLAY

| 230 | 230

| log x | 2.361727836

log 230 ≈ 2.3617

Find the antilogarithms:

49. log x = 2.6522

      ENTER   DISPLAY

      | 2.6522 | 2.6522

       | 10^x | 448.9520924

The antilogarithm 2.6522 is 448.9521.

53. log x = 4.7777

      ENTER   DISPLAY

      | 4.7777 | 4.7777

       | 10^x | 59937.68983

                       $\approx 5.99 \times 10^4$ Scientific notation.

The antilogarithm 4.7777 is approximately 59937.6898 or 5.99×10^4 in scientific notation.

Using a calculator, find the antilogarithms:

57. $ln_e\, x = 2$

      ENTER   DISPLAY

      | 2 | 2

      | e^x | 7.38906

The antilogarithm 2, base e, is approximately 7.3891.

D

61. The Slippery Rock band plays with an intensity that is 10^{12} times the least intensity that can be heard. What is the decibel level? (Use the formula: $S(i) = 10 \log \frac{i}{i_0}$ where i_0 is the least intensity that can be heard by the human ear and $S(i)$ is measured in decibels.)

Formula:

$S(i) = 10 \log \frac{i}{i_0}$

Solve:

$S(i) = 10\log \dfrac{10^{12} i_0}{i_0}$ Substitute $i = 10^{12}$

    $= 10 \log 10^{12}$

    $= 12 \cdot 10 \log 10$ Law Three of logarithms.

    $= 120 \cdot 1$ $\log 10 = \log_{10} 10 = 1$

    $= 120$

Answer:
The Slippery Rock band plays with an intensity of 120 decibels.

65. Find the pH of a mixture with a hydrogen ion concentration of 6.2×10^{-6}. (Use the formula: $pH = \log \frac{1}{H^+}$.)

Formula:

$pH = \log \dfrac{1}{H^+}$

Substitute:

$pH = \log \dfrac{1}{6.2 \times 10^{-6}}$ $H^+ = 6.2 \times 10^{-6}$

Solve:

$pH \approx \log 161290.32$ $\dfrac{1}{6.2 \times 10^{-6}} \approx 161290.32$

   ≈ 5.2 By calculator.

Answer:
The pH of the mixture is approximately 5.2.

69. The heat loss (L) from a four inch pipe with one inch of insulation that is carrying hot water is approximately $L = \dfrac{50}{\ln\left(\frac{5}{4}\right)}$. Find the heat loss (to the nearest tenth) for the given pipe. (The units are BTU per hour where a "BTU" is a British Thermal Unit.) (nearest tenth)

$L = \dfrac{50}{\ln\left(\frac{5}{4}\right)}$

$= \dfrac{50}{\ln 1.25}$ Use a calculator to evaluate $\ln\left(\frac{5}{4}\right)$ or

$= \dfrac{50}{0.22314}$ $\ln(1.25)$.

$= 224.07$

The heat loss is approximately 224.1 BTU.

73. The price (p) of a certain item depends on the quantity demanded (q_d), and is given by $p = \$150 - \ln(q_d)$. Find the anticipated price if the demand is for 500,000 of the items. (To the nearest cent.)

$p = 150 - \ln(q_d)$
$p = 150 - \ln(500,000)$ Substitute 500,000 for q_d.
$p = 150 - 13.12236$
$p = 136.88$

The anticipated price is $136.88.

STATE YOUR UNDERSTANDING

77. Use your calculator to find the y-values for the following x-values using $y = \ln x$. (3,), (4,), and (5,). Then using the first two ordered pairs, calculate the slope. Using the second two ordered pairs, calculate the slope. Are the slopes the same? What does this tell you about the graph of $y = \ln x$?

$y = \ln x$

x	y
3	1.099
4	1.386
5	1.609

Using (3, 1.099) and (4, 1.386)

$$m_1 = \frac{1.386 - 1.099}{4 - 3} = 0.287$$

Using (4, 1.386) and (5, 1.609)

$$m_2 = \frac{1.609 - 1.386}{5 - 4} = 0.223$$

Since the slopes are unequal, the graph of $y = \ln x$ is not a straight line.

CHALLENGE EXERCISES

Using the properties of logarithms, rewrite the following expressions so that the log of just one number need be taken. Then find the log of that number.

81. $2\log 5 + \log 2 - \log 10$
       $= \log 5^2 + \log 2 - \log 10$
       $= \log(5^2 \cdot 2) - \log 10$
       $= \log\left(\frac{5^2 \cdot 2}{10}\right)$
       $= \log\left(\frac{50}{10}\right)$
       $= \log 5$
       $= 0.69897$

MAINTAIN YOUR SKILLS

85.   Find the x-intercepts of the graph of $y = x^2 - 9x - 111$.

$$0 = x^2 - 9x - 111 \qquad \text{Use the quadratic formula.}$$
$$\text{Set } y = 0.$$

$$x = \frac{9 \pm \sqrt{9^2 - 4(-111)}}{2}$$

$$= \frac{9 \pm \sqrt{525}}{2}$$

$$= \frac{9 \pm 5\sqrt{21}}{2}$$

So the x-intercepts are $\left\{\left(\frac{9 \pm 5\sqrt{21}}{2}, 0\right)\right\}$.

89. Write the equation for $G^{-1}(x)$ if:

$$G(x) = x^3 + 2$$
$$y = x^3 + 2$$ Replace $G(x)$ with y.
$$x = y^3 + 2$$ Interchange y and x.

$$y^3 + 2 = x$$ Solve for y.
$$y^3 = x - 2$$

$$y = \sqrt[3]{x - 2}$$

$$G^{-1}(x) = \sqrt[3]{x - 2}$$ Replace y with $G^{-1}(x)$.

EXERCISES 10.5   EXPONENTIAL EQUATIONS, LOGARITHMIC EQUATIONS, AND FINDING LOGARITHMS, ANY BASE

A

Solve these exponential and logarithmic equations without using tables:

1. $4^x = 32$

$(2^2)^x = 2^5$ Rewrite both sides using 2 as the base.

$2^{2x} = 2^5$

$2x = 5$ If $a^x = a^y$, then $x = y$.

$x = \dfrac{5}{2}$

5. $\log_3 2 + \log_3 x = 2$

$\log_3 (2x) = 2$ Law One of logarithms.

$3^2 = 2x$ Write in exponential form.

$9 = 2x$

$x = \dfrac{9}{2}$

9. $5^{x+1} = 125^x$

$5^{x+1} = (5^3)^x$ Rewrite both sides using 5 as a base.

$5^{x+1} = 5^{3x}$

$x + 1 = 3x$

$1 = 2x$

$x = \dfrac{1}{2}$

13.
$$4^{2x+1} = 32^{x+5}$$
$$2^{2(2x+1)} = 2^{5(x+5)}$$

Rewrite both sides using 2 as the base.

$$2(2x + 1) = 5(x + 5)$$
$$4x + 2 = 5x + 25$$
$$-23 = x$$

If $a^x = a^y$, then $x = y$.

B

Use the formulas of this section and a log table or calculator to find each of the logarithms correct to four decimal places:

17. $\log_7 25$

Formula:
$$\log_a x = \frac{\log_b x}{\log_b a}$$

Substitute:
$$\log_7 25 = \frac{\log_{10} 25}{\log_{10} 7}$$

Substitute $x = 25$, $b = 10$, and $a = 7$.

$$\approx \frac{1.39794}{0.845098}$$

By calculator.

$$\approx 1.1654175$$

$$\log_7 25 \approx 1.6542$$

21. $\log_{13} 6$

Formula:
$$\log_a x = \frac{\log_{10} x}{\log_{10} a}$$

Substitute:
$$\log_{13} 6 = \frac{\log_{10} 6}{\log_{10} 13}$$

$$\approx \frac{0.7781513}{1.1139434}$$

By calculator.

$$\approx 0.6985555$$

$$\log_{13} 6 \approx 0.6986$$

Use tables (base 10) or a calculator to solve this exponential equation (round to four decimal places):

25. $25^x = 50$

        $\log 25^x = \log 50$ Take the logarithm, base 10, of each side.

        $x \log 25 = \log 50$ Property Seven of logarithms.

$$x = \frac{\log 50}{\log 25}$$

$$\approx \frac{1.6989700}{1.3979400} \qquad \text{By calculator.}$$

$$\approx 1.2153383$$

So $x \approx 1.2153$.

Solve this logarithmic equation:

29. $\log_2 x - \log_2 (x - 2) = 2$

$$\log_2 \frac{x}{x - 2} = 2 \qquad \text{Property Six of logarithms.}$$

$$2^2 = \frac{x}{x - 2} \qquad \text{Rewrite in exponential form.}$$

$$4 = \frac{x}{x - 2}$$

$$4(x - 2) = x \qquad \text{Multiply both sides by}$$

$$4x - 8 = x \qquad (x - 2).$$

$$3x = 8$$

$$x = \frac{8}{3}$$

C

Use the formulas of this section and a log table or calculator to find each of the logarithms correct to four decimal places:

33. $\log_{2.5} 45$

$$\log_a x = \frac{\log_b x}{\log_b a} \qquad \text{Formula for changing the base of a logarithm.}$$

$$\log_{2.5} 45 = \frac{\log_{10} 45}{\log_{10} 2.5} \qquad \text{Substitute 45 for } x, 2.5 \text{ for } a, \text{ and 10 for } b.$$

$$\approx \frac{1.6532125}{0.3979400} \qquad \text{By calculator.}$$

$$\approx 4.1544265$$

$\log_{2.5} 45 \approx 4.1544.$

37. $\log_{1/5} 0.33$

$$\log_a x = \frac{\log_b x}{\log_b a}$$

$$\log_{1/5} 0.33 = \frac{\log_{10} 0.33}{\log_{10}\left(\frac{1}{5}\right)}$$

$\frac{1}{5} = 0.2.$

$\text{Log } 0.2 \approx -0.6989700.$

$$\approx \frac{-0.4814861}{-0.6989700}$$

$$\approx 0.6888508$$

$$\log_{1/5} 0.33 \approx 0.6889$$

Use tables (base 10) or a calculator to solve this exponential equation (round to four decimal places):

41. $3^{x+1} = 4$

$\log 3^{x+1} = \log 4$

$(x + 1)(\log 3) = \log 4$

Take the log, base 10, of both sides. Property Seven.

$x\log 3 + \log 3 = \log 4$

$x\log 3 = \log 4 - \log 3$

$$x = \frac{\log 4 - \log 3}{\log 3}$$

$$x \approx \frac{0.6020600 - 0.4771213}{0.4771213}$$

By calculator.

$$\approx \frac{0.1249387}{0.4771213}$$

$$\approx 0.2618595$$

So $x \approx 0.2619$.

Solve these logarithmic equations:

45. $\log_3 (x + 6) + \log_3 x = 3$

$\log_3 [x(x + 6)] = 3$

Law One.

$3^3 = x^2 + 6x$

Write in exponential form.

$x^2 + 6x - 27 = 0$

$(x + 9)(x - 3) = 0$

$x = -9 \text{ or } x = 3$

$x = 3$

Powers restricted to positive values.

49. $\log_2 x + \log_2 (x - 2) \ = 3$

       $\log_2 [x(x - 2)] \ = 3$ Property Five.

                $2^3 \ = x(x - 2)$ Write in exponential form.

                $8 \ = x^2 - 2x$

       $x^2 - 2x - 8 = 0$ Standard form.

    $(x - 4)(x + 2) = 0$

  $x = 4 \text{ or } x = -2$ Reject -2 as log x is
                                     defined only for x > 0.

  So x = 4.

D

53. How long will it take for \$50 to triple in value? (Use the formula in the application.)

Formula:

$$V = P\left(1 + \frac{r}{4}\right)^{4y}$$

Substitute:

$$150 = 50\left(1 + \frac{0.08}{4}\right)^{4y}$$

Substitute 50 for P, 8% or 0.08 for r (from Exercise 51) and 150 for V since the value will triple 50.

$$3 = (1 + 0.02)^{4y}$$
$$\log 3 = 4y \log (1.02)$$

Divide both sides by 50. Take the log of each side.

$$4y = \frac{\log 3}{\log 1.02}$$

$$y = \frac{\log 3}{4\log 1.02}$$

$$\approx 13.86953$$

Using a calculator.

It takes approximately 13.87 years for \$50 to triple.

57. The amount (A) of radioactive material present at time t is given by the formula $A = A_0 2^{-t/k}$, where A_0 is the initial amount present, and k is the materials half-life. How many years will it take 100 mg of a radioactive substance to decay to 25 mg if its half-life is 1000 years?

$$A = A_0 2^{-t/k}$$

$$25 = 100(2^{-t/1000})$$

Substitute 100 for A_0, 25 for A, and 1000 for k.

$$0.25 = 2^{-t/1000}$$

Divide on both sides by 100.

$$\log 0.25 = \log 2^{-t/1000}$$

$$\log 0.25 = -\frac{t}{1000}\log 2$$

$$\frac{1000\log 0.25}{\log 2} = -t$$

$$2000 \approx t$$

It will take approximately 2000 years for the substance to decay to 25 mg..

61. Show that $\ln x \approx 2.3026 \log x$

$$\ln x = \frac{\log x}{\log e}$$ Change of base formula.

$$\approx \frac{\log x}{0.4343}$$

$$\approx \frac{1}{0.4343}(\log x)$$

$$\approx 2.3026(\log x)$$

STATE YOUR UNDERSTANDING

65. What are the differences in the approaches for solving the following exponential and logarithmic equations: $3^x = 5$ and $\log_x 9 = 2$?

In the first "exponentially" stated equation we could find the logarithm of both sides. Whereas, in the second "logarithmically" stated equation we can convert to exponential notation.

Solve: (answers to the nearest hundredth)

69. $\log_5 x - \log_5 (x - 4) = \log_5 (x - 6)$

$\log_5 \left(\dfrac{x}{x - 4}\right) = \log_5 (x - 6)$ Rewrite the left side as a quotient.

$\dfrac{x}{x - 4} = x - 6$ If $\log_b x = \log_b y$, then $x = y$.

$x = (x - 6)(x - 4)$

$x = x^2 - 10x + 24$

$x^2 - 11x + 24 = 0$

$(x - 8)(x - 3) = 0$

$x = 8$ or $x = 3$ Reject $x = 3$ since in $\log_5 (3 - 4)$ the argument is negative.

The solution set is {8}.

73. $5^x = (4)^{2x}$

$\dfrac{5^x}{2^x} = 4$ Divide both sides by 2^x.

$\left(\dfrac{5}{2}\right)^x = 4$

$(2.5)^x = 4$

$(\log)(2.5)^x = \log 4$ Take the log of both sides.

$x\log(2.5) = \log 4$

$x = \dfrac{\log 4}{\log(2.5)}$

$x \approx 1.51$

The solution set is {1.51}.

Evaluate:

77. $\begin{vmatrix} 6 & 0 & 2 \\ 0 & 4 & 8 \\ 1 & 2 & 5 \end{vmatrix}$ Use the first row and expand by minors.

$= 6(20 - 16) - 0 + 2(0 - 4)$

$= 6(4) - 8$

$= 16$

Solve by any method:

81. $\begin{cases} 4y = 3x & (1) \\ 2x^2 - 3xy + 4 = 0 & (2) \end{cases}$

$4y = 3x$ Solve (1) for y.

$y = \dfrac{3x}{4}$

$2x^2 - 3x\left(\dfrac{3x}{4}\right) + 4 = 0$ Replace $\dfrac{3x}{4}$ for y in (2).

$8x^2 - 9x^2 + 16 = 0$

$x^2 = 16$

$x = \pm 4$

$4y = 3(4)$ $4y = 3(-4)$ Replace x in (1) with 4

$y = 3$ $y = -3$ and -4.

The two solutions are $\{(4, 3), (-4, -3)\}$.

CHAPTER 11

FUNCTIONS OF COUNTING NUMBERS

EXERCISES 11.1   SEQUENCES AND SERIES

A

Write the first five terms of the sequence defined by the following:

1. $c(n) = n + 6$
      $c(1) = 1 + 6 = 7$ Replace n with 1, 2, 3, 4,
      $c(2) = 2 + 6 = 8$ and 5. Simplify.
      $c(3) = 3 + 6 = 9$
      $c(4) = 4 + 6 = 10$
      $c(5) = 5 + 6 = 11$

Write a rule for a sequence function given the first five terms:

5.   3, 5, 7, 9, 11
      $2(1) + 1 = 3$ The terms are consecutive
      $2(2) + 1 = 5$ odd numbers. To generate
      $2(3) + 1 = 7$ the set of even numbers,
      $2(4) + 1 = 9$ the rule would be 2n, so
      $2(5) + 1 = 11$ to generate the set of odd
         $2n + 1.$ use 2n + 1.

      The sequence rule is $c(n) = 2n + 1$.

9.   3, 6, 9, 12, 15
      $3(1) = 3$ The terms are every third
      $3(2) = 6$ number. To generate the
      $3(3) = 9$ set, multiply n by 3.
      $3(4) = 12$
      $3(5) = 15$

      The sequence rule is $c(n) = 3n$.

Write in expanded form, and find the sum:

13. $\displaystyle\sum_{n=1}^{3} (2n + 7)$ Expand by replacing n with
                                                   1, 2, and 3.
      $= [2(1) + 7] + [2(2) + 7] + [2(3) + 7]$
      $= 9 + 11 + 13$ Simplify.
      $= 33$

17. $\displaystyle\sum_{n=1}^{5} (-2n - 5)$ Expand by replacing n with 1, 2, 3, 4, and 5.

$= [-2(1) - 5] + [-2(2) - 5] + [-2(3) - 5]$
$\quad\quad + [-2(4) - 5] + [-2(5) - 5]$
$= (-7) + (-9) + (-11) + (-13) + (-15)$
$= -55$

B

Write a rule for a sequence function given the first five terms and find c(8):

21. $1, \dfrac{1}{4}, \dfrac{1}{9}, \dfrac{1}{16}, \dfrac{1}{25}$

$\dfrac{1}{1^2} = 1$ The terms are the reciprocals of the squares of consecutive counting numbers.

$\dfrac{1}{2^2} = \dfrac{1}{4}$

$\dfrac{1}{3^2} = \dfrac{1}{9}$

$\dfrac{1}{3^2} = \dfrac{1}{16}$

$\dfrac{1}{4^2} = \dfrac{1}{16}$

$\dfrac{1}{5^2} = \dfrac{1}{25}$

$c(n) = \dfrac{1}{n^2}$

$c(8) = \dfrac{1}{8^2} = \dfrac{1}{64}$

Write in expanded form, and find the sum:

25. $\displaystyle\sum_{n=1}^{6} (4n - n^2)$

$= (4 \cdot 1 - 1^2) + (4 \cdot 2 - 2^2) + (4 \cdot 3 - 3^2)$
$\quad\quad + (4 \cdot 4 - 4^2) + (4 \cdot 5 - 5^2) + (4 \cdot 6 - 6^2)$
$= 3 + 4 + 3 + 0 + (-5) + (-12)$
$= 10 - 17$
$= -7$

29. $\displaystyle\sum_{n=1}^{9} (n^3 - n^2)$

$\quad = (1^3 - 1^2) + (2^3 - 2^2) + (3^3 - 3^2) + (4^3 - 4^2) +$
$\quad\quad (5^3 - 5^2) + (6^3 - 6^2) + (7^3 - 7^2) + (8^3 - 8^2) +$
$\quad\quad (9^3 - 9^2)$

$\quad = 0 + 4 + 18 + 48 + 100 + 180 + 294 + 448 + 648$

$\quad = 1740$

Write the following series in summation notation:

33. $1 + \dfrac{1}{8} + \dfrac{1}{27} + \dfrac{1}{64} + \dfrac{1}{125} + \cdots$

$1 = \frac{1}{1}$. The denominators are cubes of successive counting numbers.

$\quad c(n) = \dfrac{1}{n^3}$

$\quad \displaystyle\sum_{n=1}^{\infty} \dfrac{1}{n^3}$

The ellipsis (three dots) indicates there is no boundary for n.

c

Write a rule for a sequence function given the first five terms and find c(9). Write the summation notation for the first 10 terms.

37. $\dfrac{1}{7}, \dfrac{2}{8}, \dfrac{3}{9}, \dfrac{4}{10}, \dfrac{5}{11}$

$\quad c(n) = \dfrac{n}{n + 6}$

The numerators are successive counting numbers. The denominators are respectively six more than the numerators.

$\quad c(9) = \dfrac{9}{9 + 6} = \dfrac{9}{15} = \dfrac{3}{5}$

Replace n with 9, and reduce.

$\quad \displaystyle\sum_{n=1}^{10} \dfrac{n}{n + 6}$

41. 0, 2 log 2, 3 log 3, 4 log 4, 5 log 5

$\quad c(n) = n \log n$

0 = 1 log 1 since log 1 = 0.

$\quad c(9) = 9 \log 9$

$\quad \displaystyle\sum_{n=4}^{10} n \log n$

45. $5x^3$, $4x^6$, $3x^9$, $2x^{12}$, x^{15}

$c(n) = (6 - n)x^{3n}$ The coefficients of the x terms are expressed 6 - n. The powers on the x terms are multiples of 3. Replace n with 9, and simplify.

$c(9) = (6 - 9)x^{3 \cdot 9}$
$ = -3x^{27}$

$$\sum_{n=1}^{10} (6 - n)x^{3n}$$

Write in expanded form, and find the sum:

49.
$$\sum_{n=4}^{7} (n^2 - 5n)$$

$= (4^2 - 5 \cdot 4) + (5^2 - 5 \cdot 5)$ Replace n with 4, 5, 6, and 7 successively and
$ + (6^2 - 5 \cdot 6) + (7^2 - 5 \cdot 7)$ simplify.
$= (-4) + 0 + 6 + 14$
$= 16$

53.
$$\sum_{n=10}^{14} (-2n + 5)$$
     Replace n with 10, 11, 12, 13, and 14 successively and simplify.

$= (-2 \cdot 10 + 5) + (-2 \cdot 11 + 5) + (-2 \cdot 12 + 5) + (-2 \cdot 13 + 5)$
$ + (-2 \cdot 14 + 5)$
$= (-15) + (-17) + (-19) + (-21) + (-23)$
$= -95$

D

57. For every year that strawberries are planted in the same field, the yield is reduced by 8 tons. If the yield for the first year is 91 tons, what is the yield in the fourth year?

$c(n) = 91 - 8(n - 1)$ Write the rule for the yearly yield.

$c(4) = 91 - 8(4 - 1)$ Replace n with 4.
$ = 91 - 8(3) = 91 - 24$ Simplify.
$ = 67$

The yield in the fourth year is 67 tons.

61. A basketball dropped from 10 ft rebounds 5 ft on the first bounce and 2.5 ft on the second bounce. If it continues to rebound each time to half the previous height, how high will it rebound on the eighth bounce?

$$c(n) = \frac{10}{2^n}$$ Write a rule for the sequence function.

$$c(8) = \frac{10}{2^8} = 0.0390625$$ Now find c(8).

$$\text{or } \frac{10}{2^8} = \frac{10}{256} = \frac{5}{128}$$

On the eighth bounce, the ball will rebound 0.0390625 ft or $\frac{5}{128}$ ft.

65. The value of a sheet of first issue postage stamps was $7 in the first year it was issued. The value increased to $14 the second year and $28 the third year. What is the expected value in the sixth year?

The sequence is: 7, 14, 28, 56, 112, 224, ...

$$c(n) = 7(2^{n-1})$$ Write a rule for the sequence function.

$$c(6) = 7(2^{6-1})$$ Evaluate c(6).
$$= 7(2^5)$$
$$= 224$$

The expected value in the sixth year is $224.

STATE YOUR UNDERSTANDING

69. What "trick" can be used to get the signs to alternate in a sequence?

Use $(-1)^n$ as a factor. For even values of n, the expression is positive, while for odd values of n, the expression is negative.

In the following problem, you are given the first four terms of a sequence; (a) write the rule for the sequence; and (b) find the seventh term of the sequence.

73.   -3, 1, 5, 9, ...

$$4(1) - 7 = -3$$
$$4(2) - 7 = 1$$
$$4(3) - 7 = 5$$
$$4(4) - 7 = 9$$

Multiply by 4 since the difference between consecutive integers is 4.

$$c(n) = 4n - 7$$
$$c(7) = 4(7) - 7 = 21$$

Find the seventh term.

(a) The rule for the sequence is $c(n) = 4n - 7$.
(b) The seventh term of the sequence is 21.

77. $\frac{4}{1}, \frac{9}{2}, \frac{16}{3}, \frac{25}{4}, \ldots$

$$\frac{(1 + 1)^2}{1} = \frac{4}{1}$$

The numerators are squares, and the denominators are consecutive numbers.

$$\frac{(2 + 1)^2}{2} = \frac{9}{2}$$
$$\frac{(3 + 1)^2}{3} = \frac{16}{3}$$
$$\frac{(4 + 1)^2}{4} = \frac{25}{4}$$
$$c(n) = \frac{(n + 1)^2}{n}$$
$$c(7) = \frac{(7 + 1)^2}{7} = \frac{64}{7}$$

Find the seventh term.

(a) The rule for the sequence is $c(n) = \frac{(n + 1)^2}{n}$.

(b) The seventh term is $\frac{64}{7}$.

81. Write the slope of the line perpendicular to the graph of
    $3x + 5y = 8$.

$$3x + 5y = 8$$
$$5y = -3x + 8$$

Find the slope of the given line by solving for y.

$$y = -\frac{3}{5}x + \frac{8}{5}$$

Slope-intercept form.

$$m_1 = -\frac{3}{5}$$

Slope of the given line.

$$m_2 = \frac{5}{3}$$

Write the negative reciprocal of the slope of the given line.

The slope of the line perpendicular to the given line is $\frac{5}{3}$.

85. Write the coordinates of the y-intercept of the graph of
    $y = 2x^2 - 28x + 108$.

$$y = 2x^2 - 28x + 108$$
$$y = 2(0)^2 - 28(0) + 108$$
$$y = 108$$

Replace x with 0 to find the y-intercept.

The y-intercept is at (0, 108).

EXERCISES 11.2 ARITHMETIC PROGRESSIONS (SEQUENCES)

A

Write the first five terms of the arithmetic progression and find the indicated term:

1. $a_1 = 7$, $d = 3$. Find a_{10}

$$a_n = a_1 + (n - 1)d$$

Formula for the nth term of an arithmetic progression.

$$a_1 = 7$$
$$a_2 = 7 + (2 - 1)3 = 10$$

Substitute n = 2, 3, 4, and 5 in the formula.

$$a_3 = 7 + (3 - 1)3 = 13$$
$$a_4 = 7 + (4 - 1)3 = 16$$
$$a_5 = 7 + (5 - 1)3 = 19$$

$$a_{10} = 7 + (10 - 1)3 = 34$$

Substitute a = 10 in the formula.

5. $a_1 = 7$, $a_2 = 19$. Find a_{11}

    $d = 12$ Find d, the difference between two successive terms. $19 - 7 = 12$.

    $a_n = a + (n - 1)d$ Formula.

    $a_1 = 7$ First term.

    $a_2 = 19$ Given.

    $a_3 = 7 + (3 - 1)12 = 7 + 24 = 31$ Substitute 3, 4, and 5.

    $a_4 = 7 + (4 - 1)12 = 7 + 36 = 43$

    $a_5 = 7 + (5 - 1)12 = 7 + 48 = 55$

    $a_{11} = 7 + (11 - 1)12 = 127$ Substitute $n = 11$.

9. $a_1 = 2$, $a_4 = 14$. Find a_6

    $n = 1$, $a_1 = 2$ Given information.

    $n = 4$, $a_4 = 14$

    $a_n = a_1 + (n - 1)d$ Formula for general term.

    $14 = 2 + (4 - 1)d$ Substitute 14 for a_n, 2 for a_1 and 4 for n.

    $14 = 2 + 3d$

    $12 = 3d$

    $4 = d$ Solve for d.

    $a_1 = 2$

    $a_2 = 2 + 1(4) = 6$

    $a_3 = 2 + 2(4) = 10$

    $a_4 = 2 + 3(4) = 14$

    $a_5 = 2 + 4(4) = 18$

    $a_6 = 2 + 5(4) = 22$

Find the number of terms in each of the following finite arithmetic progressions:

13. $17, 13, \ldots, -23$

    $a_1 = 17$; $d = 13 - 17 = -4$

    $a_n = -23$

    $a_n = a_1 + (n - 1)d$ Formula for n^{th} term.

    $-23 = 17 + (n - 1)(-4)$ Substitue $a_n = -23$,

    $-23 = 17 - 4n + 4$ $a_1 = 17$, and $d = -4$.

    $-23 = 21 - 4n$ Solve for n.

    $-44 = -4n$

    $11 = n$

B

Find all the terms between the given terms of the following
arithmetic progressions:

17. $a_1 = 10$, $a_6 = 0$

$a_n = a_1 + (n - 1)d$	Formula.
$0 = 10 + (6 - 1)d$	Substitute 0 for a_n, 10
$0 = 10 + 5d$	for a_1, and 6 for n.
$-10 = 5d$	
$-2 = d$	Solve for d.

$a_1 = 10$	Generate the following
$a_2 = 10 - 2 = 8$	terms by subtracting 2.
$a_2 = 8 - 2 = 6$	
$a_4 = 6 - 2 = 4$	
$a_5 = 4 - 2 = 2$	

21. $a_{10} = 22$, $1_{15} = 37$

$n = 10$, $a_{10} = 22$	Given information.
$n = 15$, $a_{15} = 37$	

$a_n = a_1 + (n - 1)d$	
	Write two equations with two variables to find a_1 and d.
$22 = a_1 + 9d$ (1)	Substitute n = 10 and $a_{10} = 22$ to get equation (1).
$37 = a_1 + 14d$ (2)	Substitute n = 15 and $a_{15} = 37$ to get equation (2).
$15 = 5d$	Solve the system by subtracting (1) from (2).
$3 = d$	

$a_{10} = 22$	Start with a_{10} and add 3
$a_{11} = 25$	to get the terms that are
$a_{12} = 28$	between a_{10} and a_{15}.
$a_{13} = 31$	
$a_{14} = 34$	

So the missing terms are 25, 28, 31, and 34.

Find the sum of the indicated number of terms of the arithmetic progressions.

25. 15, 9, 3, ... 10 terms.
     $a_1 = 5$, $d = -6$, $n = 10$

     $\begin{aligned} a_{10} &= 15 + (10 - 1)(-6) \\ &= 15 + 9(-6) \\ &= -39 \end{aligned}$ Write the tenth term.

     $S_n = \dfrac{n(a_1 + a_n)}{2}$ Formula for the sum of n terms of an arithmetic progression.
     Substitute $n = 10$, $a_1 = 15$, and $a_n = -39$.

     $\begin{aligned} &= \dfrac{10[15 + (-39)]}{2} \\ &= 5(-24) \\ &= -120 \end{aligned}$

29. $-2, -\dfrac{3}{2}, -1, \ldots$ 9 terms
     $a_1 = -2$, $d = \dfrac{1}{2}$, $n = 9$

     $a_9 = -2 + (9 - 1)\dfrac{1}{2}$ Write the ninth term.
     $d = -\dfrac{3}{2} + 2 = \dfrac{1}{2}.$

     $\begin{aligned} &= -2 + 8\left(\dfrac{1}{2}\right) \\ &= -2 + 4 \\ &= 2 \end{aligned}$

     $S_n = \dfrac{n(a_1 + a_n)}{2}$

     $S_9 = \dfrac{9(-2 + 2)}{2}$

     $= 0$

Write the first five terms of the arithmetic progression and find the indicated term:

33. $a_1 = 5$, $a_2 = 3.5$. Find a_{15}.
    $d = 3.5 - 5 = -1.5$

    $a_1 = 5$
    $a_2 = 5 - 1.5 = 3.5$
    $a_3 = 3.5 - 1.5 = 2$
    $a_4 - 2 - 1.5 = 0.5$
    $a_5 = 0.5 - 1.5 = -1$

    $a_n = a_1 + (n - 1)d$

    $a_{15} = 5 + (15 - 1)(-1.5)$
    $\phantom{a_{15}} = 5 + 14(-1.5)$
    $\phantom{a_{15}} = 5 + (-21)$
    $\phantom{a_{15}} = -16$

    So $a_{15} = -16$.

37. $a_5 = 4$, $a_{15} = -11$. Find a_{31}.

    $a_n = a_1 + (n - 1)d$

    $4 = a_1 + 4d \qquad (1)$

    $-11 = a_1 + 14d \quad (2)$
    $-15 = 10d$
    $-\dfrac{3}{2} = d$

    $4 = a_1 + 4\left(-\dfrac{3}{2}\right)$
    $ = a_1 - 6$
    $10 = a_1$

    $a_{31} = a_1 + (n - 1)d$
    $\phantom{a_{31}} = 10 + (31 - 1)\left(-\dfrac{3}{2}\right)$

    $\phantom{a_{31}} = 10 + 30\left(-\dfrac{3}{2}\right)$
    $\phantom{a_{31}} = 10 - 45$
    $\phantom{a_{31}} = -35$

Find a_1 and d by writing two equations in two variables, a_1 and d. Substitute 4 for a_5 and 5 for n for equation (1). Substitute -11 for a_{15} and 15 for n for equation (2).

Subtract equation (1) from equation (2).

Substitute $d = -\dfrac{3}{2}$ in equation (1) and solve for a_1.

Formula for general term. Substitute 31 for n and $-\dfrac{3}{2}$ for d, 10 for a_1.

C

Find the number of terms and the sum of the following finite arithmetic progression.

41. 14.2, 12.9, ..., -14.4
    $a_1 = 144.2$, $d = 12.9 - 14.2 = -1.3$, $a_n = -14.4$

$$a_n = a_1 + (n - 1)d$$

$-14.4 = 14.2 + (n - 1)(-1.3)$ Substitute for a_n, a_1, and d.

                                           Solve for n.

$-14.4 = 14.2 - 1.3n + 1.3$

$-14.4 = 15.5 - 1.3n$

$-29.9 = -1.3n$

   $23 = n$

$$S_n = \frac{n(a_1 + a_n)}{2}$$

$$= \frac{23[14.2 + (-14.4)]}{2}$$

$$= \frac{23(-0.2)}{2}$$

$$= -2.3$$

The number of terms is 23, and the sum of the terms is -2.3.

Find the indicated sum:

45. $\displaystyle\sum_{n=1}^{18} (3n + 9)$

$a_1 = 3(1) + 9 = 12$ Generate the first and
$a_{18} = 3(18) + 9 = 63$ last terms by using the
  $n = 18$ rule given.

$$S_n = \frac{n(a_1 + a_{18})}{2}$$ Formula for the sum of an arithmetic progression.

$$= \frac{18(12 + 63)}{2}$$

$$= 9(75)$$

$$= 675$$

49. $\displaystyle\sum_{n=1}^{25} (-0.4n - 10)$

$a_1 = -0.4(1) - 10 = -10.4$ Generate the first and
$a_{25} = -0.4(25) - 10 = -20$ last terms by using the
$n = 25$ rule given.

$S_n = \dfrac{n(a_1 + a_{25})}{2}$ Formula for the sum of an arithmetic progression.

$= \dfrac{25[-10.4 + (-20)]}{2}$ Substitute $n = 25$, $a_1 = -10.4$ and $a_{25} = -20$.

$= \dfrac{25(-30.4)}{2}$

$= -380$

Find the sum of the indicated number of terms of the following progression:

53. $a_5 = -5$, $a_{15} = -30$. Find S_{36}.
$a_n = a_1 + (n - 1)d$

$-5 = a_1 + 4d \qquad (1)$ First find a_1 and d by
$-30 = a_1 + 14d \qquad (2)$ generating a system of two equations in two variables.

$-25 = 10d$
$-2.5 = d$ Subtract (1) from (2).
$-5 = a_1 + 4(-2.5)$ Solve for a_1 using
$-5 = a_1 - 10$ equation (1).
$5 = a_1$

$a_n = a_1 + (n - 1)d$ Formula for general term.
$a_{36} = 5 + 35(-2.5)$ Substitue $d = -2.5$ and $a_1 = 5$.

$= -82.5$

$S_{36} = \dfrac{36[5 + (-82.5)]}{2}$ Formula for the sum of an arithmetic progression.
$= 18(-77.5)$
$= -1395$

The sum of the first 36 terms is −1395.

D

57. Pete intends to increase the distance he runs each week by two miles. If he now runs 15 miles per week, in how many weeks will he be running 3 miles per week?

    15, 17, 19, ... 63   Write the first few terms of the progression.

    $a_1 = 15$, $d = 17 - 15 = 2$, $a_n = 63$

    $a_n = a_1 + (n - 1)d$ Formula for the nth term.
    $63 = 15 + (n - 1)2$ Substitute and solve
    $63 = 15 + 2n - 2$ for n.
    $63 = 13 + 2n$
    $50 = 2n$
    $25 = n$

    In 25 weeks, Pete will be running 63 miles per week.

61. Willy is stacking blocks. How many blocks are there in his stack if there are 18 blocks in the bottom row, 17 in the second row, 16 in the third row, and so on, until there is 1 block in the top row?

    $S_n = 18 + 17 + 16 + ... + 1$ $a = 18$, $a_n = 1$
    $\quad d = 17 - 18 = -1$
    $1 = 18 + (n - 1)(-1)$ Substitute into the
    $1 = 18 - n + 1$ formula for the general
    $1 = 19 - n$ term to find n.
    $n = 18$

    $S_{18} = \dfrac{18(18 + 1)}{2} = 171$

    There are 171 blocks.

65. Two baseball players, Davis and Mooney, sign long-term contracts with a team. Davis is more experienced, so his salary at the beginning is higher. Mooney shows a lot of promise so his increases are larger. Their salaries for the first three years of their contracts are as follows:

	Davis	Mooney
1st year	$150,000	$73,000
2nd year	$155,000	$85,000
3rd year	$160,000	$97,000

(a) In what year will their salaries be the same?
(b) How much will each earn in that year?
(c) How much will each player earn from the first year to, and including, the year that their salaries are the same?

$a_n = a_1 + (n - 1)d$	Write an expression for a_n for Davis's progression.
$a_n = 150000 + (n - 1)(5000)$	Replace a_1 with 150000 and d with 5000.
$a_n = a_1 + (n - 1)d$	Write an expression for a_n for Mooney's progression.
$a_n = 73000 + (n - 1)(12000)$	Replace a_1 with 73000, and d with 12000.

$$150000 + (n - 1)(5000) = 73000 + (n - 1)(12000)$$

Set the two expressions equal and solve for n.

$$150000 + 5000n - 5000 = 73000 + 12000n - 12000$$
$$5000n + 145000 = 12000n + 61000$$
$$-7000n = -84000$$
$$n = 12$$

(a) In the twelfth year, their salaries will be the same.

Davis:
$a_{12} = 150000 + (12 - 1)(5000)$ Evaluate a_{12}.
$\quad\ = 150000 + 11(5000)$
$\quad\ = 150000 + 55000$
$\quad\ = 205000$

Mooney:
$a_{12} = 73000 + (12 - 1)(12000)$
$\quad\ = 73000 + 11(12000)$
$\quad\ = 73000 + 132000$
$\quad\ = 205000$

(b) In the twelfth year, both Davis and Mooney will earn $205,000.

$$S_{12} = \frac{12(150000 + 205000)}{2}$$ Davis's sum.

$$= 6(355000)$$
$$= 2130000$$

$$S_{12} = \frac{12(73000 + 205000)}{2}$$ Mooney's sum.

$$= 6(278000)$$
$$= 1668000$$

(c) Davis's sum is $2,130,000 and Mooney's is $1,668,000.

CHALLENGE EXERCISES

For the following arithmetic sequences, find:
(a) the tenth term of the sequences and
(b) the sum of the first ten terms.

69. 7, 7, 7, 7, ...

$$a_n = 1_1 + (n - 1)d$$ The difference between
$$a_n = 7 + (n - 1)0$$ terms is 0. Replace d
$$a_n = 7$$ with zero ,and a_1 with 7.
$$a_{10} = 7$$ Find the tenth term.

(a) The tenth term is 7.

$$S_n = \frac{n(a_1 + a_{10})}{2}$$ Find the sum of the first
                                           ten terms.
$$S_{10} = \frac{10(7 + 7)}{2}$$ Replace n with 10, a_1 with
                                           7 and a_{10} with 7.
$$= 5(14)$$
$$= 70$$

(b) The sum of the first ten terms of the sequence is 70.

73. $\sqrt{3}$, $\sqrt{12}$, $\sqrt{27}$, $\sqrt{48}$, ...

$\sqrt{3}$, $2\sqrt{3}$, $3\sqrt{3}$, $4\sqrt{3}$, ...   Simplify radicals.
$$a_n = a_1 + (n - 1)d$$ The difference between

$$a_n = \sqrt{3} + (n - 1)\sqrt{3}$$ terms is $\sqrt{3}$.

$$a_{10} = \sqrt{3} + (10 - 1)\sqrt{3}$$ Find the tenth term.

$$= \sqrt{3} + 9\sqrt{3}$$

$$= 10\sqrt{3}$$

(a) The tenth term is $10\sqrt{3}$.

$$S_{10} = \frac{10(a_1 + a_{10})}{2}$$

$$= 5(\sqrt{3} + 10\sqrt{3})$$

$$= 5(11\sqrt{3})$$

$$= 55\sqrt{3}$$

Find the sum of the first ten terms.

(b) the sum of the first ten terms of the sequence is $55\sqrt{3}$.

MAINTAIN YOUR SKILLS

77. $\log_{1/2} 0.015625 = 6$

$\left(\frac{1}{2}\right)^6 = 0.015625$

The base is $\frac{1}{2}$, and the exponent is 6.

Write in logarithmic form:

81. $(a + b)^9 = y$
$\log_{a+b} y = 9$

The base is $(a + b)$ and the exponent (log) is 9.

EXERCISES 11.3   GEOMETRIC PROGRESSIONS (SEQUENCES)

A

Write the first four terms of the geometric progression and find the indicated term:

1. $a_1 = 1$, $r = 3$. Find a_6.

$a_1 = 1$

$a_2 = 1 \cdot 3 = 3$

$a_3 = 3 \cdot 3 = 9$

$a_4 = 9 \cdot 3 = 27$

The first term is 1, and each term after is found by multiplying the preceding term by 3.

The first four terms are 1, 3, 9, and 27.

$a_n = a\, r^{n-1}$

$a_6 = 1 \cdot 3^{6-1} = 3^5 = 243$

Formula for the general term of a geometric progression.
Substitute $a_1 = 1$, $n = 6$, and $r = 3$.

The sixth term is 243.

Find the number of terms in each finite geometric progression:

5. $6, 12, \ldots, 1536$

$r = \dfrac{12}{6} = 2$ Find r.

$a_n = a_1 r^{n-1}$ Use the formula to solve for n, where $a_1 = 6$, $r = 2$, and $a_n = 1536$.

$1536 = 6 \cdot 2^{n-1}$
$256 = 2^{n-1}$
$2_8 = 2^{n-1}$ Write each side with a base of 2.

$n - 1 = 8$ If $a^x = a^y$, then $x = y$.
$n = 9$

There are 9 terms.

Find all the terms between the given terms of the following geometric progressions:

9. $a_1 = 2, \ a_5 = 162$

$a_n = a_1 r^{n-1}$ Formula for the nth term of a geometric progression.

$162 = 2 \cdot r^4$ Substitue 162 for a_n, and
$81 = r^4$ 5 for n. Solve for r.
$\pm 3 = r$

$a_2 = 2(\pm 3)^1 = \pm 6$ Use the formula for the

$a_3 = 2(\pm 3)^2 = 18$ n^{th} term to find the terms between the first and fifth terms.

$a_4 = 2(\pm 3)^3 = \pm 54$

13. $a_2 = -1, \ a_7 = 243$

$a_n = a_1 r^{n-1}$ Formula for the nth term.

So,
$\begin{cases} -1 = a_1 r^1 & (1) \\ 243 = a_1 r^6 & (2) \end{cases}$ Write two equations with two unknowns using the data to find a_1 and r.

$-\dfrac{1}{r} = a_1$ Solve (1) for a_1.

$243 = -\dfrac{1}{r} \cdot r^6$ Substitute in (2).
$243 = -r^5$
$-243 = r^5$
$-3 = 4$

452

$a_2 = -1$
$a_3 = -1 \cdot -3 = 3$
$a_4 = 3 \cdot -3 = -9$
$a_5 = -9 \cdot -3 = 27$
$a_6 = 27 \cdot -3 = -81$

Use -3 as the common ratio.

Find the sum of the indicated number of terms of the following geometric progression:

17. 2, -6, 18, ... 6 terms

$a_1 = 2, \quad r = \dfrac{-6}{2} = -3$

$S_6 = \dfrac{2 - 2(-3)^6}{1 - (-3)}$

$\quad = \dfrac{2 - 1458}{4}$

$\quad = -\dfrac{1456}{4}$

$\quad = -364$

The sum of the first 6 terms of the given geometric progression is -364.

Write the first four terms of the geometric progression and find the indicated term:

21. $a_1 = 9, \quad a_2 = 6.$ Find a_8.

$r = \dfrac{6}{9} = \dfrac{2}{3}$

$a_n = a_1 r^{n-1}$

Formula for the nth term.

$a_1 = 9\left(\dfrac{2}{3}\right)^0 = 9$

$a_2 = 9\left(\dfrac{2}{3}\right)^1 = 6$

$a_3 = 9\left(\dfrac{2}{3}\right)^2 = 4$

$a_4 = 9\left(\dfrac{2}{3}\right)^3 = \dfrac{8}{3} = 2\dfrac{2}{3}$

$a_8 = 9\left(\dfrac{2}{3}\right)^4 = \dfrac{128}{243}$

25. $a_2 = \frac{1}{6}$, $a_5 = \frac{1}{162}$. Find a_7

$$a_n = a_1 r^{n-1}$$

Formula for the nth term of a geometric progression.

So,

$$\begin{cases} \dfrac{1}{6} = a_1 r^1 \quad (1) \\[2mm] \dfrac{1}{162} = a_1 r^4 \quad (2) \end{cases}$$

Write two equations with two unknowns using the given data to find a_1 and r.

$$\frac{1}{6r} = a_1$$

Solve (1) for a_1.

$$\frac{1}{162} = \frac{1}{6r} \cdot r^4$$

Substitute in (2).

$$\frac{1}{162} = \frac{r^3}{6}$$

$$\frac{6}{162} = r^3$$

Multiply both sides by 6.

$$\frac{1}{27} = r^3$$

Reduce the fraction.

$$\frac{1}{3} = r$$

$$\frac{1}{6} = a_1 \left(\frac{1}{3}\right)$$

Substitute $\frac{1}{3}$ for r in (1).

$$\frac{1}{6} = \frac{a_1}{3}$$

$$\frac{3}{6} = a_1$$

Multiply both sides by 3.

$$\frac{1}{2} = a_1$$

$$a_1 = \frac{1}{2}$$

$$a_2 = \frac{1}{2}\left(\frac{1}{3}\right)^1 = \frac{1}{6}$$

Use the formula for the nth substituting $\frac{1}{2}$ for a

$$a_3 = \frac{1}{2}\left(\frac{1}{3}\right)^2 = \frac{1}{18}$$

and $\frac{1}{3}$ for r.

$$a_4 = \frac{1}{2}\left(\frac{1}{3}\right)^3 = \frac{1}{54}$$

$$a_7 = \frac{1}{2}\left(\frac{1}{3}\right)^6 = \frac{1}{1458}$$

Find the number of terms and the sum of the following finite geometric progressions:

29. 80, 20, ..., $\frac{5}{64}$

$a_1 = 80$, $r = \frac{20}{80} = \frac{1}{4}$, $a_n = \frac{5}{64}$

80, 20, 5, $\frac{5}{4}$, $\frac{5}{16}$, $\frac{5}{64}$ Use the ratio to find the missing terms.

There are 6 terms.

$$S_n = \frac{a_1 - a_1 r^n}{1 - 4}$$

$$S_6 = \frac{80 - 80\left(\frac{1}{4}\right)^6}{1 - \frac{1}{4}} = \frac{80 - \frac{1}{4096}}{\frac{3}{4}}$$

$$= 106\frac{41}{64} \text{ or } 106.640625$$

The sum of the six terms is $106\frac{41}{64}$.

33. -2.1, 1.05, ..., -0.13125

$a_1 = -2.1$, $r = \frac{1.05}{-2.1} = -0.5$, $a_n = -0.13125$

-2.1, 1.05, -0.525, 0.2625, -0.13125   Use the ratio to find the missing terms.

There are five terms.

$$S_n = \frac{a_1 - a_1 r^n}{1 - r}$$

$$S_5 = \frac{-2.1 - (-2.1)(-0.5)^5}{1 - (-0.5)}$$ Substitute -2.1 for a_1, -0.5 for r, and 5 for n.

$$= \frac{-2.1 - (0.065625)}{1.5}$$

$$= \frac{-2.165625}{1.5}$$

$$= -1.44375$$

C

Find the indicated sum:

37. $\displaystyle\sum_{n=1}^{10} 10(-1)^n$

$= 10(-1)^1 + 10(-1)^2 + 10(-1)^3 + 10(-1)^4 + 10(-1)^5$
$\qquad + 10(-1)^6 + 10(-1)^7 + 10(-1)^8 + 10(-1)^9 + 10(-1)^{10}$

$= -10 + 10 - 10 + 10 - 10 + 10 - 10 + 10 - 10 + 10$

$= 0$

41. $\displaystyle\sum_{n=1}^{10} 2187(3)^{-n}$

$n = 11$ Find a_1, r, and n.
$a_1 = 2187(3)^{-1} = 729$
$a_2 = 2187(3)^{-2} = 243$
$r = \dfrac{243}{729} = \dfrac{1}{3}$

$S_n = \dfrac{a_1 - a_1 r^n}{1 - r}$ Formula.

$S_{11} = \dfrac{729 - 729\left(\dfrac{1}{3}\right)^{11}}{1 - \dfrac{1}{3}} = \dfrac{729 - \dfrac{729}{177147}}{\dfrac{2}{3}}$ Substitute.

$= 1093\dfrac{40}{81}$ or approximately 1093.4938

The sum is $1093\dfrac{40}{81}$.

45. $\displaystyle\sum_{n=1}^{5} (3)\left(-\dfrac{1}{3}\right)^n$

$n = 5$ Find n, a_1, and r.
$a_1 = (3)\left(-\dfrac{1}{3}\right)^1 = -1$
$a_2 = (3)\left(-\dfrac{1}{3}\right)^2 = \dfrac{1}{3}$
$r = \dfrac{a_2}{a_1} = \dfrac{\dfrac{1}{3}}{-1} = -\dfrac{1}{3}$

So

$$S_n = \frac{a_1 - a_1 r^n}{1 - 4}$$

$$S_5 = \frac{-1 - (-1)\left(-\frac{1}{3}\right)^5}{1 - \left(-\frac{1}{3}\right)}$$ Substitute.

$$= \frac{-\frac{244}{243}}{\frac{4}{3}}$$

$$= -\frac{61}{81}$$

The sum is $-\frac{61}{81}$ or approximately -0.7531.

49. $\displaystyle\sum_{n=1}^{5} \left(\frac{3}{2}\right)\left(\frac{1}{3}\right)^n$

$n = 5$ Find n, a_1, and r.

$$a_1 = \left(\frac{3}{2}\right)\left(\frac{1}{3}\right)^1 = \frac{1}{2}$$

$$a_2 = \left(\frac{3}{2}\right)\left(\frac{1}{3}\right)^2 = \frac{1}{6}$$

$$r = \frac{a_2}{a_1} = \frac{\frac{1}{6}}{\frac{1}{2}} = \frac{1}{3}$$

$$S_n = \frac{a_1 - a_1 r^n}{1 - r}$$

$$S_5 = \frac{\frac{1}{2} - \frac{1}{2}\left(\frac{1}{3}\right)^5}{1 - \frac{1}{3}}$$ Substitute.

$$= \frac{\frac{242}{486}}{\frac{2}{3}}$$

$$= \frac{121}{162}$$

The sum is $\frac{121}{162}$ or approximately 0.7469.

53. The price of a can of juice increases at a rate of 5% per year because of inflation. If the price of the juice is now 73¢, what will the price be four years from now? Twenty years from now? (Hint: Use r = 1.05 since the price each year is 105% of last year's price.)

$a_n = a_1 r^{n-1}$

The progression of the price of the juice in each successive year is geometric.

Formula for the nth term of a geometric progression.

$a_5 = 0.73(1.05)^5$

Substitute $a_1 = 0.73$ and r = 1.05. Use n = 5 since the fourth year represents a_5 or four additional years after a_1.

≈ 0.89

$a_{21} = 0.73(1.05)^{20}$

≈ 1.94

Substitute $a_1 = 0.73$, r = 1.05, and n = 21 for the twentieth year.

The price of the juice four years from now will be $0.89 and twenty years from now will be $1.94.

57. If you start with a single square and draw a vertical and a horizontal line through the square as shown, you will now have four smaller squares. If you divide each of those, you will have sixteen still smaller squares. If this process is continued six times, how many very small squares will there be?

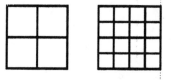

$a_n = a_1 r^{n-1}$
$a_n = 1 \cdot 4^{n-1}$

Substitute $a_1 = 1$ and r = 4.
Now replace n with 6.

$a_6 = 1 \cdot 4^{6-1} = 4^5$
$= 1024$

If the process is repeated six times, there will be 1024 very small squares.

61. A ball is dropped from a height of 40 feet and it bounces up to 70% of the previous height. How high will it bounce on the seventh bounce? What will be the total "up and down" distance that the ball has travelled after its seventh bounce (Careful)! (All answers to the nearest hundredth)

$a_n = a_1 r^{n-1}$
$a_7 = 40(0.7)^{7-1}$

Replace r with 70% or 0.7, n with 7, and a_1 with 40.

$= 40(0.7)^6$
$= 40(0.117649)$
$= 4.71$

$S_n = \dfrac{a_1 - a_1 r^n}{1 - 4}$

Find the sum.

$S_7 = \dfrac{40 - 40(0.7)^7}{1 - 0.7}$
≈ 122.35

Replace a_1 with 40, r with 0.7, and n with 7. This is the total of the distances the ball has bounced up.

$2(122.35) \approx 244.70$

Multiply by 2 to find the total distance the ball has bounced up and down.

$244.70 - 40$
≈ 204.70

Now subtract 40 since the initial distance downward was traveled before the first bounce.

The ball will bounce up approximately 4.71 ft on the seventh bounce. The total "up and down" distance the ball will have traveled after its seventh bounce is approximately 204.70 ft.

CHALLENGE EXERCISES

For each of the following geometric sequences, find:
(a) the seventh term of the sequence; and
(b) the sum of the first seven terms.

65. $4, -2, 1, -\dfrac{1}{2}, \ldots$

The ratio between terms is $-\dfrac{1}{2}$, since $\dfrac{-2}{4} = -\dfrac{1}{2}$.

$a_n = a_1 r^{n-1}$
$a_7 = 4\left(-\dfrac{1}{2}\right)^{7-1}$

Replace r with 7, and a_1 with 4.

$= 4\left(-\dfrac{1}{2}\right)^6$
$= 4(0.15625)$
$= 0.0625$

$$S_n = \frac{a_1 - r(a_n)}{1 - r}$$

Sum of a geometric progression.

$$S_7 = \frac{4 - \left(-\frac{1}{2}\right)(0.0625)}{1 - \left(-\frac{1}{2}\right)}$$

Replace n with 7, a_1 with 4, r with $-\frac{1}{2}$, and a_n with 0.0625.

$$= \frac{4.03125}{1.5}$$

$$= 2.6875$$

(a) the seventh term is 0.0625.
(b) the sum of the first seven terms is 2.6875.

69. 1, $\sqrt{5}$, 5, $\sqrt{125}$, ...

The ratio of this progression is $\sqrt{5}$ since $\frac{\sqrt{5}}{1}$ is $\sqrt{5}$.

$$a_n = a_1 r^{n-1}$$
$$a_7 = 1(\sqrt{5})^{7-1}$$
$$= 1(125)$$
$$= 125$$

$$S_n = \frac{a_1 - r(a_n)}{1 - r}$$

$$= \frac{1 - \sqrt{5}(125)}{1 - \sqrt{5}}$$

$$= \frac{1 - 125\sqrt{5}}{1 - \sqrt{5}}$$

$$= \frac{(1 - 125\sqrt{5})(1 + \sqrt{5})}{(1 - \sqrt{5})(1 + \sqrt{5})}$$

Rationalize the denominator.

$$= \frac{1 - 124\sqrt{5} - 625}{1 - 5}$$

$$= \frac{-625 - 124\sqrt{5}}{-4}$$

$$= 156 + 31\sqrt{5}$$

(a) The seventh term is 125.

(b) the sum of the first seven terms is $156 + 31\sqrt{5}$ or approximately 225.32.

MAINTAIN YOUR SKILLS

Solve without a calculator:

73. $5^{x-1} = 125$
     $5^{x-1} = 5^3$
     $x - 1 = 3$ If $b^x = b^y$, then $x = y$.
         $x = 4$

     The solution set is {4}.

Solve with the help of a calculator or logarithm tables; write the answer to the nearest ten-thousandths:

77. $\log_{17} 5 = x$
         $x = \dfrac{\log 5}{\log 17}$ Change of base formula.
         $x = \approx 0.5681$ Use a calculator.

EXERCISES 11.4   INFINITE GEOMETRIC PROGRESSIONS

A

Find the sum of each of the following infinite geometric series:

1. $8 + 4 + 2 + \ldots$

    $a_1 = 8, \ r = \dfrac{4}{8} = \dfrac{1}{2}$ Identify a_1 and r.

    $S_\infty = \dfrac{a_1}{1 - 4}$

    $ = \dfrac{8}{1 - \dfrac{1}{2}}$ Substitute 8 for a_1 and $\dfrac{1}{2}$ for r.

    $ = \dfrac{8}{\dfrac{1}{2}}$

    $ = 16$

5. $21 + 7 + \dfrac{7}{3} + \ldots$

    $a_1 = 21, \ r = \dfrac{1}{3}$

    $S_\infty = \dfrac{21}{\dfrac{2}{3}} = 31\dfrac{1}{2}$

9. $30 + 15 + \dfrac{15}{2} + \ldots$

    $a_1 = 30, \quad r = \dfrac{1}{2}$

    $S_\infty = \dfrac{30}{\dfrac{1}{2}} = 60$

13. $-100 + (-50) + (-25) + \ldots$

    $a_1 = -100$

    $r = \dfrac{-50}{-100} = \dfrac{1}{2}$

    $S_\infty = \dfrac{-100}{1 - \dfrac{1}{2}}$

       $= -200$

B

17. $15 - 3 + \dfrac{3}{5} + \cdots$

    $a_1 = 15, \quad r = \dfrac{-3}{15} = -\dfrac{1}{5}$ Identify a_1 and r.

    $S_\infty = \dfrac{15}{1 - \left(-\dfrac{1}{5} \right)}$

       $= \dfrac{15}{\dfrac{6}{5}}$

       $= \dfrac{75}{6}$

       $= \dfrac{25}{2}$

21. $\frac{2}{3} + \frac{1}{3} + \frac{1}{6} + \cdots$

$a_1 = \frac{2}{3}, \quad r = \dfrac{\frac{1}{3}}{\frac{2}{3}} = \frac{1}{2}$ Identify a_1 and r.

$S_\infty = \dfrac{\frac{2}{3}}{1 - \frac{1}{2}}$

$= \dfrac{\frac{2}{3}}{\frac{1}{1}}$

$= \frac{4}{3}$ or $1\frac{1}{3}$

25. $-22 - 8.8 - 3.52 - \cdots$

$a_1 = -22, \quad r = \dfrac{-8.8}{-22} = \frac{2}{5}$

$S_\infty = \dfrac{-22}{1 - \frac{2}{5}}$

$= \dfrac{-22}{\frac{3}{5}}$

$= -\frac{110}{3}$

29. $5 - \frac{15}{2} + \frac{45}{2} - \cdots$

$a_1 = 5, \quad r = \dfrac{-\frac{15}{2}}{5} = -\frac{3}{2}$

Since $r \leq -1$, the sum is not defined.

33. $\displaystyle\sum_{n=1}^{\infty} \left(\frac{7}{8}\right)^n = \frac{7}{8}, \frac{49}{64}, \frac{2401}{4096}, \ldots$

$a_1 = \dfrac{7}{8}, \quad r = \dfrac{7}{8}$

$S_\infty = \dfrac{\frac{7}{8}}{1 - \frac{7}{8}}$

$= 7$

c

Find the common fraction name for the following repeating decimals.

37. 0.151515 ...

0.151515... = 0.15 + 0.0015 + 0.000015 + ...

$a_1 = 0.15, \quad r = \dfrac{0.0015}{0.15} = 0.01$ Identify a_1 and r.

$S_\infty = \dfrac{0.15}{1 - 0.01}$

$= \dfrac{5}{33}$

41. 0.181818...

0.181818... = 0.18 + 0.0018 + 0.000018 + ...

$a_1 = 0.18, \quad r = \dfrac{0.0018}{0.18} = 0.01$

$S_\infty = \dfrac{0.18}{1 - 0.01}$

$= \dfrac{18}{99}$

$= \dfrac{2}{11}$

45. $0.027027027\ldots$

$0.027027027\ldots = 0.027 + 0.000027 + 0.000000027 + \ldots$

$a_1 = 0.027, \quad r = \dfrac{0.000027}{0.027} = 0.001$

$\begin{aligned} S_\infty &= \dfrac{0.027}{1 - 0.001} \\ &= \dfrac{27}{999} \\ &= \dfrac{1}{37} \end{aligned}$

49. $\displaystyle\sum_{n=1}^{\infty} \frac{3}{5} \cdot \left(\frac{4}{7}\right)^n$

Generate the first few terms of the progression.

$= \dfrac{12}{35} + \dfrac{48}{245} + \dfrac{192}{1715}$

$a_1 = \dfrac{12}{35}, \quad r = \dfrac{48}{245} \div \dfrac{12}{35} = \dfrac{4}{7}$

$\begin{aligned} S_\infty &= \dfrac{\frac{12}{35}}{1 - \frac{4}{7}} \\ &= \dfrac{4}{5} \end{aligned}$

D

53. A ball rebounds to $\frac{1}{3}$ of its prevous height with each bounce. How far has the ball traveled if it is dropped from a height of 10 feet? (Assume the ball bounces an infinite number of times.)

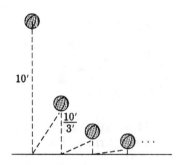

$\dfrac{10}{3} + \dfrac{10}{3} \cdot \dfrac{1}{3} + \dfrac{10}{3} \cdot \dfrac{1}{3} \cdot \dfrac{1}{3} + \cdots$

Generate the first few terms.

$\dfrac{10}{3} + \dfrac{10}{9} + \dfrac{10}{27} + \cdots$

$a_1 = \dfrac{10}{3}, \quad r = \dfrac{1}{3}$

465

$$S_\infty = \frac{\frac{10}{3}}{1 - \frac{1}{3}}$$

Sum of distances the ball bounces upward.

$$= 5$$
$$2(5) + 10 = 20$$

Multiply by 2 and add in the initial 10 ft to find the total distance the ball was dropped plus distances the ball bounced both <u>up</u> and <u>down</u>.

The ball traveled 20 ft.

57. An equilateral triangle has three equal sides, and the dimensions of three triangles in a sequence are given below:

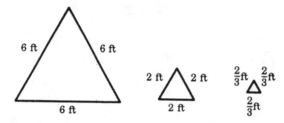

If the perimeter of each triangle is three times the length of one side, find sum of the perimeters of all of the triangles in the sequence if the sequence were to continue indefinitely.

$$18 + 6 + 2 + \ldots$$

Find the first few perimeters of the sequence.

$$a_1 = 18, \quad r = \frac{6}{18} = \frac{1}{3}$$

$$S_\infty = \frac{18}{1 - \frac{1}{3}}$$

Sum of a geometric progression.

$$= 27$$

The sum of the perimeters of all the triangles in the sequence is 27.

61. A diamond (a square which is resting on a corner) is 4 feet by 4 feet. By connecting the midpoints of the sides of the diamond, a square is formed inside. What is the length of each side of the new square? Then, another diamond is formed by connecting the midpoints of the sides of the square, and so on. If the area of each is bound by squaring one of its sides, what would be the sum of the areas of all of the diamonds and squares if the process were to continue forever?

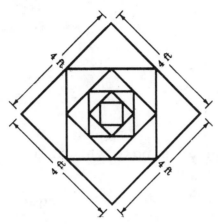

	Length of each side	Area
First diamond	4	16
Second	$\sqrt{2^2 + 2^2} = 2\sqrt{2}$	8
Third	$\sqrt{\sqrt{2}^2 + \sqrt{2}^2} = 2$	4

$a_1 = 16, \quad r = \dfrac{8}{16} = \dfrac{1}{2}$

$S_\infty = \dfrac{16}{1 - \dfrac{1}{2}}$

$= \dfrac{16}{\dfrac{1}{2}}$

$= 32$

The sum of the areas of all the diamonds and squares is 32 ft^2.

CHALLENGE EXERCISES

65. If the sum of the following geometric series is 4, find "r."

$$3 + 3r + 3r^2 + 3r^3 + \ldots$$

$$S_\infty = \frac{a_1}{1 - r}$$

$$4 = \frac{3}{1 - r}$$ Replace S_∞ with 4, and a_1 with 3.

$$4 - 4r = 3$$ Solve for r.

$$-4r = -1$$

$$r = \frac{1}{4}$$

MAINTAIN YOUR SKILLS

Write as a combination of single variables:

69. $\log\frac{2a}{b}$ = log2a - logb   Logarithm of a quotient.

     = log2 + loga - logb   Logarithm of a product.

Write as a single logarithm.

73. $\ln x^{-1} - \ln y^{-1} + \ln\frac{1}{z}$

$= \ln\frac{1}{x} - \ln\frac{1}{y} + \ln\frac{1}{z}$ Definition of negative exponent.

$= \ln\frac{y}{x} + \ln\frac{1}{z}$ Difference of logarithms.

$= \ln\frac{y}{xz}$ Sum of logarithms.

EXERCISES 11.5   BINOMIAL EXPANSION

A

Evaluate:

1. $3! = 3\cdot2\cdot1 = 6$ Write in expanded form and simplify.

5. $\frac{9!}{6!} = \frac{9\cdot8\cdot7\cdot6!}{6!} = 9\cdot8\cdot7 = 504$

Write in expanded form:

9. $(x + 1)^4$
   $n = 4, \ a = x, \ b = 1$ Identify a, b, and n.

$$\frac{4!}{4!0!}(x)^4(1)^0 \ + \ \frac{4!}{3!1!}(x)^3(1)^1$$ Use the formula to expand.

$$+ \ \frac{4!}{2!2!}(x)^2(1)^2 \ + \ \frac{4!}{1!3!}(x)^1(1)^3$$

$$+ \ \frac{4!}{0!4!}(x)^0(1)^4$$

$$= x^4 + 4x^3 + 6x^2 + 4x + 1$$ Simplify.

13. $(x + 2)^8$
    $n = 8, \ a = x, \ b = 2$ Identify a, b, and n.

$$\frac{8!}{8!0!}(x)^8(2)^0 \ + \ \frac{8!}{7!1!}(x)^7(2)^1$$ Use the formula to expand $(x + 2)$.

$$+ \ \frac{8!}{6!2!}(x)^6(2)^2 \ + \ \frac{8!}{5!3!}(x)^5(2)^3$$

$$+ \ \frac{8!}{4!4!}(x)^4(2)^4 \ + \ \frac{8!}{3!5!}(x)^3(2)^5$$

$$+ \ \frac{8!}{2!6!}(x)^2(2)^6 \ + \ \frac{8!}{1!7!}(x)^1(2)^7$$

$$+ \ \frac{8!}{0!8!}(x)^0(2)^8$$

$$= (1)(x^8)(1) \ + \ (8)(x^7)(2) \ + \ (28)(x^6)(4) \ + \ (56)(x^5)(8)$$
$$+ \ (70)(x^4)(16) \ + \ (56)(x^3)(32) \ + \ (28)(x^2)(64)$$
$$+ \ (8)(x)(128) \ + \ (1)(1)(256)$$

$$= x^8 + 16x^7 + 112x^6 + 448x^5 + 1120x^4 + 1792x^3$$
$$+ \ 1792x^2 + 1024x + 256$$

B

Evaluate:

17. $\dfrac{18!}{16!2!} = \dfrac{18 \cdot 17 \cdot 16!}{16!2!} = \dfrac{18 \cdot 17}{2 \cdot 1} = 153$

21. $\left(3x - \dfrac{y}{3}\right)^4$

$a = 3x, \; b = \left(-\dfrac{y}{3}\right), \; n = 4$ Use the formula to expand.

$\dfrac{4!}{4!0!}(3x)^4\left(-\dfrac{y}{3}\right)^0 + \dfrac{4!}{3!1!}(3x)^3\left(-\dfrac{y}{3}\right)^1$

$+ \dfrac{4!}{2!2!}(3x)^2\left(-\dfrac{y}{3}\right)^2 + \dfrac{4!}{1!3!}(3x)^1\left(-\dfrac{y}{3}\right)^3$

$+ \dfrac{4!}{0!4!}(3x)^0\left(-\dfrac{y}{3}\right)^4$

$= 1(81x^4)(1) + 4(27x^3)\left(-\dfrac{y}{3}\right) + 6(9x^2)\left(\dfrac{y^2}{9}\right)$

$+ 4(3x)\left(-\dfrac{y^3}{27}\right) + 1(1)\left(\dfrac{y^2}{81}\right)$

$= 81x^4 - 36x^3y + 6x^2y^2 - \dfrac{4}{9}xy^3 + \dfrac{1}{81}y^4$

25. $(2x - 3)^5$
$a = 2x, \; b = -3, \; n = 5$

$\dfrac{5!}{5!0!}(2x)^5(-3)^0 + \dfrac{5!}{4!1!}(2x)^4(-3)^1 + \dfrac{5!}{3!2!}(2x)^3(-3)^2$

$+ \dfrac{5!}{2!3!}(2x)^2(-3)^3 + \dfrac{5!}{1!4!}(2x)^1(-3)^4$

$+ \dfrac{5!}{0!5!}(2x)^0(-3)^5$

$= 1(32x^5)(1) + 5(16x^4)(-3) + 10(8x^3)(9) + 10(4x^2)(-27)$

$+ 5(2x)(81) + 1(1)(-243)$

$= 32x^5 - 240x^4 + 720x^3 - 1080x^2 + 810x - 243$

Find the specified term of the expanded form:

29. $(x + 1)^{10}$, seventh term
    $a = x$, $b = 1$, $n = 10$, $k = 6$ Identify a, b, n, and k.
                                             Note that $k = 7 - 1 = 6$.

$$\frac{10!}{4!6!}(x)^4 (1)^6 = 210x^4 \qquad \text{Substitute in the formula.}$$

C

Write the first four terms of the expanded form:

33. $(2x + y)^{24}$
    $a = 2x$, $b = y$, $n = 24$ Identify a, b, and n.

$$\frac{24!}{24!0!}(2x)^{24}(y)^0 + \frac{24!}{23!1!}(2x)^{23}(y)^1 \quad \text{Use the formula to expand.}$$

$$+ \frac{24!}{22!2!}(2x)^{22}(y)^2 + \frac{24!}{21!3!}(2x)^{21}(y)^3$$

$$= (1)(2x)^{24}(1) + (24)(2x)^{23}(y) + 276(2x)^{22}(y^2)$$
$$+ 2024(2x)^{21}(y^3)$$

$$= (2x)^{24} + 24(2x)^{23}y + 276(2x)^{22}y^2 + 2024(2x)^{21}y^3$$

37. $(x^2 - 2y^3)^{12}$
    $a = x^2$, $b = -2y^3$, $n = 12$

$$\frac{12!}{12!0!}(x^2)^{12}(-2y^3)^0 + \frac{12!}{11!1!}(x^2)^{11}(-2y^3)^1$$

$$+ \frac{12!}{10!2!}(x^2)^{10}(-2y^3)^2 + \frac{12!}{9!3!}(x^2)^9 (-2y^2)^3$$

$$= 1(x^{24})(1) + 12(x^{22})(-2y^3) + 66(x^{20})(4y^6)$$
$$+ 220x^{18}(-8y^9) + \ldots$$

$$= x^{24} - 24x^{22}y^3 + 264x^{20}y^6 - 1760x^{18}y^9 \ldots$$

Find the specified term of the expanded form:

41. $(x + 1)^{22}$, eleventh term
    $a = x$, $b = 1$, $n = 22$, $k = 11 - 1 = 10$

$$\frac{22!}{12!10!}(x)^{12}(1)^{10} = 646646x^{12}$$

45. $\left(\dfrac{x}{2} + 2\right)^{15}$, eighth term

   $a = \dfrac{x}{2}$, $b = 2$, $n = 15$, $k = 8 - 1 = 7$

   $$\dfrac{15!}{8!7!}\left(\dfrac{x}{2}\right)^{8}(2)^{7} = 6435\left(\dfrac{x^8}{256}\right)(128)$$

   $$= \dfrac{6435}{2}x^8$$

D

49. If a coin is tossed eight times, the number of ways that exactly four heads and four tails will show is the coefficient of the fifth term of the expansion of $(H + T)^8$. Find the number of ways.

   $a = H$, $b = T$, $n = 8$, $k = 5 - 1 = 4$

   $$\dfrac{8!}{4!4!} = \dfrac{8\cdot7\cdot6\cdot5\cdot4!}{4!4!}$$

   The variables are not needed, since the coefficient of the term is the answer to the question.

   $$= \dfrac{8\cdot7\cdot6\cdot5}{4\cdot3\cdot2\cdot1}$$

   $$= 70$$

   Answer:
   There are 70 ways you can get 4 heads and 4 tails.

53. If the number of bacteria in a culture initially containing 1000 bacteria is given by $N = 1000(1.02)^t$ (where t is the time in hours), use the binomial expansion of $(1 + 0.02)^6$ to find the number present after 6 hours.

   Expand $(1 + 0.02)^6$.

   $(1 + 0.02)^6$

   $= \dfrac{6!}{6!0!}(1)^6(0.02)^0 + \dfrac{6!}{5!1!}(1)^5(0.02)^1 + \dfrac{6!}{4!2!}(1)^4(0.02)^2$

   $\quad + \dfrac{6!}{3!3!}(1)^3(0.02)^3 + \dfrac{6!}{2!4!}(1)^2(0.02)^4$

   $\quad + \dfrac{6!}{1!5!}(1)^1(0.02)^5 + \dfrac{6!}{0!6!}(1)^0(0.02)^6$

   $= 1 + 6(0.02) + 15(0.0004) + 20(0.000008)$
   $\quad + 15(0.00000016) + 6(0.0000000032) + 0.000000000064$

   $= 1.126$

   Round to the nearest thousandth.

$$N = 1000(1.02)^6$$
$$= 1000(1 + 0.02)^6$$
$$= 1000(1.126)$$

Replace $(1 + 0.02)^6$ with 1.126

$$= 1126$$

The approximate number of bacteria present after 6 hours is 1,126.

57. Of the 12 member nations of an international treaty group, how many subgroups can be formed containing representatives from 4 nations?

$$\frac{n!}{(n - k)!k!}$$

Formula used to calculate the number of ways to select k-elements from a set containing n-elements. Replace n with 12, an k with 4.

$$\frac{12!}{8!4!} = \frac{479001600}{(40320)(24)}$$
$$= 495$$

The number of subgroups that can be formed is 495.

STATE YOUR UNDERSTANDING

61. Why is the sum of the coefficients of the expansion $(a - b)^5$ equal to zero?

From Pascal's triangle, the coefficients for expanding $(a + b)^5$ are 1, 4, 6, 4, and 1. Since in $(a - b)^5$ the term -b will have a positive value for even powers and a negative value for odd powers, the signs of the coefficients will alternate: -1, 4, -6, 4, -1. The sum of these coefficients is zero.

65. Use the binomial expansion to find the sixth term of $\left(x + \dfrac{1}{2y}\right)^9$.

$$\frac{n!}{(n-k)!k!}\ a^{n-k}b^b$$

$a = x,\quad b = \dfrac{1}{2y},\quad n = 9,\ \text{and}$ Identify a, b, n, and k.
$k = 6 - 1 = 5$

$$\frac{9!}{4!5!}\ x^4\left(\frac{1}{2y}\right)^5$$

$$= \frac{9\cdot 8\cdot 7\cdot 6}{4\cdot 3\cdot 2\cdot 1}\ x^4\left(\frac{1}{32y^5}\right)$$

$$= \frac{126\ x^4}{32\ y^5}$$

$$= \frac{63\ x^4}{16\ y^5}$$ Sixth term.

MAINTAIN YOUR SKILLS

Find the logarithm to four decimal places:

69. ℓn 5950

ENTER	DISPLAY
5950	5920

ENTER	DISPLAY
$\ell n\ x$	8.6861

73. $\ell n\ x = -2.3112$

ENTER	DISPLAY
-2.3112	-2.3112

ENTER	DISPLAY
e^x	0.0991